Space Shuttle Legacy

How We Did It and What We Learned

Space Shuttle Legacy

How We Did It and What We Learned

Roger D. Launius
Smithsonian Institution
John Krige
Georgia Institute of Technology
James I. Craig
Georgia Institute of Technology

Ned Allen, Editor-in-Chief
Lockheed Martin Corporation
Bethesda, Maryland

Published by
American Institute of Aeronautics and Astronautics, Inc.
1801 Alexander Bell Drive, Reston, VA 20191-4344

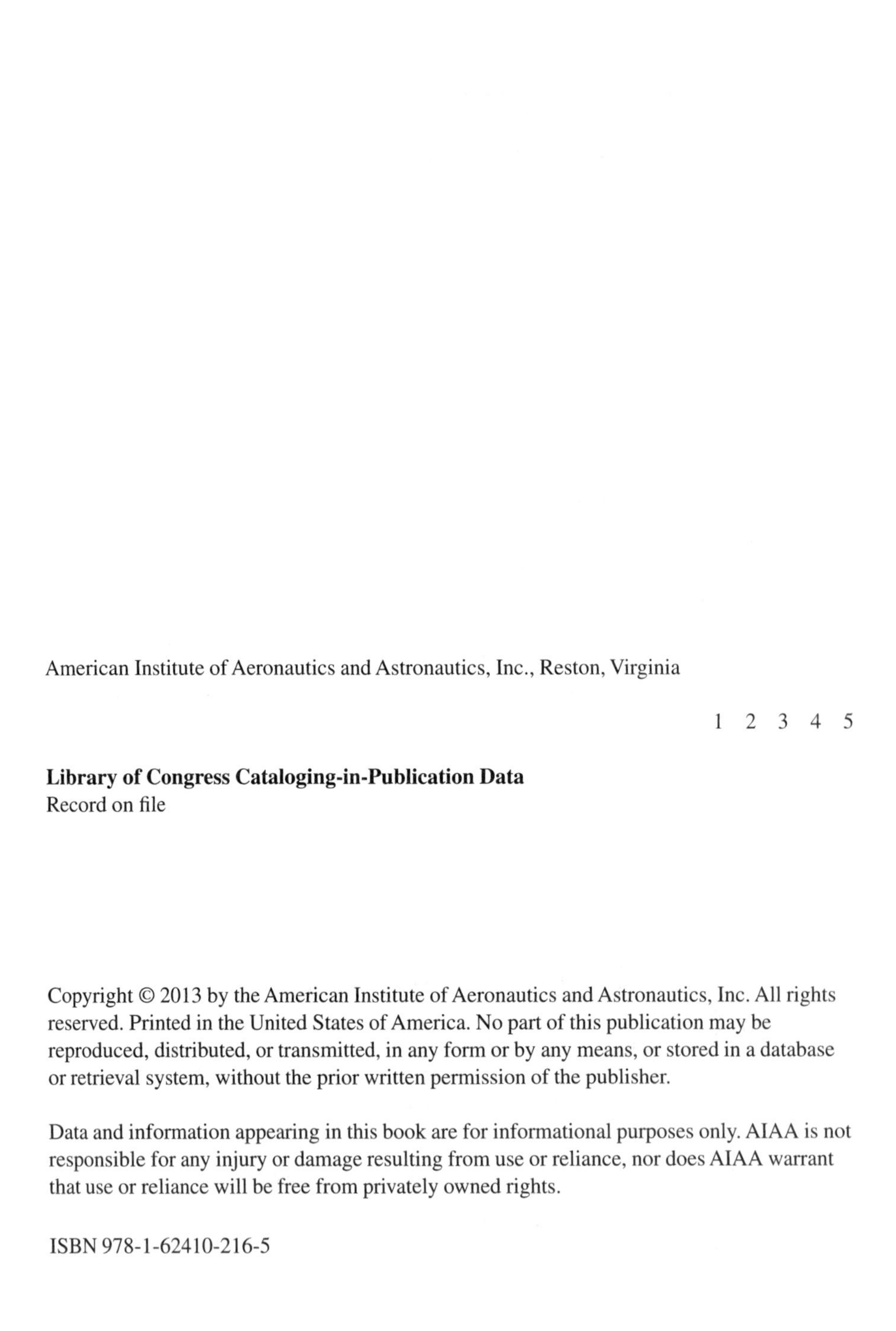

American Institute of Aeronautics and Astronautics, Inc., Reston, Virginia

1 2 3 4 5

Library of Congress Cataloging-in-Publication Data
Record on file

ISBN 978-1-62410-216-5

Dedication

To all those who worked to make the Space Shuttle Program a success.

Note from the Series Editor

This book, more than any other, distills and captures the historical essence of the Space Shuttle program from its post-Apollo inception to its winding down and retirement in 2011. The Shuttle program is the single most significant, and among the most inspirational, science and technology programs in human history to date. In *Space Shuttle Legacy* Roger D. Launius, John Krige, James I. Craig and an eminent group of colleagues assess the program's historic impact upon our technoculture as the world's premiere "big science" and "high technology" program, both in terms of its stunning and unique benefits and its immeasurable and troubling opportunity costs – both inspirational and practical.

This book is significant in documenting and commenting upon perhaps the most important case study in history of large-scale internationally collaborative public science and technology. Many of the lessons learned documented here have already been passed on, for better or for worse, to shape the programs of such ambitious and crucial endeavors as the Large Hadron Collider (LHC) project at CERN or the International Thermonuclear Experimental Reactor (ITER) in Cadarache, France. These lessons will surely continue to influence the conception and gestation of future engineering efforts aimed at managing and preserving a livable ecosystem for our spaceship, Earth, and someday transporting us to terra-formed planets in the stars. Thus, this book more than earns a place in the Library of Flight.

The Library of Flight is part of the comprehensive portfolio of information services from the American Institute of Aeronautics and Astronautics. It extends the Institute's offerings with the best material in a growing variety of topics in aerospace from policy, to histories, to law, management, and beyond. The Library of Flight documents the crucial role of aerospace in enabling, facilitating, and accelerating our technological society and its global commerce, communication, and defense.

Ned Allen
Bethesda, Maryland
September 2013

Contents

PREFACE

In July 2011, NASA completed the 135th and final successful mission of the Space Shuttle program. On many occasions during the Space Shuttle's 30-year career, it had demonstrated remarkable capabilities, but the cost and complexity of flying this first-ever reusable space transportation system always ensured controversy and difference of opinion. NASA moved to retire the shuttle before the vehicle's 30th anniversary because of its age and cost of operations and because the space agency planned a new mission to move beyond Earth orbit and return to the moon.

It is difficult to overestimate the significance of the Space Shuttle to the cause of human spaceflight in the United States, and perhaps even in the world; certainly, few recognized how important the program would become to American life during the late twentieth century. Indeed, since the Space Shuttle's first flight in 1981 until its retirement in 2011, it was an important symbol of this nation's technological capability, and both the American people and the larger international community have universally recognized it as such. After a career spanning a third of a century, the Space Shuttle left an indelible mark on the pursuit of space activities in Earth orbit, making various scientific pursuits, grand engineering activities, and national security endeavors possible, and enabling commercial and other types of actions that were previously unattainable. The shuttle allowed NASA to extend its capabilities from the pioneering experiences of Mercury, Gemini, and Apollo into an era of more routine operations, in the process turning human spaceflight from an exceptional experience into one that became strikingly conventional—and, for the public, mundane.

The shuttle also defined the space agency in much the same way as the moon landing program had during NASA's infancy, although in not quite the same unblemished manner. This was true in no small part for the Space Shuttle program, for all of its very real successes, because of the public memory of desperate, tragic moments. The two most important were the *Challenger* and *Columbia* accidents, which each resulted in the loss of their seven-member crews, but several other failures also came close to costing the lives of astronauts and others working on the Shuttle program.

After 30 years of Space Shuttle operations—40 years, if one includes the research and development period of the 1970s—what is the legacy of the Space Shuttle? That is the question we seek to answer in this work, *Space Shuttle Legacy: How We Did It/What We Learned*, which focuses on the history of the Space Shuttle specifically by delving deeply into various aspects of the program's evolution over its long duration.

This book opens new vistas in our understanding of the program, its place in the larger history of the Space Age, and the arc of national history in the late twentieth and early twenty-first centuries. The book's chapters, each written by a singular expert, broadly analyze major aspects of the Space Shuttle's history. The book's purpose is to consider key aspects of the Space Shuttle program, its successes and its failures, and what lessons individuals who are engaged in future programs might gain from understanding this experience. Each chapter explores the specific history of an individual topic, laying out in narrative form its broad contours. Each also explicitly considers in depth three lessons learned relative to these specific issues with a view to ensuring the chapter passes on a strong message that is useful to future practitioners in the space community.

This book was the brainchild of Vigor Yang, chair of the School of Aerospace Engineering at the Georgia Institute of Technology, and James I. Craig, emeritus professor in the School, who has a special interest in the place of the Space Shuttle in the development of space technology. With administrative support and funding from NASA, the Boeing Company, and Georgia Tech, they hosted a symposium, "The Space Shuttle: An Engineering Milestone," in 2011, to recognize the technological accomplishments of the many individuals involved in the program and to involve aerospace engineering students in the commemoration. The symposium offered a beginning point for assessing the legacy of the Space Shuttle over its more than 30-year career, and a variety of contributors engaged in first-rate proceedings that focused on the technical lessons learned from the program. Yang and Craig followed this up in 2012 by recruiting the historians Roger D. Launius and John Krige to design, and along with Craig to edit, a book to undertake a broader evaluation of the program. The writers published in this book were chosen for both their broad knowledge and their disciplines; they include historians, political scientists, public administrators, engineers, and scientists. All intend to highlight lessons learned in the Space Shuttle program, with the objective of informing NASA officials, national executives, the aerospace community, and the broader public.

Throughout this book, the authors highlight useful analogies with the objective of informing future decision-making. Of course, all individuals in any organization use history to make decisions every day; most just do not realize they are doing so, and often they do so implicitly and without rigor. This book seeks to help address this issue and to offer methodologies that might be useful in employing historical insights both more explicitly and effectively. One outcome will be to enable practitioners to use the history of the Space Shuttle program effectively in dealing with engineering, scientific, and programmatic questions by determining the applicability of analogies, testing presumptions, precisely reviewing issues, and accurately diagnosing the nature of problems.

We intend the book to fulfill a threefold purpose: (1) to provide to a broad readership a basic and accessible history of the Space Shuttle program over its long history, (2) to offer insights from the Space Shuttle program that may be useful to space professionals in future endeavors, and (3) to serve as a catalyst for additional study of the Space Shuttle and its legacy. In this last context, this book is intended as a work that offers important insights about themes that investigators in the future might productively explore and as an opportunity to energize scholars to pursue this important topic more effectively. Although there are other books on the Space Shuttle and its history, none seek to do anything of this sort, and this one, therefore, fills a critically important niche in the historiography of this critical human space program.

All 14 chapters, each written by specialists, cover in broad strokes the complex and lengthy history of the Space Shuttle era in the United States.

In Chapter 1, Roger D. Launius traces the long shadow of the quest for an American spaceplane from the 1920s to the realization of the Space Shuttle in the 1970s. He asks: What is it about the concept of a spaceplane that was so inviting to American engineers throughout the twentieth century? A winged, reusable space vehicle for human flight beyond Earth dominated thinking about the task before the Space Age, but to compete with the Soviet Union in human space spectaculars beginning in the late 1950s, the United States opted for ballistic capsules that were easier to build and fly. No sooner had that competition ebbed, however, than NASA returned to the pursuit of a spaceplane, resulting in building the winged, reusable Space Shuttle and flying it for 30 years.

Launius explores how the longstanding quest for the spaceplane in the United States represents a striking example of the social construction of technology—a theme in the history of technology that suggests that some engineering decisions are not decided through hard-headed analysis but instead by other conventions and priorities. He concludes that this ideal of a spaceplane has enthralled Americans for nearly 100 years and has significantly influenced decisions about technological direction in the human exploration and development of space.

In Chapter 2, Launius investigates the decision to build the Space Shuttle, a decision made in 1972 after a difficult debate inside Washington over the next-generation human spaceflight vehicle. The relations of NASA, the aerospace industry, the presidency, Congress, and the American public fundamentally affected how this decision proceeded. Additionally, the budgetary priorities of the nation's leaders, especially austerity in domestic spending during the Vietnam era, affected at a fundamental level the design of the Space Shuttle. NASA compromised on many of its most fundamental goals in pursuing the shuttle as reflected in this situation, leading to a creature of compromise that couldn't fully meet its requirements. To reach even this end, the system had to be oversold and its capabilities overstated. This necessitated NASA's compromise with the Department of Defense (DoD) and other potential users on the shuttle's design and capabilities; however, it also forced the space agency to overpromise results and underbid costs. The result was a vehicle that failed to meet expectations.

In Chapter 3, political scientist W. Henry Lambright offers a fascinating account of the debate over management approaches for the Space Shuttle program. At the conclusion of the Apollo program, NASA officials realized that they would never again have the resources available for the moon landings and that they had to find another means of accomplishing their projects without such a broad effort. Accordingly, NASA adopted for the Space Shuttle program a "lead center" style of management that devolved the program from the NASA headquarters to the various centers, with "work packages" to each. In theory, it allowed center directors more flexibility to complete their tasks free from excessive oversight and internal coordination. They had control of discrete elements of the organizational structure, budget, contractor management, cost/schedule/reliability issues, and the like. Lambright traces how and why this approach to management unfolded throughout the Shuttle program, and how successive generations of NASA executives altered, reformed, and modified the management structure.

The three most difficult technical issues to be resolved in the Space Shuttle program involved the Space Shuttle Main Engine (SSME), the thermal protection system (TPS), and the advanced software employed on the orbiter. Engineers Robert E. Biggs, Dennis R. Jenkins, and Nancy G. Leveson have contributed Chapters 4–6, which detail the processes followed in conquering these engineering challenges.

Likewise, in Chapter 7, N. Wayne Hale analyzes the manner in which the Space Shuttle was prepared for launch, completed its operational mission, and safely returned to Earth.

In Chapter 8, Matthew H. Hersch explains how the Space Shuttle, which began flying in 1981, became all things to all people, especially a low-cost, reusable spaceplane to ferry humans, cargo, and various technologies into outer space for a variety of purposes, as well as to return objects from space. The Space Shuttle was also intended to be a laboratory in orbit for scientific research on a broad range of questions and involving a wide array of research specialties. It performed many of these tasks effectively, if not inexpensively, as Hersch demonstrates.

All are familiar at some level with the *Challenger* (1986) and *Columbia* (2003) Space Shuttle accidents. In Chapter 9, Stephen P. Waring seeks to understand and explain these accidents, as well as several near disasters during spaceflight. Through the lens of horrific accidents, Waring analyzes the shuttle's dark shadows to illuminate causes and offer comparisons over time. Through this historical and comparative process, he offers broad lessons relating to vehicle design, program architecture, research and development, and the challenges of maintaining cost, schedule, and reliability.

Two critical aspects of the Space Shuttle program are discussed in Chapters 10 and 11. First, Howard E. McCurdy considers the importance of the shuttle for the building of the International Space Station (ISS), occupied for the first time in 2000. This chapter focuses on the role of the Space Shuttle in building the station on orbit between 1998 and 2011. In essence, it could not have been constructed in the way it was without the Space Shuttle. John Krige's chapter follows and emphasizes efforts to internationalize the Space Shuttle program by inviting European allies into the construction program, only to have those plans nixed by technocrats and diplomats in the White House and at the State Department. There remained a core international component to the Space Shuttle program, but none of the larger collaborative desires were realized. Instead, Europeans built Spacelab to fly inside the space shuttle's payload bay and joined the program with astronauts. It represents a cautionary tale in the internationalization of space activities.

Amy E. Foster discusses in Chapter 12 how and why the practices of astronauts on the Space Shuttle have evolved over the life of the program. She details the history of the astronaut corps in this era, its evolution in terms of make-up and skill set, and how it carried out the tasks before it during the program. Most importantly, she highlights how NASA was able to diversify the astronaut corps and what that meant, the evolving procedures for orbital operations and their antecedents and results, and the structuring of astronaut crews and support teams for optimal operations.

The two final chapters, on the Space Shuttle as a cultural icon by Linda Billings and on its retirement by John M. Logsdon, deal with larger meanings for the program in modern American life. From first flight to the present, the Space Shuttle served as an important symbol of U.S. technological capability, universally recognized as such by both the American people and the larger international community. Billings shows how the Space Shuttle remained until retirement one of the most highly visible symbols of American excellence worldwide. The decision to retire the Space Shuttle in the aftermath of the *Columbia* accident, originally set for 2010 but extended a year, and all that resulted from it has profoundly influenced the course of human space activities in the United States, as Logsdon makes clear. This is especially true because of the story of shuttle replacement and the various efforts undertaken to move to the next human space flight system.

Lastly, the epilogue seeks to cast the Space Shuttle in a larger context as a benchmark in American aerospace engineering, as a vehicle for the successful opening of low-Earth orbit (LEO), and as a program that served as a useful follow-on to the more famous but not necessarily more significant Apollo moon landings of the space race era.

In sum, at least five core legacies seem to deserve consideration in any assessment of the Space Shuttle. First, the Space Shuttle has a reputation as a mistake resulting from a policy failure that should never have been pursued. Second, the Space Shuttle has been criticized as a program that prohibited other paths for the U.S. space program. Third, and more positively, the Space Shuttle provided three decades of significant human spaceflight capability and stretched the nature of what could be accomplished in Earth orbit much beyond where it had previously been. Fourth, the Space Shuttle served as a relatively flexible platform for scientific activities. Finally, and perhaps most significantly because the American human spaceflight program has always been focused on national prestige, the Space Shuttle served well as a symbol of American technological verisimilitude.

When the Shuttle's 30-year career ended, it had racked up an impressive record of achievement. As a scientific platform, as a construction vehicle, as a cargo and logistics carrier, and for a host of other activities, the Space Shuttle was a remarkable vehicle. With 135 flights over 30 years, it was most appropriate that NASA gave the vehicle an honorable retirement, after an important career as the primary vehicle carrying U.S. astronauts, as well as others, into orbit. Its legacy is assured.

ACKNOWLEDGMENTS

Whenever scholars take on a project such as this, they stand squarely on the shoulders of earlier investigators and incur a good many intellectual debts. The editors and authors would like to acknowledge the assistance of the institutions and individuals who aided in the preparation of this book. Our greatest debt is to the School of Aerospace Engineering at Georgia Tech, and especially Vigor Yang and Jim Craig, who conceived of this project. We gratefully acknowledge the support provided by Georgia Tech. We also express our deep gratitude to the many people at the Smithsonian Institution and at NASA who supported this endeavor. This book was planned at Georgia Tech; this was also where the editors lined up the authors, and where Yang and Craig organized a workshop to discuss the authors' first drafts. Foremost, we thank the individuals who participated in the project as authors, editors, and others involved at various stages.

For their many contributions in completing this project, we also wish to thank Jane Odom and her staff archivists at the NASA History Division, who helped track down information and correct inconsistencies, in addition to Bill

Barry and Nadine Andreassen at NASA; the staffs of the NASA Library and the Scientific and Technical Information Program, who provided assistance in locating materials; Marilyn Graskowiak and her staff at the NASM Archives; and many archivists and scholars throughout several other organizations. Patricia Graboske, head of publications at the National Air and Space Museum, provided important guidance for this project. Our deep thanks are due to all of these fine people.

In addition to these individuals, we wish to acknowledge the following individuals who aided in a variety of ways: Jeff Bingham, Lynn Cline, Tom D. Crouch, Gen. John R. Dailey, Dwayne A. Day, Lori B. Garver, G. Michael Green, Richard P. Hallion, Roger Handberg, James R. Hansen, David A. Hounshell, Peter Jakab, Sylvia K. Kraemer, Alan M. Ladwig, Jennifer Levasseur, Elaine Liston, Valerie Lyons, W. Patrick McCray, Jonathan C. McDowell, Karen McNamara, Valerie Neal, Allan A. Needell, Michael J. Neufeld, Frederick I. Ordway III, Scott Pace, Dominick Pisano, Robert Poole, Tony Springer, Bert Ulrich, Margaret Weitekamp, Peter Westwick, and Joni Wilson. Several interns provided assistance at various stages of this project, and to them we offer our sincere thanks: Jonathan Cohen, Anna Olkovsky, and Bryn Pernot. We also thank Rodger Williams and the staff of AIAA for their efforts in seeing this book through to publication.

These chapters are offered in the spirit of scholarly debate. Not all will agree with everything in them, but we envision that this volume will consolidate and bring to the fore current understandings on this important subject.

Roger D. Launius
John Krige
James I. Craig
August 2013

INTRODUCTION

When NASA began work on what became known as the Space Shuttle in the late 1960s, few recognized how important a part of American life the program would become over the next 40 years. It has become the symbol of American technological prowess, recognized worldwide as such for both its triumph and its tragedy. With the shuttle program, NASA intended to lower the cost of space flight so that it could conduct an aggressive space exploration effort. The shuttle was intended to provide the United States with low-cost, routine access to space [1].

On January 5, 1972, President Nixon announced the decision to build a space shuttle. What emerged in the early 1970s consisted of three primary elements: a delta-winged orbiter spacecraft with a large crew compartment, a 15×60-ft cargo bay, and three main engines; two solid rocket boosters (SRBs); and an external fuel tank that housed the liquid hydrogen and oxidizer that burned in the main engines. The orbiter and the two SRBs were reusable. The shuttle was designed to transport approximately 45,000 tons of cargo into near-Earth orbit, 115 to 250 statute miles above the Earth. It could also accommodate a flight crew of up to 10 persons for a basic space mission of 7 days. During a return to Earth, the orbiter was designed so that it had a cross-range maneuvering capability of 1265 statute miles to meet requirements for liftoff and landing at the same location after only one orbit.

With much public excitement, *Columbia*, the first orbiter that could be flown in space, took off from Kennedy Space Center, Florida, on April 12, 1981. After 36 orbits during 2 days in space, *Columbia* landed like an aircraft at Edwards Air Force Base in California, successfully accomplishing the first such landing of an orbital vehicle in the history of the Space Age. The first flight had been an enormous success, and with it, the United States had embarked on a new era of human space flight. Through 30 years of operational life, until the end of the Space Shuttle program in 2011, the vehicle came to symbolize the muscular but positive nature of American technological verisimilitude [2].

To summarize, at least five core legacies deserve consideration in any assessment of the Space Shuttle. First, the Space Shuttle has a reputation as a mistake resulting from a policy failure that should never have been pursued. Second, it has been criticized as a program that prohibited other paths for the U.S. space program. Third, and more positively, the Space Shuttle provided three decades of significant human spaceflight capability and stretched the nature of what could be accomplished in Earth orbit much beyond where it had previously been. Fourth, the Space Shuttle served as a relatively flexible platform for scientific activities. Finally, and perhaps most significantly because the American human spaceflight program has always been focused on national prestige, the Space Shuttle served well as a symbol of American technological prowess [3].

One of the most unfortunate common mischaracterizations of the Space Shuttle program is that it resulted from a "policy failure" made in planning for the post-Apollo space program and that it was, in essence, a mistake relentlessly pursued by NASA for more than a generation. For example, a well-publicized statement of this belief came in 1985, when Alex Roland, a former member of the NASA History Division who went on to become a professor at Duke University, offered a thoughtful and reasoned criticism of the shuttle in the era before the *Challenger* accident [4]. Roland's arguments about the shuttle as a mistake continue to be believed even now. He summarized his critique in testimony before the U.S. Senate in the aftermath of the 2003 *Columbia* accident [5]:

> Briefly stated, NASA made two mistakes in shuttle development in the late 1960s and early 1970s. First, it traded development costs for operational costs. Second, it convinced itself that a recoverable launch vehicle would be inherently more economical than an expendable [one]. NASA promised savings of 90%, even 95%, in launch costs. In practice, it costs more to put a pound of payload in orbit aboard the Shuttle than it did aboard the Saturn launch vehicle that preceded it. These mistakes produced a program that cannot work. NASA could conceivably operate the Shuttle safely and reliably, but it dares not admit what it would cost. The evidence for this was abundant before the *Challenger* accident. Instead of listening to the data, NASA consistently allowed its judgment to be clouded by its hopes and predictions for human activities in space. The agency cares about astronaut safety, but it is trapped by its own claims about Shuttle costs. And, unlike expendable launch vehicles, the Shuttle grows more dangerous and more expensive to fly with each passing year.

The cost that Roland noted is incorrect, however. The cost per pound to orbit was $21,300 per pound for the Shuttle and $36,000 per pound for Saturn (combined I/IB/V)—all assuming a full payload for each launch.

Other aerospace analysts, including the dean of the space policy community, John M. Logsdon, criticized the Space Shuttle as a flawed decision that was unable to meet expectations [6].

Since that point, many others have criticized the Space Shuttle effort as ill conceived, politically suspect, and poorly executed. The result has been a steady stream of critiques of the Shuttle program, especially coming whenever there is a public failure. Virtually all of these appraisals emphasize the convoluted history of the Space Shuttle's origins, its evolution, its operation, and the continuing challenge of space access. Sometimes the criticisms are well reasoned and temperate, whereas at other times they are muckraking and outrageous; sometimes they are also erroneous. Always they emphasize missteps, policy reversals, organizational inertia, and leadership failings, as well as seemingly impossible technical challenges, which combined to prevent NASA from realizing the vision it had set out for itself in advocating the Space Shuttle [7].

Ironically, although there is no question that the Space Shuttle is a creature of compromise and that it does not enjoy a universally positive perception, its faults may have been exaggerated over the years. There is a public perception that Apollo was an unmitigated success worthy of celebration at every level; there is a public perception that the Space Shuttle was a "mistake," and in 2005, the NASA administrator even characterized it as such [8]. Both perceptions are incorrect. Apollo had failures—one catastrophic—that blemish the project, and the Space Shuttle's successes far outweighed the failures of the program. Because the Space Shuttle did not live up to all expectations, many criticized NASA for failing to meet the promises it made to gain approval for the Shuttle program [9]. This criticism has served the useful purpose, however, of inviting a range of questions that deserve consideration. What are the most effective ways to lower the cost of space access? Is reusability an effective way? Are "big, dumb boosters" the way? Is the launch vehicle for efficient operations? Is the solution something else altogether, or several "something elses"? Perhaps the solution is a combination of these and many other, more sublime factors. Whatever the answer, it is important to take into careful consideration the legacies of these earlier research and development efforts in assessing the Space Shuttle and its contributions to future technologies.

A second powerful legacy of the Shuttle program points to the possibilities of the paths not taken. Because NASA pursued the Shuttle program, it was unable to undertake other projects that might have been more fruitful. There is no question that this is true, and several in the spaceflight community have noted this in recent articles and presentations. There is, unfortunately, no evidence—because NASA did not pursue those other paths—that these other avenues would have been better than the Shuttle. Usually critiques along these lines take one of three forms. The first is a criticism that NASA spent the last 30 years in Earth orbit and that it could have—indeed, should have—used the same funding that it received for the Shuttle to return to the moon or to explore Mars.

Robert Zubrin, a persistent advocate for a mission to Mars, made this case in testimony before the U.S. Senate in 2003 when he said:

> In today's dollars, NASA['s] average budget from 1961–1973 was about $17 billion per year. This is only 10% more than NASA's current budget. To assess the comparative productivity of the Apollo Mode with the Shuttle Mode, it is therefore useful to compare NASA's accomplishments between 1961–1973 and 1990–2003, as the space agency's total expenditures over these two periods were equal.

During that earlier period, Zubrin asserted, NASA was able to undertake Apollo as well as a broad array of scientific probes and other missions. In the later Shuttle era, it spent the majority of its resources on this space access vehicle. He concluded: "Comparing these two records, it is difficult to avoid the conclusion that NASA's productivity in *both* missions accomplished *and* technology development during its Apollo Mode was at least ten times greater than under the current Shuttle Mode" [10].

Despite the validity of Zubrin's facts, he makes two assumptions that are important to consider explicitly. First, the Shuttle was a program approved through the democratic process in place in the United States. Other programs, especially continued lunar exploration and a human mission to Mars, were explicitly rejected in the policy process despite their recommendation in the Space Task Group report of 1969 [11]. There is no reason to believe that the nation's leaders would have allowed NASA to reprogram funds from one endeavor to another even if its leaders had wanted to do so. Second, these alternative missions might have been successful, but there is no evidence to believe this because we are engaging in a counterfactual consideration [12]. It never happened; the assumption that it would have worked as intended is problematic. That is not to deny that Zubrin might be correct and that the unfolding of space history might have been dramatically different, and presumably better, had another course been pursued. We simply will never know.

A second criticism of "paths not taken" comes from representatives of the scientific community and usually involves questioning the role of humans in space altogether. Many scientists have questioned the viability of the Shuttle for scientific activities and suggested that its budget could have more usefully been applied to expendable systems and robotic probes that promised higher scientific returns on investment. University of Iowa astrophysicist and discoverer of the radiation belt that surrounds Earth and that bears his name, James A. Van Allen, never believed that the Space Shuttle was worth the expense. In 2004, he remarked, "Risk is high, cost is enormous, science is insignificant. Does anyone have a good rationale for sending humans into space?" [13]

Undoubtedly, large numbers of scientific missions could have been developed had funding used for the Shuttle been used instead to fund other types of scientific missions. But it is not a zero-sum game. Again, as in the case with Zubrin's argument, there is very little reason to believe that the political process would have put the kind of funds used to support the Shuttle into scientific missions. Daniel S. Goldin, NASA administrator between 1992 and 2001, observed in 1992 that this money would more likely be put into some other congressional priority rather than space science. "NASA's budget is $14 billion," he said. "Our federal budget is $1.5 trillion. We could take the whole NASA budget and, in a feeding frenzy in the U.S. Congress, vaporize that budget in two hours" [14].

Finally, regarding "paths not taken," we might consider what the exploration of space might have been like if NASA had not pursued the Space Shuttle and if it instead had taken another path in developing launch vehicles. Would the $90+ billion spent to build and fly the Space Shuttle between 1972 and 2011 have been better spent on a different combination of launchers? Some certainly thought that with the investment made in Apollo technology, it should not have been abandoned in favor of the Space Shuttle, it was a waste of both money and a fully reliable technology, and the costs of moving in an entirely different technological direction at the conclusion of Apollo far outweighed the benefits that might accrue. This "minority position" on the Space Shuttle created a scandal between 1969 and 1971. Iconoclastic aerospace engineer Robert C. Truax, for one, suggested that abandoning Apollo technology was ill advised. He thought there was no need to build a new winged space vehicle. Instead, the approach taken in Apollo would do just as well and be both recoverable and reusable. That would cut down development costs drastically, but because capsules were "inelegant," he thought, NASA was committed to a winged spacecraft that "could be an unparalleled money sponge" [15].

And so, why did NASA turn away from effective launch vehicle technology in favor of creating a new launch system from scratch? Although the timetable of the Apollo project, tied as NASA was to Cold War public policy concerns, certainly drove the agency to exploit ballistic missile technology, the budget pressures in the post-Apollo era replaced policy concerns. Given that the era of virtually unlimited budgets that Congress gave NASA in the 1960s had ended, again one might think that adapting existing technology would have been attractive. As things turned out, it took almost a decade between the political announcement to build the Shuttle and its first flight; it therefore ended up costing much more than anticipated for development alone [16]. NASA certainly could have stayed with a variant of Saturn launch technology. Would that have been better? Opinions differ wildly, and it is impossible to reach any definitive answers.

As for the third legacy of the Space Shuttle, there is no question about its overwhelming complexity and its role as a driver for the technology of spaceflight. It is indubitably a magnificent machine. The Space Shuttle has proven itself one of the most flexible space vehicles ever flown. Most assuredly, the range of possibilities for operations in orbit expanded dramatically with the launch of *Columbia* in 1981. When making any effort to develop a follow-on system, we need to consider the Space Shuttle's ability to carry a diversity of payloads, to accomplish a myriad of tasks on orbit, and to deploy and retrieve satellites. The United States, as well as its partners in the International Space Station (ISS), will certainly miss the Space Shuttle should the planned Crew Exploration Vehicle (CEV) or some other vehicle be unable to provide comparable capabilities in the future. In this regard, it is important to realize that the CEV as currently conceived replaces not the Shuttle but the next human spaceflight vehicle. It has little up-mass or down-mass past its crew, it cannot easily conduct independent extra-vehicular activities (EVAs), it cannot supply a large amount of cargo to the ISS, and it remains to be seen whether it will be more economical, although that is clearly an important goal of the program.

No flights demonstrate the Shuttle's flexibility more effectively than the five Hubble Space Telescope servicing missions. After the telescope was launched in 1990, many believed that a spherical aberration in its mirror would cripple the instrument. In December 1993, because of the difficulties with the mirror, NASA launched the shuttle *Endeavour* on a dramatic repair mission to insert new components to correct for the aberration and to service its other instruments. During a weeklong mission, *Endeavour*'s astronauts conducted five spacewalks and successfully completed all programmed repairs to the spacecraft. The first reports from the Hubble spacecraft afterward showed that the images were more than an order of magnitude crisper than those that had been obtained before. For this outstanding effort, NASA's Hubble Space Telescope Recovery Team received the Robert J. Collier Trophy "for outstanding leadership, intrepidity, and the renewal of public faith in America's space program by the successful orbital recovery and repair of the Hubble Space Telescope" [17]. Four additional servicing missions extended the capabilities of the telescope into the first decade of the 21st century, and perhaps another mission will extend the life of the telescope yet another decade.

Through the end of the program in 2011, there had been 135 Space Shuttle missions, and the range of activities on each of these had been impressive: ISS construction and logistics missions; pure science missions; DoD missions; collaborative flights with Russian cosmonauts to the Mir Space Station; and satellite deployments, retrievals, and repairs.

Undoubtedly, and this is a significant aspect of the Shuttle's flexibility, its size and capability greatly expanded the opportunity for human spaceflight. A crew of 3 flew on the Apollo missions, whereas the Shuttle routinely flew 7, and by the end of 2005, the number of astronauts flown aboard shuttles stood at more than 250. Accordingly, among other notable developments, the Shuttle allowed NASA to expand the astronaut corps beyond the white male test pilots who had exclusive domain during the Mercury, Gemini, and Apollo eras. The Shuttle enabled an expansion of the astronaut complement to nonpilots and to women and minorities. As all know, in June 1983, Sally K. Ride, a NASA scientist-astronaut, became the first American woman to fly in space, aboard *STS-7*, and in August 1983, Guion S. Bluford became the first African American astronaut to fly, on *STS-8*. The Shuttle era also saw flights by people who were not truly astronauts. NASA inaugurated both a payload specialist program to fly individuals associated with specific experiments as well as a "Space Flight Participant Program" aimed at allowing nonscientists and nonengineers to experience orbital flight. The first person was a teacher, Christa McAuliffe, who died in the *Challenger* accident in January 1986, but a journalist and perhaps a poet were also possibilities for future missions. In addition, astronauts from many other nations flew aboard the Shuttle, including astronauts from Russia, Israel, Saudi Arabia, Canada, France, Germany, Italy, Japan, and Switzerland. This democratization of human spaceflight was a major attribute of the Shuttle era and the result of its flexibility as a space vehicle.

As for the Space Shuttle's fourth legacy, throughout its three-decade career, it served as a versatile platform for scientific inquiry. Although the program was not conceptualized as a science effort—rather, it was a technology demonstrator and workhorse for space access—it has been used as an exemplary platform for all manner of microgravity and space science experiments. President Nixon, announcing the decision to build the Space Shuttle in 1972, minimized its scientific role. Instead, he argued that it was "the right step for America to take, in moving out from our present beach-head in the sky to achieve a real working presence in space—because the Space Shuttle will give us routine access to space" [18].

Even so, the Space Shuttle has been a useful instrument in the hands of scientists. All of the operational flights undertook a range of scientific experiments, including the deployment of important space probes to other planets, the periodic flight of the European-built Spacelab science module, and a dramatic set of Earth observations made over a 25-year period. One example of a valuable science experiment, among others that might be offered, is the flight of the Italian Tethered Satellite System in 1992 and again in 1996. This was designed to investigate new sources of spacecraft power and ways to study Earth's upper atmosphere. It demonstrated that tethered systems might be used to generate thrust to compensate for atmospheric drag on orbiting platforms such as the ISS. Deploying a tether towards Earth could place movable science platforms in hard-to-study atmospheric zones. Tethers also could be used as antennas to transmit extremely low-frequency signals that could penetrate land and sea, providing for communications not possible with standard radio equipment. In addition, nonelectrical tethers could be used to generate artificial gravity and to boost payloads to higher orbits [19].

Additionally, a stunning science experiment occurred with the flight of the Shuttle Radar Topography Mission (SRTM) in 2000. This flight on *Endeavour* obtained elevation data on a near-global scale to generate the most complete high-resolution digital topographic data ever created. It consisted of a specially modified radar system that flew during an 11-day shuttle mission. Virtually the entire land surface between +/ – 60 degrees latitude was mapped by SRTM, and it has been an enormously significant data set for land use scientists [20].

The Space Shuttle also deployed scientific probes to other planets' satellites for Earth observation and beyond. It was long used as a workhorse of space exploration. Perhaps most important, since the *Challenger* accident in 1986, seven major scientific payloads required the unique capabilities of the Shuttle. It launched the Magellan spacecraft to Venus, the Galileo spacecraft to Jupiter, and the Ulysses spacecraft to study the sun. The Shuttle also deployed the Compton Gamma Ray Observatory, the Hubble Space Telescope, the Upper Atmosphere Research Satellite, and the Chandra X-Ray Telescope [21].

In addition, 16 Spacelab and Spacehab missions could not have been conducted by any other existing spacecraft. These Shuttle missions allowed significant scientific work to be accomplished, and Shuttle advocates often cite these activities as important contributions that justified the cost of the program. Spacelab, a sophisticated laboratory built by the European Space Agency, fit into the Shuttle's cargo bay and enabled a complex array of scientific experiments to be completed [22]. Others noted the more than 2000 life sciences experiments flown aboard the Space Shuttle, as well as numerous other demonstration science activities on orbit [23].

Once again, this is a mixed legacy. The Hubble Space Telescope, for one, had to undergo repeated alterations to fly on the Shuttle, had to be launched on something else after the *Challenger* accident, and then had to be reconfigured for Shuttle deployment. All of this added to the schedule and the cost of the Hubble program [24]. This happened in other spacecraft research and development (R&D) projects as well. Could issues such as these have been handled more effectively? Probably, but project managers did not have access to hindsight at the time. Even so, the significance of the Shuttle as a platform for scientific investigation is secure.

Finally, from first flight to the present, the Space Shuttle has been an important symbol of U.S. technological capability, universally recognized as such by both the American people and the larger international community. NASA's Space Shuttle remains after more than 30 years one of the most highly visible symbols of American excellence worldwide—and a positive one, unlike the nation's enormous military might.

No other nation on the face of the Earth had the technological capability to build such a sophisticated reusable vehicle during the 1970s. Few could do so today. A massively complex system—with more than 200,000 separate components that must work in synchronization with each other and to specifications more exacting than any other technological system in human history—the Space Shuttle must be viewed as a triumph of engineering. As such, it has been an enormously successful program. The research, development, and operation of the Space Shuttle represent a worthy follow-on to the positively viewed Apollo program of the 1960s and early 1970s.

Indeed, if there is one hallmark of the American people, it is their enthusiasm for technology and what it can help them to accomplish. Historian Perry Miller wrote of the Puritans of New England that they "flung themselves in the technological torrent, how they shouted with glee in the midst of the cataract, and cried to each other as they went headlong down the chute that here was their destiny" as they used technology to transform a wilderness into their "City upon a hill" [25]. Since that time, the United States has been known as a nation of technological system builders who could use this ability to create great machines of wonder and the components of their operation.

Perceptive foreigners might be enamored with American political and social developments, with democracy, and with pluralism, but they are more taken with U.S. technology. The United States is not just the nation of George Washington, Thomas Jefferson, Abraham Lincoln, Frederick Douglass, and Elizabeth Cady Stanton; it is also of Thomas Edison, Henry Ford, the Tennessee Valley Authority, Apollo, and the Manhattan Project. These reinforced the belief throughout the globe that America was *the* technological giant of the world. Until the loss of *Challenger* and *Columbia*, NASA and its accomplishments symbolized more than any other institution America's technological creativity. That symbolism, misplaced as it might have been all along, accounts for the difficulties the agency has felt in the recent past. Every NASA failure raises the question of American technological virtuosity in the world, and the questioning of much American capability in so many other areas is already underway, making setbacks in this one area all the more damaging to the American persona. Doubts about America increased with every perceived failure in the space program [26].

Everyone seems to agree on this symbolism. Conservative columnist Tony Blankley opined in 2005 [27]:

> The Shuttle is the most complex moving machine ever built by man. Conceived in the 1960s, the current machines are up to 25 years old and weigh 4.5 million pounds at launch.... Despite the age of the machines and the technology, no other people on the planet yet have had the skill, wealth and will to build such a thing. It is a triumph of engineering to assemble millions of parts into the necessary complexity that permits the machine to function with only two failures in a quarter of a century under the extreme pressures of launch, space, reentry and re-use.

Even critics of the program, such as journalist Greg Easterbrook, acknowledge this. As he wrote eloquently in *Time* just after the *Columbia* accident [28]:

> A spacecraft is a metaphor of national inspiration: majestic, technologically advanced, produced at dear cost and entrusted with precious cargo, rising above the constraints of the earth. The spacecraft carries our secret hope that there is something better out there—a world where we may someday go and leave the sorrows of the past behind. The spacecraft rises toward the heavens exactly as, in our finest moments as a nation, our hearts have risen toward justice and principle.

Easterbrook appropriately characterized the sense of wonder and awe that the Space Shuttle has evoked around the world. Because of its technological magnificence, it has become an overwhelmingly commanding symbol of American technological virtuosity. Ask individuals outside the United States what ingredients they believe demonstrate America's superpower status in the world, and almost everyone will quickly mention the Space Shuttle as a constant reminder of what Americans can accomplish when they set their minds to it.

When the Space Shuttle ended its 30-year career, it had racked up an impressive record of achievement. As a scientific platform, as a construction vehicle, as a cargo and logistics carrier, and for a host of other activities, the Space Shuttle was a remarkable vehicle. With 135 flights over 30 years, it was most appropriate that NASA gave the vehicle an honorable retirement after an important career as the primary vehicle carrying U.S. astronauts, as well as others, into orbit. Its legacy is assured.

REFERENCES

[1] Jenkins, D. R., *Space Shuttle: The History of the National Space Transportation System, the First 100 Missions*, 3rd ed., Cape Canaveral, FL, 2001; Launius, R. D., "A Baker's Dozen of Key Historical Books about the Space Shuttle," 19 Oct. 2011, http://launiusr.wordpress.com/2011/10/19/a-baker's-dozen-of-key-historical-books-about-the-space-shuttle/ [retrieved 27 May 2013].

[2] Neal, V., "Framing the Meanings of Spaceflight in the Shuttle Era," *Societal Impact of Spaceflight*, edited by S. J. Dick and R. D. Launius, NASA SP-2007-4801, 2007, Washington, DC, pp. 67–88.

[3] Launius, R. D., "Assessing the Legacy of the Space Shuttle," *Space Policy,* Vol. 22, no. 4, Nov. 2006, pp. 226–234.

[4] Roland, A., "The Shuttle: Triumph or Turkey?" *Discover*, Vol. 6, no. 11, Nov. 1985, pp. 14–24.

[5] Roland, A., "Statement before the Subcommittee on Science, Technology, and Space of the Senate Committee on Commerce, Science, and Transportation," 2 April 2003, http://history.nasa.gov/columbia/Troxell/Columbia%20Web%20Site/Documents/Congress/Senate/FEBRUA~1/roland_statement.html [retrieved 2 Feb. 2006].

[6] Logsdon, J. M., "The Space Shuttle Program: A Policy Failure," *Science,* Vol. 232, no. 7454, 30 May 1986, pp. 1099–1105.

[7] Gehman, H. L., Gehman, chair, H. L., Turcotte, S. A., Barry, J., Hess, K. W., Hallock, J. N., Wallace, S. B., Deal, D., Scott Hubbard, G., Tetrault, R. E., Widnall, S. E., Osheroff, D. D., Ride, S. K. and Logsdon, J. M., *Columbia Accident Investigation Board Report,*

Volume 1, Government Printing Office, Washington, DC, Aug. 2003; Vaughan, D., *The Challenger Launch Decision: Risky Technology, Culture and Deviance at NASA*, University of Chicago Press, Chicago, IL, 1996; Brewer, G. D., "Perfect Places: NASA as an Idealized Institution," *Space Policy Reconsidered*, edited by R. Byerly Jr., Westview, Boulder, CO, 1989, pp. 155–169; McCurdy, H. E., *Inside NASA: High Technology and Organizational Change in the U.S. Space Program*, The Johns Hopkins University Press, Baltimore, MD, 1993; Launius, R. D., "After Columbia: The Space Shuttle Program and the Crisis in Space Access," *Astropolitics: The International Journal of Space Power and Policy,* Vol. 2, no. 3, July–Sept. 2004, pp. 277–322; Logsdon, J. M., "'A Failure of National Leadership': Why No Replacement for the Space Shuttle?" *Critical Issues in the History of Spaceflight*, edited by S. J. Dick and R. D. Launius, NASA SP-2006-4702, Washington, DC, 2006, Chap. 9; Hall, J. L., "*Columbia* and *Challenger*: Organizational Failure at NASA," *Space Policy,* Vol. 19, No. 4, Nov. 2003, pp. 239–247; Cabbage, M., and Harwood, W., *Comm Check...: The Final Flight of Shuttle Columbia*, Free Press, New York, 2004; Klerkx, G., *Lost in Space: The Fall of NASA and the Dream of a New Space*, Pantheon Books, New York, 2004; McConnell, M., *Challenger: A Major Malfunction*, Doubleday and Co., Garden City, NY, 1987; Trento, J. T., *Prescription for Disaster: From the Glory of Apollo to the Betrayal of the Shuttle*, reporting and editing by S. B. Trento, Crown Publishers, New York, 1987; Easterbrook, G., "The Case Against NASA," *New Republic*, 8 July 1991, pp. 18–24.

[8] Watson, T., "NASA Administrator says Space Shuttle was a Mistake," *USA Today*, 28 Sept. 2005, p. 1A.

[9] Logsdon, J. M., "The Space Shuttle Program: A Policy Failure," *Science*, Vol. 232, no. 7454, 30 May 1986, pp. 1099–1105; Launius, R. D., *NASA: A History of the U.S. Civil Space Program*, Krieger Publishing Company, Malabar, FL, 1994, pp. 114–115.

[10] Testimony of Robert Zubrin to the Senate Commerce Committee, 29 Oct. 2003, p. 2, http://history.nasa.gov/columbia/Troxell/Columbia%20Web%20Site/Documents/Congress/Senate/OCTOBE~1/Zubrin's%20Testimony.pdf [retrieved 28 May 2013].

[11] "The Post-Apollo Space Program: A Report for the Space Task Group," NASA, Washington, DC, Sept. 1969.

[12] Ferguson, N. (ed.), *Virtual History: Alternatives and Counterfactuals*, Basic Books, New York, 1999, pp. 2–32.

[13] Van Allen, J. A., "Is Human Spaceflight Obsolete?" *Issues in Science and Technology,* Vol. 20, Summer 2004, http://www.issues.org/20.4/p_van_allen.html [retrieved 3 Aug. 2004].

[14] Goldin, D. S., speech at California Institute of Technology, 4 Dec. 1992, NASA Historical Reference Collection.

[15] Truax, R. C., "Shuttles—What Price Elegance?" *Astronautics & Aeronautics*, Vol. 8, No. 6, 8 June 1970, pp. 22–23.

[16] Heppenheimer, T. A., *Development of the Space Shuttle, 1972–1981 (History of the Space Shuttle, Vol. 2)*, Smithsonian Institution Press, Washington, DC, 2002.

[17] Tatarewicz, J. N., "The Hubble Space Telescope Servicing Mission," *From Engineering Science to Big Science: The NACA and NASA Collier Trophy Research Project Winners*, Washington, DC, edited by P. Mack, NASA SP-4219, 1998, pp. 365–396.

[18] White House Press Secretary, "The White House, Statement by the President," 5 Jan. 1972, Richard M. Nixon Presidential Files, NASA Historical Reference Collection.

[19] Laue, J. L., "Tethered Satellite System Project Overview," *Appl. of Tethers in Space,* Vol. 1, no. 16, pp. 11–37, Mar 1, 1985, pp. 11–37; Vallerani, E., Bevilacqua, F., and Giani, F., *Europe/United States Space Activities*, Univelt, Inc., San Diego, CA, 1985, pp. 161–173; Bilén, S. G., Gilchrist, B. E., Bonifazi, C., and Malchioni, E., "Transient Response of an Electrodynamic Tether System in the Ionosphere: TSS-1 First Results," *Radio Science,*

Vol. 30, No. 5, Pages 1519–1535, Sept.–Oct. 1995; Fuhrhop, K. R., and Gilchrist, B. E., "Electrodynamic Tether System Analysis Comparing Various Mission Scenarios," 41st AIAA/ASME/SAE/ASEE Joint Propulsion Conference & Exhibit, AIAA-2005-4435, 10–13 July 2005.

[20] Werner, M., "Shuttle Radar Topography Mission (SRTM)—Mission Overview," *Proceedings of the 3rd European Conference on Synthetic Aperture Radar*, Munich, Germany, 23–25 May 2000, pp. 209–212.

[21] Siddiqi, A. A., *Deep Space Chronicle: Robotic Exploration Missions to the Planets*, NASA SP-2002-4524, Washington, DC, 2002; Harland, D. M., *The Story of the Space Shuttle*, Springer-Praxis, Chichester, United Kingdom, 2004.

[22] Sebesta, L., *Spacelab in Context*, ESA Publications Division, ESA HSR-21, Noordwijk, The Netherlands, October 1997; *Science in Orbit: The Shuttle and Spacelab Experience, 1981–1986*, NASA, Washington, DC, 1988; Shapland, D., and Rycroft, M., *Spacelab: Research in Earth Orbit*, Cambridge University Press, Cambridge, UK, 1984.

[23] *Life into Science: Space Life Sciences Experiments, Ames Research Center, Kennedy Space Center, 1991–1998*, edited by K. Souza, G. Etheridge, and P. X. Callahan, NASA SP-2000-534, Washington, DC, 2000; *Get Away Special Experimenter's Symposium*, edited by C. R. Prouty, NASA, Washington, DC, 1984; *Get Away Special...: The First Ten Years*, Goddard Space Flight Center, Greenbelt, MD, 1989; *Biological* and *Medical Experiments on the Space Shuttle, 1981–1985*, edited by T. W. Halstead and P. A. Dufour, NASA, Washington, DC, 1986.

[24] Smith, R. W., *The Space Telescope: A Study of NASA, Science, Technology, and Politics*, revised ed., Cambridge University Press, New York, 1989, 1994.

[25] Miller, P., "The Responsibility of a Mind in a Civilization of Machines," *The American Scholar*, Vol. 31, No. 1, Winter 1961–1962, pp. 51–69.

[26] Hughes, T. P. *American Genesis: A Century of Invention and Technological Enthusiasm, 1870–1970*, Viking, New York, 1989, p. 2.

[27] Blankley, T., "Space Shuttle America," 5 Aug. 2005, http://townhall.com/columnists/tonyblankley/2005/08/03/space_shuttle_america/page/full/ [retrieved 28 May 2013].

[28] Easterbrook, G., "The Space Shuttle Must Be Stopped," *Time*, 2 Feb. 2003, http://www.time.com/time/magazine/article/0,9171,1004201,00.html [retrieved 28 May 2013].

Chapter 1

Defining the Shuttle: The Spaceplane Tradition

Roger D. Launius

Introduction

What is it about the concept of a space plane that was so inviting to American engineers throughout the 20th century? A winged, reusable space vehicle for human flight beyond Earth dominated thinking about the task before the Space Age, but to compete with the Soviet Union in human space spectaculars beginning in the late 1950s, the United States opted for ballistic capsules that were easier to build and fly. No sooner had that competition ebbed, however, than NASA returned to the pursuit of a spaceplane, resulting in building the winged, reusable Space Shuttle and flying it for 30 years.

This essay reviews the more than 50-year history of the quest for a spaceplane that eventually found fruition with the operations of the Space Shuttle and that continued on as a large number of potential replacements also offered spaceplane dreams into the future. Three major lessons concerning this quest come to the fore and are highlighted in this chapter: (1) the power of an idea to push engineering reality, (2) the delta between technological knowledge and potential, and (3) the social construction of technology—how critical design elements are fostered not only by hardheaded engineering analysis but also by other conventions and priorities. An analogy to this story is employed in presenting a discussion of how the spaceplane ideal continues to overshadow other concepts in the 21st century, and the continuing saga of how best to place humans into space. This quest for the spaceplane remains powerful to the present, even as the United States moves forward with the replacement of the Space Shuttle and the capsule concepts that offer the easiest technology for the future.

The Power of an Idea

For nearly 40 years before the beginning of human spaceflight in 1961, leading U.S. experimenters and advocates for the human exploration and development of space envisioned a future in which astronauts flew into orbit aboard reusable, efficient, winged vehicles and then came back to Earth and landed on runways. Their model for that effort was the emerging airline industry. This belief dominated thinking before the Space Age, but in the rush to place humans into space during the Cold War of the late 1950s, the United States abandoned these dominant ideas about reusable spaceplanes in favor of the expedience of ballistic capsules, a technology already well underway to ensure the reentry of nuclear missile warheads. During the 1960s Projects Mercury, Gemini, and Apollo employed these ballistic concepts even as the dream of a spaceplane continued unabated and as studies persisted [1].

No sooner had the Moon landing in 1969 ensured victory in the space race than NASA leaders returned to the idea of the spaceplane for the future. NASA built the Space Shuttle to supplant those earlier human spaceflight capsules, an important step toward a spaceplane but not truly the vehicle previously envisioned. A succession of projects designed to replace the Space Shuttle with more advanced spaceplanes dominated much of the research and development for human spaceflight vehicles undertaken by NASA throughout the 1980s and 1990s. Only in the 21st century has NASA shifted its focus away from the spaceplane concept to return to a ballistic flight approach, although other possibilities for new spaceplanes remain in the post-2012 period. There are, of course, always multiple ways of constructing space vehicles, and any given approach is the result of a creative process of give and take that reflects a range of technical possibilities. The quest for an orbital spaceplane represents a unique story of the social construction of technology, one in which spaceflight professionals pursued a technological path dictated not so much by well-defined engineering considerations but by other, nontechnical concerns [2].

Winged Flight into Space and the Social Construction of Technology

In recent years, the "social construction of technology" has illuminated understanding and helped to explain seemingly difficult issues in the development of spaceflight. The social construction of technology offers tantalizing prospects for studying the orbital spaceplane. It suggests that the nature of technological choice is sometimes made not for clear-cut technological reasons but for broader nontechnical reasons. What perhaps should be suggested is that a complex web or system of ties between various people, institutions, and interests bring forward any space system, and that each fundamentally affects the direction taken. The process whereby those decisions are made and implemented offers an object lesson for current engineers and policy-makers involved in making difficult decisions [3].

Social constructionism is present throughout the story of this quest for a winged, reusable human orbital capability. In three major ways, the spaceplane ideal has dominated the thinking of spaceflight advocates until the present. First, it represented a strategic approach to flying into, through, and back from space in the era before the advent of the Space Age. Second, it remained a dream of spaceflight engineers and advocates even as the expediency of ballistic capsules became the norm during the space race, and it led to a return to the effort when the Moon landing program ended. Third, it governed the direction of replacement efforts for the Space Shuttle throughout the 1980s and 1990s, with a shift back to capsules coming only in the aftermath of losing the Space Shuttle *Columbia* in 2003. Although the follow-on human space vehicle for the United States might be a capsule, there is renewed commitment to new designs that extend the spaceplane approach in the aftermath of the Space Shuttle's retirement. Without question, the spaceplane concept has demonstrated remarkable resilience as the dominant approach pursued by the United States in the overarching trajectory of human spaceflight throughout the 20th century and into the 21st.

In terms of spaceplanes, there has been an infatuation with winged rockets since Robert H. Goddard stated in his classic study, *A Method of Reaching Extreme Altitudes* (published in early January 1920 but with a 1919 copyright), that rockets—and only rockets—could be used to reach beyond Earth [4]. This flew in the face of earlier conceptions of space travel, which relied on cannon shots, flywheels, or exotic methods of flight such as lodestones or antigravity. Goddard's ideas in favor of rockets were ridiculed on the editorial page of the *New York Times* but also set off a media debate worldwide over the role of rockets in space exploration. He fundamentally reshaped both technical conceptions of how spaceflight might be accomplished and fictional visions of it. By the middle part of the 1920s, rocketry had become the accepted mode of reaching space [5].

Linking rocketry to visions of human flight in space to visions beginning in the 1920s—added to the reality of atmospheric flight with airplanes—suggested the necessity of winged vehicles that bore a strikingly close visual relationship to the emerging technological sophistication of aircraft. This became a powerful icon of spaceflight, although in two of the three different regimes in which the vehicle must perform, the features of a spaceplane are unnecessary or even a detriment. During orbital operations, the spaceplane design is superfluous to the vehicle's performance, and during launch, a spaceplane's wings are an unnecessary drag on the vehicle. Only during reentry and landing are the unique features of the spaceplane significant components of the vehicle's performance. In such a situation, social construction represents a useful way of analyzing the decision-making process that led to a winged, reusable vehicle as the chosen path for American spaceflight by humans [6].

Spaceplanes of the Imagination

Since the point in the 1920s when it became conventional wisdom that rockets were the only effective means of reaching Earth orbit, concepts for winged reusable spaceplanes have dominated the thinking of American engineers focused on space exploration. Spaceplanes, for example, were principal technologies in the science fiction of Buck Rogers. His Patrol Ship was able to take off horizontally like an airplane and accelerate to a high enough speed to go beyond the pull of a planet's gravity into orbit. Buck Rogers first appeared in the August 1928 issue of *Amazing Stories* but became a successful, long-running syndicated comic strip that premiered in newspapers on January 7, 1929. In addition, the Buck Rogers radio program commenced in 1932, airing four times each week through World War II. This sparked a series of films, beginning with a 10-minute short that premiered at the 1933 World's Fair in Chicago and a 12-part serial released by Universal Pictures in 1939 starring Buster Crabbe [7].

In every case, Buck Rogers's winged reusable patrol ship served as the means of travel for the hero and his sidekicks. Although the patrol ship was never a true "character" in these Buck Rogers stories in the same way that the starship Enterprise was in *Star Trek*, fans accepted at face value that it should be a winged vehicle that could undertake airplane-like operations. This popular culture icon paralleled the advancement of space technology in the 20th century and introduced Americans to outer space as a familiar environment for swashbuckling adventure. The ideas expressed in the series informed popular conceptions of what a human spaceflight vehicle might look like even as advances in technology informed the artwork used in science fiction. As former NASA Deputy Administrator Fred Gregory recalled in 2001 about the Space Shuttle program, "We had such great ideas about where we were going to go and what we were going to do; I was caught up in it. But, of course, I grew up with Buck Rogers" [8].

Soon thereafter, serious scientific studies on the feasibility of spaceplanes also began, and these captured the imagination of engineers and enthusiasts alike. The theoretical rocketplane studies of the 1930s by Austrian aerospace designer Eugen Sänger, for instance, proved influential in laying the foundations for the modern concept of a spaceplane. Sänger's basic concept—the Silbervogel (Silverbird)—was a cross between a powered booster rocket and an aerodynamic glider. Inspired by Hermann Oberth's 1923 book, *By Rocket into Planetary Space*, Sänger changed his studies at the Viennese Polytechnic Institute from civil engineering to aeronautics. Beginning in 1929, Sänger conceptualized a reusable, rocket-powered spaceplane with straight wings.

Sänger, in collaboration with mathematician Irene Bredt, whom he later married, pursued a spaceplane for the next 30 years [9]. By 1938, the Sänger/ Bredt collaboration had yielded a design—propelled by a liquid rocket engine—that would boost to lift-off velocity via a rocket-propelled sled, eventually reaching Mach 24 [10].

Americans were keenly interested in Sänger's ideas. In 1936, at Caltech, William Bollay gave a lecture on what was known about these ideas beyond Nazi Germany, and this lecture seemingly kick started rocket research efforts at the university. Because of a newspaper report on Bollay's lecture, several spaceflight enthusiasts gathered at Caltech around another graduate student, Frank Malina, to form the research team that eventually established the Jet Propulsion Laboratory in World War II [11]. After World War II, the ideas of Sänger and Bredt made their way from Germany to the U.S. Navy's Bureau of Aeronautics (BuAer) and succeeded in changing the perspective of many about the possibility of spaceplanes. Plans for a rocket-powered hypersonic aircraft could be built with only minor advances in technology, they convincingly argued. At the same time, many of Sänger's technical studies were translated and thoroughly studied by American aerospace engineers in the late 1940s.

In August 1961, Sänger, working for Junkers in the Federal Republic of Germany, wrote an influential report that offered his spaceplane ideas in the most detail ever. He commented [12]:

> It is my firm opinion that for civil use of the aerospace transporter the catapult start by means of steam rockets and the main propulsion by liquid-hydrogen liquid-oxygen high pressure rocket engines is the best initial approach. Later on the main stage may be powered by thermal nuclear fission rocket engines. The total launching weight should initially be chosen between 100 and 1,000 tons, and the use of a single stage may be justified if catapults are applied for launching.

Sänger's powerful advocacy for a single-stage-to-orbit spaceplane that would be catapulted through the atmosphere toward orbital velocity clearly enamored many American aerospace engineers.

Sänger's most important contribution, given that none of his designs for a spaceplane were ever built, was to focus attention on the potential of winged hypersonic flight, ultimately leading to such vehicles as the X-15 and the Space Shuttle. His ideas remained in vogue for nearly 20 years, anticipating the Space Shuttle by some 30 years, even though they were essentially "paper airplanes" that never got beyond the level of engineering studies [13].

By the early 1950s, winged-reusable spaceplane concepts had matured sufficiently that Darrell C. Romick's Goodyear Aircraft Corporation design for the three-stage Manned Earth-Satellite Terminal Evolving from Earth-To-Orbit Ferry Rockets (METEOR) met with approval by the American Rocket Society when unveiled in 1954. This spaceplane was to stand 142 feet tall and carry 35 tons while weighing roughly 18 million pounds and producing 32 to 38 million pounds of thrust at lift-off. Each stage of the vehicle was to have been piloted and had a delta wing design with retractable landing gear. In this way, each individual stage could be flown as a glider to its landing site on Earth. The METEOR's purpose was to bring cargo and people back and forth between Earth and a space station.

Romick got into spaceplane design when he began working on Goodyear's experimental MX-778, a 100-mile-range missile, and he realized that ideas such as those pursued by Sänger had potential. "The one thing that was missing in these studies," he later recalled, was a "practical launch vehicle that you could run like an airliner—the transport of cargo on a regular basis" [14]. Scrapping the catapult idea, Romick came up with another answer: a reusable booster that in its most elaborate form involved a three-stage launch vehicle with oversized delta fins. The multiple stages allied the METEOR concept with ballistic missile ideas being pursued elsewhere. Its major difference was the spaceplane's reusability because each stage boasted a crew to fly that segment back to a landing on a runway. Romick aggressively hawked his METEOR concept, even appearing on television and radio. METEOR was never built, but Romick was convinced for the rest of his life that it was simply "too ambitious" for its time but that it still represented a reasonable approach to reaching into space [15].

Throughout the 1950s, until the actual beginning of the Space Age in 1957, the spaceplane concept remained the dominant approach offered by those who imagined an expansive human future in space. For these advocates, it seemed more credible than ever before. The popular weekly news magazine *Collier's* featured a reusable spaceplane on the cover of its first special issue devoted to the prospects of human spaceflight, on March 22, 1952, thereby promoting the idea to a broad audience [16]. The charismatic German émigré Wernher von Braun, technical director of the V-2 ballistic missile effort in World War II, led the *Collier's* issue with an impressionistic article. It described human possibilities in space made possible by the development of a winged reusable spacecraft that could travel to and from Earth orbit and that ultimately could reach Mars, again with spaceplanes that could land on Mars. As von Braun wrote, "Imagine the size of this huge three-stage rocket ship: it stands 265 feet tall, approximately the height of a 24-story building. Its base measures 64 feet in diameter. And the overall weight of this monster rocket ship is 14,000,000 pounds, or 7,000 tons—about the same weight of a light cruiser" [17].

Following close on the heels of the *Collier's* series, Walt Disney Productions contacted von Braun seeking assistance in the production of spaceflight shows for Disney's weekly television series (Fig. 1.1). Two of these, "Man in Space" and "Man and the Moon," premiered on Disney's weekly television show in 1955 with an estimated audience of 42 million. They depicted a spaceplane

Fig. 1.1 The von Braun–Disney Spaceplane: Wernher von Braun (right) and Walt Disney (left), 1954. They met on the set of a Walt Disney television program that helped to build public support for an ambitious space program. The spaceplane depicted here was one of the ideas popularized on television. (NASA)

supporting a wheel-like space station as a launching point for a mission to the Moon. Von Braun appeared on camera to explain his concepts for human spaceflight, whereas Disney's characteristic animation illustrated the basic principles and ideas with wit and humor [18]. Media observers noted the public's favorable response to the Disney shows and concluded that "the thinking of the best scientific minds working on space projects today" went into them, "making the picture[s] more fact than fantasy" [19].

A Dose of Reality

Although the public discussion of technologies for human spaceflight in the United States emphasized spaceplanes, the Cold War pressures of the 1950s impinged on those ideas. The DoD abandoned spaceplanes in favor of ballistic capsules that could be placed atop launchers being developed to deliver nuclear warheads to the Soviet Union, and NASA pursued a similar approach after its creation in 1958. The decision to pursue a capsule instead of the spaceplane ideal came only after aerodynamicists realized that the need to orbit a human vehicle before the Soviet Union outweighed the creation of the more elegant spaceplane solution. Although research remained a critical component of the spaceplane concept, National Advisory Committee for Aeronautics (NACA) engineer John V. Becker recalled in 1968, "The exciting potentialities of these rocket-boosted aircraft could not be realized without major advances in technology in all areas of aircraft design. In particular, the unprecedented problems of aerodynamic heating and high-temperature structures appeared to be so formidable that they were viewed as 'barriers' to hypersonic flight" [20].

As early as 1954, engineers at the DoD and the NACA began to wonder if, in the interest of expediency, the spaceplane would have to be tabled in favor of the capsule to assure the ability to achieve human orbital flight within the next decade. By early 1957, various proposals for human space missions using ballistic capsules similar to warhead designs circulated among various government organizations. The most well developed was the U.S. Air Force's Man-in-Space Soonest (MISS) concept [21]. This program called for a four-phase capsule orbital process that would first use instruments, then primates, and then a pilot, with the final objective being to land humans on the Moon. MISS was initially discussed before the launch of Sputnik 1 in October 1957, and afterwards, the Air Force invited Edward Teller and several other leading members of the scientific/technological elite to study the issue of human spaceflight and to make recommendations for the future. Teller's group concluded that the Air Force could place a human in orbit within two years using the proposed capsule and urged that the department pursue this effort. Teller understood, however, that there was essentially no military reason for undertaking this mission. He chose not to tie his recommendation to any specific rationale, falling back on a basic belief that the first nation to do so would accrue national prestige and advance science and technology in a general manner [22].

Soon after the new year, Lieutenant General Donald L. Putt, the USAF Deputy Chief of Staff for Development, informed NACA Director Hugh L. Dryden of the Air Force's intention to pursue aggressively "a research vehicle program having as its objective the earliest possible manned orbital flight which will contribute substantially and essentially to follow-on scientific and military space systems" [23]. This eventually led to Project Mercury, an effort to place an American in orbit at "the earliest possible" time, contributing to "manned orbital flight which will contribute substantially and essentially to follow-on scientific and military space systems." Implicit in this effort was the abandonment of the spaceplane [24].

The remainder of the 1960s followed the accelerated timetable of the Cold War space race, including the Apollo Moon landings, driving NASA to exploit ballistic missile technology despite desires to emphasize the spaceplane ideal. In the end, expediency froze out more elegant solutions to the prospect of flying humans to and from space. But even as Apollo was being pursued, NASA still undertook research on the technology. As the Moon landing objective was achieved, NASA leaders moved back toward a winged, reusable spaceplane as the post-Apollo technology for reaching orbit. As soon as

Apollo was completed, NASA chose to retire that ballistic technology, despite its genuine serviceability, in favor of the Space Shuttle [25].

This begs the following question: Had there not been the crisis of the Cold War and the Apollo commitment that flowed from it, might NASA have pursued reusable spaceplane concepts as the launcher of choice from the beginning? The answer to this counterfactual question, of course, can never be known with certainty, but with all of the *sturm und drang* associated with spaceplanes even while the capsule era of the 1960s reigned, it seems logical that NASA would have done so.

Spaceplanes: More than Side Trips in the 1960s

Aggressive spaceplane research continued throughout the Mercury-Gemini-Apollo programs and yielded tangible results even during the space race between the United States and the Soviet Union. For example, a concerted group of engineers at the NACA continued to argue that a spaceplane would be far superior to a nonlifting capsule. In fact, at the NACA Conference on High-Speed Aerodynamics in March 1958, John Becker presented a concept for a piloted 3060-pound winged orbital satellite. According to Becker, this concept created more industry reaction—almost all of it favorable—than any other study he had previously written. What ruled out acceptance of his proposal was that the 1000 pounds of extra weight (compared to the capsule design) was beyond the capability of any launch vehicle then available [26].

In addition, between 1959 and 1968, the most celebrated experimental vehicle ever flown—the X-15—operated at the edge of space. Although the X-15 was not an orbital spaceplane, the concepts for it emphasized spaceplane ideas [27]. In all, it made 199 flights divided among 3 aircraft, established many records, and yielded over 765 research reports. It was, according to one engineer, a spaceplane program that "returned benchmark hypersonic data for aircraft performance, stability and control, materials, shock interaction, hypersonic turbulent boundary layer, skin friction, reaction control jets, aerodynamic heating, and heat transfer" [28]. At a fundamental level, this was spaceplane research that was critical to all future efforts in this technology.

Likewise, during the same period the U.S. Air Force pursued the X-20 Dyna-Soar, a military spaceplane to be launched atop a newly developed launcher (Fig. 1.2). It is, without question, one of the most memorable vehicles never flown. Dyna-Soar was officially designated System 620A on November 9, 1959. After several reviews and much political infighting, the Air Force, on April 27, 1960, awarded a contract to a Boeing-Vought team. With 10 Dyna-Soar gliders contracted for, the procurement schedule called for 2 vehicles to be delivered during 1965, 4 in 1966, and 2 during 1967. An additional two spaceplanes were to be used for static tests beginning in 1965 [29]. At the

Fig. 1.2 Dyna-Soar: An artist's conception of the Dyna-Soar X-20 military spaceplane pursued by the United States in the early 1960s. This program underwent development between 1958 and 1963. (USAF)

same time, the Glenn L. Martin Company was selected to develop a human-rated version of the Titan launch vehicle.

The Air Force believed that the X-20 would provide long-range bombardment and reconnaissance capability by flying at the edge of space and skipping off the Earth's atmosphere to reach targets anywhere in the world. However, several problems were immediately apparent and never were solvable with the technology available. Because of funding issues, political infighting, and confusion over the mission, on December 10, 1963—after spending $410 million ($3.8 billion in 2012 inflation-adjusted dollars) on its development, with another $373 million needed to attain an orbital test flight—Secretary of Defense Robert S. McNamara cancelled Dyna-Soar over the objections of many of his senior advisors. This ended the first serious attempt to build a spaceplane in the 1960s [30]. Had Dyna-Soar flown, it could have provided valuable information on entry flight control and heating, something that was seriously lacking during the development of the Space Shuttle 10 years later [31].

The Delta between the Ideal and Technological Reality

The dream of the orbital spaceplane did not die with Dyna-Soar. As the United States completed its major capsule programs in the 1960s—Mercury,

Gemini, and Apollo—most individuals involved in space advocacy still envisioned a future in which humans would venture into space aboard winged, reusable vehicles. Advocates asserted that the most expeditious, inexpensive, and reliable method for humans to reach Earth orbit was to use a reusable spaceplane. This became the raison d'être for NASA as the Apollo program ended [32].

In general, NASA's engineers abandoned their proven ballistic capsule technology in favor of a spaceplane as they worked toward developing the Space Shuttle, in large measure because of the agency's organizational culture. In effect, the space agency reverted to its heritage as the NACA. Just as all the early astronauts (at least until the Space Shuttle era) had been trained as military jet pilots, most astronautical engineers at that time were trained initially as aeronautical engineers and had an affinity for airplanes and atmospheric flight. For many years, few involved in NASA's spaceflight efforts appreciated the sharp delineation between air and space and largely viewed spaceflight as an extension of flight in the atmosphere, despite striking differences [33].

Equally significant, NASA's organizational identity as a cutting-edge aerospace research and development institution fostered an affinity for complicated technology to accomplish the task. The Moon landing program, never attempted anywhere before, invited the pursuit of the most complex technological solutions that one could imagine. NASA engineers boasted about the complexity of the effort, commenting that the Saturn V/Apollo launch combination were the most sophisticated technologies ever built. Such boasts made other engineers cringe; complexity is never the objective in a technical system if robustness, reliability, and cost are priorities. Robert Truax, a former U.S. Navy captain with much experience in rocketry, has argued concisely that in pursuing the spaceplane, NASA was too concerned with finding an "elegant" solution. In his view, splashing down in the ocean was certainly inelegant, but developing a spaceplane so that astronauts could fly vehicles to runway landings was impractical because of its technological overreach [34]. Truax, revisiting the criticism of a spaceplane in 2002, emphasized that the Space Shuttle concept was "illogical," even as he "lauded its makers for their execution—for their ability to make such a ludicrous concept work at all" [35]. This commitment to "elegance," despite Truax's criticism of spaceplanes, also speaks to the issue of technological style among various organizational cultures [36].

NASA officials in the early 1970s did not see the spaceplane as a technological overreach—in fact, quite the opposite. With the Shuttle program, NASA announced that it intended to lower the cost of spaceflight so that it could conduct an aggressive space exploration effort. To do so, NASA officials declared, "Efficient transportation to and from the earth is required." This could be best provided, they said, with "low-cost access by reusable chemical and nuclear rocket transportation systems" [37]. Some NASA officials even compared the older method of using expendable launch vehicles to operating a railroad and throwing away the locomotive after every trip. A spaceplane, they claimed, would provide the United States with low-cost, routine access to space by creating the same type of economies present in air transportation. NASA moved forward with planning for what became the Space Shuttle, emphasizing that this spaceplane concept promised to offer less expensive and more flexible space access [38]. George M. Low, NASA's deputy administrator at the time, said in a memorandum to other NASA leadership on January 27, 1970, "I think there is really only one objective for the Space Shuttle program, and that is 'to provide a low-cost, economical space transportation system.' To meet this objective, one has to concentrate both on low development costs and on low operational costs" [39].

The task proved more challenging than they believed. Truax had been right; as elegant as the vehicle that resulted might have been, it was essentially a technological overreach that failed to achieve what had been promised for it. Over a career of more than 30 years of orbital operations, the Space Shuttle program has flown 135 missions and has left an important but divided legacy. In sum, it is a striking example of the social construction of technological choice, leading NASA to pursue the spaceplane path in human spaceflight based more on desire than the simplest solutions to technical challenges.

Despite the mounting difficulties in the experience of flying the Space Shuttle as the program proceeded, NASA remained wedded to the spaceplane as the preferred technology for future human space operations. The case study by Walter G. Vincenti appears to apply to NASA's continuing commitment to the spaceplane. As NASA did with the spaceplane, Vincenti documents how Northrop failed to adopt retractable landing gear on aircraft in the 1920s and 1930, despite evidence that this offered a good solution to the problem at hand [40].

An Analogy: Persisting in Belief

For more than two decades after the Space Shuttle's first flight in 1981, NASA relentlessly pursued shuttle replacement programs that took as their starting point a spaceplane approach. Overall, these became large, ambitious programs that were overhyped, and each failed because of insurmountable technical challenges in the face of budgetary pressures. These programs typically have blurred the line, which should be bright, between revolutionary, high-risk, high-payoff R&D efforts and low-risk, marginal-payoff evolutionary efforts to improve operational systems. Three examples demonstrate the continued quest for a spaceplane despite both its technical difficulties and high cost.

The first of these was the National Aero-Space Plane (NASP), originated in the early 1980s as a full-fledged spaceplane that could replace the Shuttle.

Fueled by the realization that the Space Shuttle could never live up to its early expectations, aerospace leaders argued for the development of this single-stage-to-orbit (SSTO) hypersonic spaceplane that could take off and land on runways. With the beginning of Ronald Reagan's administration and its associated military buildup, the Defense Advanced Research Projects Agency (DARPA) began work on a hypersonic vehicle powered by a hybrid integrated engine of scramjets and rockets as a "black" program code named "Copper Canyon."

After several years of classified work, the Reagan administration unveiled NASP in 1986. It was also designated the X-30, as "a new Orient Express that could, by the end of the next decade, take off from Dulles Airport and accelerate up to 25 times the speed of sound, attaining low-Earth orbit or flying to Tokyo within 2 hours" [41]. With this announcement of NASP, the hypersonic aerospace plane had returned [42]. The NASP program initially proposed to build two research craft, at least one of which could achieve orbit by flying in a single stage through the atmosphere at speeds up to Mach 25. It would use a multicycle engine that shifted from jet to ramjet and to scramjet as the vehicle ascended through the atmosphere [43]. NASA loved the idea and partnered with DARPA on the X-30 as a potential successor to the Space Shuttle. The program, however, never achieved anything approaching flight status.

NASP finally fell victim to budget cuts in 1992, in part because the Cold War ended. The program also ended because of its technological overstretch. For instance, NASA futurist and longstanding advocate of SSTO Ivan Bekey called NASP "the biggest swindle ever to be foisted on the country" because it "was full of dubious aerodynamic claims and engine performance claims and thermal claims" [44]. By the time NASP was cancelled in 1992, the government had admitted to making a $1.7 billion investment in NASP, but parts of the R&D were highly classified and there probably were some additional expenditures [45].

With the demise of NASP, NASA undertook two additional Space Shuttle replacement efforts. One, the X-34, also known as the Reusable Small Booster Program, was a spaceplane intended to demonstrate technologies and operations useful to smaller reusable vehicles launched from an aircraft. Among these were autonomous ascent, reentry, and landing; composite structures; reusable liquid oxygen tanks; rapid vehicle turnaround; and thermal protection systems [46].

The other, the X-33 spaceplane, known also as the Advanced Technology Demonstrator Program, was far more challenging both technologically and politically. Among the operations and technologies it would demonstrate were reusable composite cryogenic tanks, graphite composite primary structures, metallic thermal protection materials, reusable propulsion systems, autonomous flight control, and electronics and avionics [47]. NASA began the hypersonic X-33 program in 1995, and the agency's leadership expressed high hopes that this small suborbital vehicle would demonstrate the technologies required for an operational SSTO launcher. The X-33 project, undertaken in partnership with Lockheed Martin, had an ambitious timetable to fly by 2001. Once the technology reached some maturity, Lockheed vowed to scale up the X-33 into a human-rated vehicle, VentureStar™, which could serve as the Space Shuttle's replacement [48].

Both the X-33 and X-34 programs became mired down in seemingly inscrutable technological problems and bureaucratic challenges; NASA officials lost faith in them and terminated both efforts in 2001. Thereafter, NASA officials expressed a deeper understanding that the technical hurdles for a spaceplane were more daunting than had been anticipated. Any spaceplane, and both the X-33 and X-34 perpetuated this pattern, would require breakthroughs in a number of technologies. Nevertheless, NASA engineers remained enamored with the ideal of the spaceplane [49].

Few at the space agency questioned the commitment to the spaceplane concept, and officially, NASA remained enthralled with the elegance of flying on wings to and from space. Although NASA spent billions of dollars and more than 20 years pursuing reusable spaceplane technology, the emphasis on this approach guaranteed tardiness in fielding a shuttle replacement because of the strikingly difficult technological challenges. Critics—and there were some—referred to these spaceplanes as being built from "unobtainium" and thought the United States should instead pursue more conventional space access technologies. Whether the quest for a spaceplane was appropriate or not for human space access in the long run, they insisted that thus far, this quest had proven a detriment to the cause.

The spaceplane ideal remained the norm into the 21st century. Only after the loss of the *Columbia* Space Shuttle in 2003 did this begin to change, albeit slowly and certainly incompletely. In the fall of 2005, the NASA administrator of some six months, but a longtime member of the space community, famously called the Space Shuttle a "mistake." Michael D. Griffin commented that NASA had pursued the wrong path with these spaceplanes and that it persisted down this path long after its flaws had been discovered. "It is now commonly accepted that was not the right path," Griffin told *USA Today* in an interview that appeared as a front-page story on September 28, 2005. "We are now trying to change the path while doing as little damage as we can" [50]. Griffin's assertion that the Space Shuttle and subsequent spaceplanes had been the "wrong path" set off a firestorm of debate within the spaceflight community [51]. It also triggered not a little soul searching about the importance of the spaceplane in both the history of space exploration and in the larger context of future efforts [52].

On January 14, 2004, President George W. Bush decided to retire the Space Shuttle as a human launcher and to replace it with something else, and he directed NASA to focus on a new Moon/Mars exploration agenda. In the aftermath of the *Columbia* accident, questions about the importance of a spaceplane were swirling within the U.S. human spaceflight community, and this Vision for Space Exploration plan affected these questions. The plan required a new vehicle, one that could be used to support operations in Earth orbit, but also it had to go farther. What NASA came up with was the Constellation program, a presumed reuse of much of the existing Space Shuttle technology. The program was to build a new Ares I crew launch vehicle, consisting of a Space Shuttle solid rocket booster as a first stage and an external tank as the beginning point for a second stage. A human space capsule, Orion, was to sit atop this system. A proposed second rocket, the Ares V cargo launch vehicle, would provide the heavy lift capability necessary to journey back to the Moon or to go beyond. Ares I was intended to carry a crew of up to six astronauts to low-Earth orbit in the Orion spacecraft, with the capability for expanding its use to send four astronauts to the Moon. Ares V was intended to serve as the agency's primary vehicle for delivering large-scale hardware and cargo to support this expansive Vision for Space Exploration. Abandoning a spaceplane concept in favor of returning to capsules was controversial but expedient, and that expediency assured that most spaceflight advocates supported the decision [53].

Such was not true when President Barak Obama's administration ended the Constellation program because of cost and technical challenges. Obama hit the reset button on human space vehicles, but would the follow-on be another capsule, à la Orion, or a spaceplane? Clay Dillow, writing in *Popular Science*, voiced the position of many human spaceflight enthusiasts when he argued for a spaceplane:

> A reusable space plane design was the cheaper and safer way to move crews to and from the ISS; its "blended lifting body" allows it to move from its orbital trajectory as it re-enters and place its point of landing where the pilot wishes. Capsules, of course, come screaming through the atmosphere more or less at their orbital trajectory and rely on parachutes to soften the "splash down" and a recovery crew to locate and pick up the crew.

Dillow expressed well the dilemmas present in the proposed way forward for human space access: "Both capsules and space planes have their advantages, and neither has a spotless safety record. But it will be interesting to see which mode NASA eventually selects for the next generation of ISS missions" [54].

Nothing about a replacement for the Space Shuttle was resolved as the second decade of the 21st century began. John M. Logsdon, a longstanding space policy expert, commented that in 40 years of watching NASA, he had not seen such an unsettled situation [55]. What accounted for the disarray in

NASA's human spaceflight efforts? Was it because of the desire for a new spaceplane or because of the contention over not having a replacement for the shuttle regardless of the type of vehicle? There is some reason to believe that at least some of the unrest revolved around the capsule/spaceplane debate. Many have argued that a spaceplane remains by far the preferred technology, and that capsules are both an inelegant solution to sending astronauts into space and a step backward from the technologies demonstrated in the Space Shuttle program. "Spaceplanes with low-risk technologies and built-in growth potential and aircraft-like operations provide the most return on investment," advocates have long insisted. Would it not, therefore, be better to build on shuttle concepts rather than to resurrect earlier capsule technologies? [56]. The power of the spaceplane ideal has not abated despite its incomplete realization.

MEANINGS

There seems to be every reason to believe that the spaceplane concept will continue with enthusiasm into the future. Eighty-plus years of orientation to that approach have proven difficult to overturn. This is especially the case when it appears that, for the most part, the space community holds the view that a spaceplane is elegant. Even though a return to the capsule concept used in the 1960s with Mercury, Gemini, and Apollo might be an expedient approach to future human spaceflight, few express much enthusiasm for the idea. An engineer working on NASA's lifting body research and development program, Weneth D. Painter of NASA's Flight Research Center, demonstrates a measure of this enthusiasm. In 1965, Painter drew a two-paneled cartoon; in one panel, a Gemini spacecraft was bobbing in the ocean as its crew awaited rescue at sea, and in another panel, a spaceplane was landing on a runway. The caption read, "Don't be rescued from outer space, fly back in style" (Fig. 1.3). It captured the key difference between space capsule splashdowns at sea and spaceplane landings on a runway. It expressed well the elegance of a spaceplane, an approach that was incompletely realized with the Space Shuttle but still something that has remained an objective of human spaceflight ever since those first shuttle flights in the 1980s. Both approaches to spaceflight work; the engineers involved in building human spaceflight vehicles persist in its pursuit to the present. Hence, there remains a continual desire to pursue spaceplane technologies such as the Dream Chaser and the X-37 as the replacement for the Space Shuttle.

At an essential level, this persistence of vision represents a case study in the social construction of technology. In this instance, the longstanding desire for a winged spaceplane that combines elements of atmospheric flight with space technology shaped the ideology of human space access. With all

Fig. 1.3 This 1965 cartoon by Weneth Painter, who then worked at NASA's High Speed Flight Test Center in the Mojave Desert of California, captured the essential difference between space capsule splashdowns at sea and more elegant runway landings by a spaceplane. (NASA)

of the failure that has resulted in the pursuit of this path to human space access, it is impossible to explain the primacy of this technology without understanding this deeply embedded social context. The persistence of support for this technology through more than 80 years of theoretical analysis, R&D, testing, and operations—in the last case with the Space Shuttle—has not happened merely because it was "the best," the most appropriate technological choice. It has much more to do with the belief system of those pursuing these technologies: that flight in the atmosphere is the normative approach to any attempt to reach beyond the atmosphere. Unique aspects of spaceflight have been trumped by commonalities that interlock air and space technologies.

Central to this concept is not that the design of space vehicles is dictated solely by one best way to do something, but that there are a range of options and a trading space for decisions in which pros and cons are always present. As often as not, critical decisions rest not specifically on technical considerations but on other, more aesthetic, rationales. The lack of one best way to accomplish human spaceflight is an important insight arising from a consideration of the technology's social construction. Over time, as the spaceplane concept found succeeding generations of advocates, its popularity seemed to close off discussion of other possibilities. Certainly, this seemed to be the case in the Space Shuttle era, during which NASA pursued a series of spaceplanes as replacement options for the shuttle—to the exclusion of any type of capsule. During this period, it seems that the community viewed the problem of space access as being solved, at least in its broad contours, and felt no need to pursue alternative designs. At the same time, as spaceplane technology advanced to the stage of flight with the Space Shuttle, flying on wings through the atmosphere and landing on a runway like an airplane provided the elegant solution to spaceflight that had motivated engineers from the early part of the 20th century. Why do anything different in the future? NASP, X-33, X-34, X-37, and a host of other projects fell in line until each ran into cost and technological problems and were cancelled. This eventually forced a reintroduction of design flexibility, and recent machinations to move beyond the Space Shuttle to a new vehicle reflect that uncertainty [57].

This new round of debate about the superiority of one type of technology over another has not yet led to consensus, and advocates of both capsules and spaceplanes continue to argue for their preferred choices. This may yet result in the adoption of a spaceplane because there is considerable enthusiasm for it; however, capsules might also become the norm for the next generation of human space access vehicles. What is obvious is that the technology of the spaceplane emerged in no small measure from a unique set of circumstances having to do with the individuals involved in developing space technology—especially their training in and close relationship to aeronautics—and their beliefs that flight in space was an extension of atmospheric flight. The persistence of this belief in the face of all manner of technological challenges attests to the power of these ideas.

Spaceplanes have demonstrated their power as a major element of the space faring vision in the United States and seem destined to continue their primacy into the future. Now that the Shuttle program has ended, it seems that longstanding ideas about efficient, safe, winged, reusable spaceplanes remain the ultimate answer for many people for long-term Earth-to-orbit space access. The recognition that a spaceplane might be the best human space access vehicle represents an inability to think "outside the box." It signifies a case of longstanding path dependency; once someone is persuaded that the spaceplane is the best option—the capsule being a lesser, short-term solution—it is exceptionally difficult to back away after the investment of time, resources, and knowledge that made it a reality with the Space Shuttle. Such thinking effectively closes all other possibilities [58].

REFERENCES

[1] Swenson, Jr., L. S., Greenwood, J. M., and Alexander, C. C., *This New Ocean: A History of Project Mercury*, Washington, DC, NASA SP-4201, 1966, pp. 75–166.

[2] Jenkins, D. R., *Space Shuttle: The History of the National Space Transportation System*, The First 100 Missions, 3rd ed., Cape Canaveral, FL, 2001.

[3] Law, J., "*Technology and Heterogeneous Engineering: The Case of Portuguese Expansion*," pp. 111–134; and MacKenzie, D., "*Missile Accuracy: A Case Study in the Social Processes of Technological Change*," pp. 195–222; in Bijker, W. E., Hughes, T. P., and Pinch, T. J. (eds.), *The Social Construction of Technological Systems: New Directions in the Sociology and History of Technology*, The MIT Press, Cambridge, MA, 1987.

[4] Goddard, R. H., *A Method of Reaching Extreme Altitudes*, Smithsonian Miscellaneous Collections, Vol. 71, No. 2, Washington, DC, 1919, p. 54.

[5] "Topics of the Times," *New York Times,* 18 Jan. 1920, p. 12; Skoog, Å. I. (ed.), "The Silent Revolution: For R. H. Goddard Helps Start the Space Age," *Proceedings of the Thirty-Eighth History Symposia of the International Academy of Astronautics*, Vancouver, BC, Canada, AAS History Series, Univelt, Inc., San Diego, CA, pp. 3–54; "The Misunderstood Professor," *Air & Space Smithsonian*, April/May 2008, pp. 56–59.

[6] Garber, S. J., "*Why Does the Space Shuttle Have Wings? A Historical Case Study in the Social Construction of Space Technology*," M.A. Thesis, History Dept., Virginia Institute of Technology, Virginia Beach, VA, 2002.

[7] Dille, R. C. (ed.), *The Collected Works of Buck Rogers in the 25th Century*, A & W Publishers, New York, 1977; Horrigan, B., "Popular Culture and Visions of the Future in Space, 1901–2001," in Sinclair, B. (ed.), *New Perspectives on Technology and American Culture*, American Philosophical Society, Philadelphia, 1986, pp. 49–67.

[8] Gregory, F., transcript of remarks at *Looking Backward, Looking Forward: Forty Years of U.S. Human Space Flight Symposium*, 8 May 2001, NASA Historical Reference Collection.

[9] Sänger, E., and Bredt, I., "The Silver Bird Story: A Memoir," edited by Hall, R. C. (ed.), *Essays of the History of Rocketry and Astronautics: Proceedings of the Third Through Sixth History Symposia of the International Academy of Astronautics*, Vol. 1, NASA, 1977, pp. 195–228; Sänger, E., Rocket Flight Engineering, from the 1933 German Raketenflugtechnik, translated by NASA as TT F-223, NASA, 1965; Ley, W., *Rockets, Missiles, and Space Travel*, The Viking Press, New York, 1957, pp. 429–434.

[10] Sänger, E., *Recent Results in Rocket Flight Technique*, translated by the NACA from a 1934 German paper, TM-1012, NACA, 1942; Sänger E., and Bredt, I., *A Rocket Drive for Long-Range Bombers*, from the German *Über einen Raketenantrieb für Fernbomber*, translated by the Naval Technical Information Branch, Bureau of Aeronautics as CGD-32, U.S. Navy, 1952.

[11] Hunley, J. D., *Preludes to U.S. Space-Launch Vehicle Technology: Goddard Rockets to Minuteman III*, University Press of Florida, Gainesville, 2008, pp. 94–95.

[12] Sänger, E., "Preliminary Proposals for the Development of a European Space Vehicle," Aug. 1961, quoted in Sänger, Bredt, "The Silver Bird Story," p. 225.

[13] Hallion, R. P. (ed.), *The Hypersonic Revolution: Case Studies in the History of Hypersonic Technology,* Vol. I, From Max Valier to Project PRIME (1924–1967), Air Force History and Museum Program, Bolling AFB, Washington, DC, 1998, pp. xiv–xx.

[14] Collins, M. (ed.), *After Sputnik: 50 Years of the Space Age*, HarperCollins, New York, 2007, p. 28.

[15] Romick, D. C., Knight, R. E., and Van Pelt, J. M., "A Preliminary Study of a Three Stage Satellite Ferry Rocket Vehicle with Piloted Recoverable Stages," *9th American Rocket Society Annual Meeting*, 1954, ARS paper 186–54; Romick, D. C., Knight, R. E., and

Black, S., "Meteor Jr., A Preliminary Design Investigation of a Minimum Sized Ferry Rocket Vehicle of the Meteor Concept," *8th International Astronautical Congress*, Barcelona, Spain, 1957; Stine, G. H., "All Aboard for Outer Space!" *Mechanix Illustrated*, Jan. 1956, pp. 56–61, 198; Liebermann R., and Romick, D. C., "Darrell C. Romick: America's Space Visionary of the 1950s," in Ciancone, M. L. (ed.), *History of Rocketry and Astronautics: Proceedings of the Thirty-Sixth History Symposium of the International Academy of Astronautics (IAA), Houston, Texas, 2002*, American Astronautical Society History Series, Vol. 33, Univelt, Inc., San Diego, CA, 2010, pp. 111–120.

[16] Lieberman, R., "The *Collier's* and Disney Series," in Ordway III, R. I. and Lieberman, R. (eds.), *Blueprint for Space: Science Fiction to Science Fact*, Smithsonian Institution Press, Washington, DC, 1992, pp. 135–144; Miller, R., "Days of Future Past," *Omni*, Oct. 1986, pp. 76–81.

[17] Von Braun, W., "Crossing the Last Frontier," *Collier's*, 22 March 1952, pp. 24–29, 72–73, quote from p. 27; von Braun, W., with Ryan, C., "Can We Get to Mars?" *Collier's*, 30 April 1954, pp. 22–28.

[18] Lieberman, R., "The *Collier's* and Disney Series," in Ordway III, R. I. and Lieberman, R. (eds.), *Blueprint for Space: Science Fiction to Science Fact*, Smithsonian Institution Press, Washington, DC, 1992, pp. 144–146; Smith, D. R., "They're Following Our Script: Walt Disney's Trip to Tomorrowland," *Future*, No. 2, May 1978, pp. 59–60; Wright, M., "The Disney–von Braun Collaboration and Its Influence on Space Exploration," in Schenker, D., Hanks, C., and Kray, S. (eds.), *Inner Space, Outer Space: Humanities, Technology, and the Postmodern World*, Southern Humanities Press, Huntsville, AL, 1993, pp. 151–160; Ley, W., *Rockets, Missiles, and Space Travel*, Viking Press, New York, 1961 ed., p. 331.

[19] *TV Guide*, 5 March 1955, p. 9.

[20] Becker, J. V., "The X-15 Program in Retrospect," 3rd Eugen Sänger Memorial Lecture, *1st Annual Meeting, Deutsche Gesellschaft für Luft- und Raumfahrt*, Bonn, Germany, 4–5 Dec. 1968, NASA Historical Reference Collection, NASA History Division, pp. 1–3.

[21] Spires, D. N., *Beyond Horizons: A Half Century of Air Force Space Leadership*, Air Force Space Command, Peterson AFB, CO, 1997, p. 75.

[22] Swenson, L. S., Jr., Greenwood, J. M., and Alexander, C. C., *This New Ocean: A History of Project Mercury*, NASA SP-4201, 1966, pp. 33–97, 73–74.

[23] Lt. Gen. Putt, D. L., USAF Deputy Chief of Staff, Development, to Dryden, H. L., NACA Director, 31 Jan. 1958, NASA Historical Reference Collection.

[24] NACA to USAF Deputy Chief of Staff, Development, "Transmittal of Copies of Proposed Memorandum of Understanding between Air Force and NACA for joint NACA–Air Force Project for a Recoverable Manned Satellite Test Vehicle," NASA Historical Reference Collection, 11 April 1958; memorandum for Silverstein, "Assignment of Responsibility for ABMA Participation in NASA Manned Satellite Project," NASA Historical Reference Collection, 12 Nov. 1958; Silverstein, A. to Lt. Gen. Wilson, R. C., USAF Deputy Chief of Staff, Development, NASA Historical Reference Collection, 20 Nov. 1958; Dryden, H. L., Deputy Administrator, NASA, memorandum for Emme, E. for NASA Historical Files, "The "Signed" Agreement of 11 April 1958, on a Recoverable Manned Satellite Test Vehicle," NASA Historical Reference Collection, 8 Sept. 1965.

[25] Heppenheimer, T. A., *The Space Shuttle Decision: NASA's Search for a Reusable Space Vehicle*, NASA SP-4221, Washington, DC, 1999; Launius, R. D., "NASA and the Decision to Build the Space Shuttle, 1969-72," *The Historian*, Vol. 57, Autumn 1994, pp. 17–34.

[26] Hansen, J. R., *Engineer in Charge*, Washington, DC, NASA, 1988, pp. 377–381.

[27] Thompson, M. O., *At the Edge of Space: The X-15 Flight Program*, Smithsonian Institution Press, Washington, DC, 1992; McDowell, J., "The X-15 Spaceplane,"

Quest: The Magazine of Spaceflight History, Vol. 3, No. 2, Spring 1994, pp. 4–12; Hallion, R. P., *On the Frontier: Flight Research at Dryden, 1946-1981*, Washington, DC, NASA SP-4303, 1984, pp. 106–129; Gubitz, M. B., *Rocketship X-15: A Bold New Step in Aviation*, Julian Messner, New York, 1960; Walker, J. A., "I Fly the X-15," *National Geographic*, Sept. 1962, pp. 428–450; Becker, J. V., "The X-15 Project," *Astronautics & Aeronautics*, Feb. 1964, pp. 52–61; Stillwell, W. H., *X-15 Research Results*, NASA SP-60, 1965; Houston, R. S., Hallion, R. P., and Boston, R. G.,"Transiting from Air to Space: The North American X-15," in Hallion, R. P. (ed.), *The Hypersonic Revolution: Eight Case Studies in the History of Hypersonic Technology*, two volumes, Special Staff Office, Aeronautical Systems Division, Wright-Patterson AFB, OH, 1987, Vol. 1, pp. 1–183; Kay, W. D., "The X-15 Hypersonic Flight Research Program: Politics and Permutations at NASA," in Mack, P. E. (ed.), *From Engineering Science to Big Science: The NACA and NASA Collier Trophy Research Project Winners*, NASA SP-4219, Washington, DC, 1998, Chap. 6.

[28] Iliff, K. W., and Shafer, M. F., *Space Shuttle Hypersonic Aerodynamic and Aerothermodynamic Flight Research and the Comparison to Ground Test Results*, NASA Technical Memorandum 4499, 1993, quotation on p. 2; Iliff, K. W., and Shafer, M. F., "A Comparison of Hypersonic Flight and Prediction Results," 31st AIAA Aerospace Sciences Meeting & Exhibit, 11–14 Jan., AIAA paper 93-0311, Reston, VA, 1993.

[29] Houchin, R. A., *U.S. Hypersonic Research and Development: The Rise and Fall of Dyna-Soar*, 1944–1963, Routledge, New York, 2006; Geiger, C. J., "Strangled Infant: The Boeing X-20 Dyna-Soar," Case II of Hallion, R. P. (ed.), *The Hypersonic Revolution: Case Studies in the History of Hypersonic Technology, Vol. 1, From Max Valier to Project PRIME (1924–1967)*, U.S. Air Force Histories and Museums Program, Bolling, AFB, Washington, DC 1998, pp. 188, 189; TWX, RDZSXB-31253-E, Headquarters USAF to Headquarters BMD, 10 Nov. 1059; TWX, AFDAT-90938, Headquarters USAF to Commander, ARDC, 17 Nov. 1959.

[30] TWX, AFCVC-1918/63, Headquarters USAF to All Commands, 10 Dec. 1963; News Briefing, Secretary of Defense, "Cancellation of the X-20 Program," 10 Dec. 1963.

[31] Houchin III, R. F., "Air Force Office of the Secretary of Defense Rivalry: The Pressure of Political Affairs in the Dyna-Soar (X-20) Program, 1957–1963," *Journal of the British Interplanetary Society* 50, May 1997, pp. 162–168; Bacon, M., "The Dyna-Soar Extinction," *Space* 9, May 1993, pp. 18–21; Houchin III, R. F., "Why the Air Forçe Proposed the Dyna-Soar X-20 Program," *Quest: The Magazine of Spaceflight* 3, Winter 1994, pp. 5–11; Smith, T., "The Dyna-Soar X-20: A Historical Overview," *Quest: The Magazine of Spaceflight* 3, Winter 1994, pp. 13–18; Houchin III, R. F., "Interagency Rivalry: NASA, the Air Force, and MOL," *Quest: The Magazine of Spaceflight* 4, Winter 1995, pp. 40–45; Pealer, D., "Manned Orbiting Laboratory (MOL), Part 1," *Quest: The Magazine of Spaceflight* 4, Fall 1995, pp. 4–17; Pealer, D., "Manned Orbiting Laboratory (MOL), Part 2," *Quest: The Magazine of Spaceflight* 4, Winter 1995, pp. 28–37; Pealer, D., "Manned Orbiting Laboratory (MOL), Part 3," *Quest: The Magazine of Spaceflight*, Vol. 5, No. 2, 1996, pp. 16–23.

[32] Launius, R. D., "NASA and the Decision to Build the Space Shuttle, 1969–72," *The Historian*, Vol. 57, Autumn 1994, pp. 17–34.

[33] Fries, S. D., *NASA Engineers and the Age of Apollo*, NASA SP-4104, Washington, DC, 1992; see also, McCurdy, H. E., *Inside NASA: High Technology and Organizational Change in the U.S. Space Program*, Johns Hopkins University Press, Baltimore, MD, Vol. 7, 1993, pp. 90–131.

[34] Truax, R. C., "Shuttles—What Price Elegance?" *Astronautics & Aeronautics*, June 1970, pp. 22–23.

[35] Truax, R. C., "Charging Down the Wrong Track," in Ciancone, M. L. (ed.), *History of Rocketry and Astronautics: Proceedings of the Thirty-Sixth History Symposium of the*

International Academy of Astronautics, Houston, Texas, 2002, Univelt, Inc., AAS History Series, Vol. 33, San Diego, CA, 2010, pp. 91–98.

[36] Hughes, T. P., "The Evolution of Large Technological Systems," in Bijker, W. E., Hughes, T. P., and Pinch, T. J. (eds.), *The Social Construction of Technological Systems: New Directions in the Sociology and History of Technology*, MIT Press, Cambridge, MA, 1987, pp. 51–82.

[37] NASA, The Post-Apollo Space Program: A Report for the Space Task Group, Sept. 1969, p. 1, p. 6.

[38] Heppenheimer, T. A., *The Space Shuttle Decision: NASA's Search for a Reusable Space Vehicle*, NASA SP-4221, 1999; Logsdon, J. M., "The Decision to Develop the Space Shuttle," *Space Policy*, Vol. 2, May 1986, pp. 103–119; Logsdon, J. M., "The Space Shuttle Decision: Technology and Political Choice," *Journal of Contemporary Business*, Vol. 7, 1978, pp. 13–30; Launius, R. D., "NASA and the Decision to Build the Space Shuttle, 1969–72," *The Historian*, Vol. 57, Autumn 1994, pp. 17–34.

[39] Low, G. M., and Myers, D. D., "Space Shuttle Objectives," George M. Low Collection, NASA Historical Reference Collection, 27 Jan. 1970.

[40] Vincenti, W. G., "The Retractable Airplane Landing Gear and the Northrop 'Anomaly': Variation-Selection and the Shaping of Technology," *Technology and Culture*, Vol. 35, No. 1, Jan. 1994, pp. 1–33.

[41] Reagan, R., "State of the Union Address," 4 Feb. 1986.

[42] Hallion, R. P., "Yesterday, Today, and Tomorrow: From Shuttle to the National Aero-Space Plane," in Hallion, R. P. (ed.), *The Hypersonic Revolution*, Vol. 2, p. 1334, p. 1337, pp. 1340–1341, p. 1345, pp. 1362–1364; Heppenheimer, T. A., *The National Aerospace Plane*, Pasha Market Intelligence, Arlington, VA, 1987, p. 14; Schweikart, L. E., "The Quest for the Orbital Jet: The National Aerospace Plane Program, 1983–1995," manuscript, NASA Historical Reference Collection, pp. I.30–I.31; Becker, J. V., "Confronting Scramjet: The NASA Hypersonic Ramjet Experiment," in Hallion, R. P. (ed.), *The Hypersonic Revolution*, Vol. 2, pp. vi–xi.

[43] Schweikart, L. E., "Command Innovation: Lessons from the National Aerospace Plane Program," in Launius, R. D. (ed.), *Innovation and the Development of Flight*, Texas A&M University Press, College Station, TX, 1999, pp. 299–322.

[44] Butrica, A. J., "The Quest for Reusability," in Launius, R. D., and Jenkins, D. R. (eds.), *To Reach the High Frontier: A History of U.S. Launch Vehicles*, University Press of Kentucky, Lexington, KY, 2002, p. 453.

[45] Schweikart, L. E., "Quest for the Orbital Jet," pp. ii.37–iii.38, pp. iii.41–iii.42, manuscript in possession of author; Schweikart, L. E., "Command Innovation," pp. 299–322, in Launius, R. D., *Innovation and the Development of Flight*, Texas A&M University Press, College Station, 1999.

[46] Cole, J. W., "X-34 Program," in "X-33/X-34 industry briefing," NASA Historical Reference Collection, 19 Oct. 1994, especially p. 1A-1216.

[47] X-33 announcement in *Commerce Business Daily*, 29 Sept. 1994, in File 276, X-33 Archive, NASA Historical Reference Collection.

[48] Sietzen, F., "VentureStar Will Need Public Funding," *Space Daily Express*, NASA Historical Reference Collection, 16 Feb. 1998.

[49] David, L., "NASA Shuts Down X-33, X-34 Programs," Space.com, 1 March 2001, http://www.space.com/missionlaunches/missions/x33_cancel_010301.html [retrieved 28 March 2003].

[50] Watson, T., "NASA Administrator Says Space Shuttle Was a Mistake," *USA Today*, 28 Sept. 2005, p. 1A.

[51] NASA Press Release, "NASA Memo: Griffin Point Paper on USA Today Article, 9/28/05," NASA Historical Reference Collection, 29 Sept. 2005.

[52] Pielke, R. A., Jr., "Griffin: The Space Shuttle was a Mistake," *Prometheus*, 28 Sept. 2005, available online at http://sciencepolicy.colorado.edu/prometheus/archives/space_policy/000586griffin_the_space_s.html [retrieved 2 Feb. 2006].

[53] David, L., "The Next Shuttle: Capsule or Spaceplane?" Space.com, 21 May 2003, http://www.space.com/businesstechnology/technology/osp_debate_030521.html [retrieved 19 April 2004]; David, L., "NASA's Orbital Space Plane Project Delayed," Space.com, 26 Nov. 2003, http://www.space.com/businesstechnology/technology/space_plane_delay_031126.html [retrieved 19 April 2004]; White House Press Release, "Fact Sheet: A Renewed Spirit of Discovery," 14 Jan. 2004, http://www.whitehouse.gov/news/releases/2004/01/20040114-1.html [retrieved 4 April 2004].

[54] Dillow, C., "Jumping Into the New Space Race, Orbital Sciences Unveils Mini-Shuttle Spaceplane Design," *Popular Science*, 16 Dec. 2010, http://www.popsci.com/technology/article/2010-12/jumping-new-space-race-orbital-sciences-unveils-mini-shuttle-spaceplane-design [retrieved 27 Feb. 2011].

[55] Logsdon, J. M., "A New U.S. Approach to Human Spaceflight?" *Space Policy* 27, Issue 1, Feb. 2011, pp. 15–19.

[56] Mehta, U. B., and Bowles, J. V., "A Two-Stage-to-Orbit Spaceplane Concept with Growth Potential," NASA/TM-2001-209620, Feb. 2001, p. 19.

[57] Pinch, T. J., and Bijker, W. E., "The Social Construction of Facts and Artefacts: Or How the Sociology of Science and the Sociology of Technology Might Benefit Each Other," *Social Studies of Science,* Vol. 14, No. 3, pp. 399–441; Russell, S., "The Social Construction of Artefacts: Response to Pinch and Bijker," *Social Studies of Science,* Vol. 16, No. 2, pp. 331–346; Winner, L., "Upon Opening the Black Box and Finding it Empty: Social Constructivism and the Philosophy of Technology," *Science Technology & Human Values,* Vol. 18, No. 3, Summer 1993, pp. 362–378.

[58] Krige, J., email to author, "Your Spaceplane Chapter," 4 Oct. 2012.

Chapter 2

Designing the Shuttle: Living within the Political System

Roger D. Launius

Introduction

The decision to build the Space Shuttle, made in 1972 after a difficult debate inside Washington over the next-generation human spaceflight vehicle, became a viable program for the U.S. space agency in the late 1960s as the political climate, the economic situation, the technological base, and other factors came together to press it to the fore. The relations of NASA, the aerospace industry, the presidency, Congress, and the American public fundamentally affected how this decision proceeded [1]. Additionally, the budgetary priorities of the nation's leaders, especially austerity in domestic spending during the Vietnam era, affected at a fundamental level the design of the Space Shuttle. NASA compromised on many of its most fundamental goals in pursuing the shuttle as reflected in this situation, leading to a creature of compromise that couldn't meet fully its requirements. To reach even this end, the system had to be oversold and its capabilities overstated. Three key lessons from this situation are (1) the politics of adoption in big technology, (2) the wages of overpromising, and (3) the importance—and hazards—of coalition building. A key analog compares this story to earlier major space decisions such as Apollo.

The birthing process for the Space Shuttle was a difficult one, and it endured numerous permutations during the late 1960s and early 1970s as many design proposals were studied for cost effectiveness, technical feasibility, and political appropriateness. This process led to the decision to build the

Space Shuttle in essentially the form that it took at the time of its first operational mission in 1981. Numerous modes of operation, each involving different designs, were presented on how best to achieve the goal of a reusable spacecraft. Some were impressively logical, whereas some read like science fiction or at least a Jack Kerouac novel. All, however, had champions who were both vocal and persistent. There was an impasse of more than two years concerning the best method of pursuing a shuttle—single stage, fully reusable; two stage, fully reusable; two stage, partially reusable; and a host of variations on those themes. This impasse was finally broken in the fall of 1971. It paved the way for President Richard Nixon to announce the shuttle decision on January 5, 1972 (Fig. 2. 1).

This chapter reviews the early evolution of the Shuttle program, describing the process by which major proposals lived and sometimes died, as well as the political and economic ramifications of each shuttle concept. Because of its technological perspective, steeped as its engineers were in the longstanding quest for a reusable system, NASA leaders developed ideas for a fully reusable shuttle to the exclusion of almost any other concept. NASA's agenda, in essence, called for the fully reusable shuttle to serve as the nation's method of delivery from the Earth to an orbiting space station. Other governmental agencies, with other priorities and agendas, had different perspectives on what would be acceptable for space flight. The Office of Management and Budget (OMB), for example, sought to hold the line on NASA's budget and saw the development of a fully reusable shuttle as a hole in the sky through which the Federal Government poured money, and the OMB advocated an

Fig. 2.1 President Richard Nixon meets with NASA Administrator James C. Fletcher to approve the Space Shuttle as a program, January 5, 1972. (NASA)

expendable launch system or a small, partially reusable shuttle. These competing factors were resolved only in 1971, when independent analyses reviewed all viable proposals and offered recommendations for a system that met both financial constraints and technological progress. The result was a compromise system with an expendable liquid fuel booster, two reusable solid rocket boosters, and a reusable orbiter. This compromise was really quite different from what had been envisioned by the various agencies and personalities involved in the process of defining the shuttle mode. The bobs and weaves, the ebbs and flows, of this shuttle design debate inform this discussion. In reality, this chapter analyzes the manner in which competing technological priorities are negotiated and how political compromise is attained so that some worthwhile goal can be more readily accomplished in a complex democracy.

The Politics of Adoption in Big Technology

When NASA went to the moon with Project Apollo, it had a well-defined national objective aimed at enhancing national prestige and demonstrating both American technological capabilities and national resolve. It was the result of a top-down decision made by President John K. Kennedy in May 1961, and it committed the United States to pulling out all stops to complete a human lunar mission by the end of the decade [2]. From at least the mid-1960s until the shuttle decision, one of the central questions related to the space program was defining the U.S. goal for putting a human in space beyond the moon landing. President Lyndon B. Johnson's commitment to the Apollo program was unwavering, but as budgets mushroomed during his presidency because of his expensive Great Society program domestically and because of the mounting expenditures for the Vietnam War, a serious financial squeeze began. Apollo was a crash program to meet a stringent timetable; Johnson accepted it as such and was willing to spend accordingly, but he saw no reason to continue accelerated space efforts beyond Apollo. He assumed, at least by 1965, that after the successful lunar missions, NASA's budget would shrink to near a pre-Apollo level [3].

Johnson first showed concern about the space program's future on January 30, 1964, when he asked NASA Administrator James E. Webb to "review our future space exploration plans" in an effort to match "hardware and development programs to prospective missions" [4]. Webb's response called for no new start programs and "assume[d] that resources on the order of those currently programmed ($5.25 billion per year) [would] continue to be available" [5]. Johnson countered, in a post-Apollo arena: "We must moderate our efforts in certain space projects" [6].

Johnson's successor, Richard Nixon, showed even less interest in setting aggressive goals for the American space program [7]. NASA Administrator

Thomas O. Paine cautioned Nixon: "The impact and positive image of your leadership would be seriously downgraded in the eyes of the nation, the Congress, and the public, in my view, if the U.S. were once again placed in the position of reacting to Soviet initiatives in space" [8]. In January 1970, Paine met with the president to discuss this problem, and Nixon "understood these were very severe [cuts] and he had done it most reluctantly, but the cuts were necessary in view of the overall budget situation—the reduced revenues and inflation" [9].

The Space Shuttle commitment that emerged from this situation never experienced the consensus of the Apollo moon landing; NASA leaders constantly fended off attacks on the program, modified the parameters of the technology to reduce costs, and brought the Shuttle program to operational status through years of development that had to respond to diverse and far-reaching permutations. Four basic attributes dominated this history: (1) a decoupling of human space endeavors from broader national security/Cold War concerns, (2) increased antitechnology bias as NASA and the DoD's war-making were linked in the public mind, (3) budgetary concerns during the inflationary 1970s that prompted retrenchment in government activities, and (4) the equating of NASA's efforts with Great Society programs, which prompted concerns about "budget-busting" on both sides of the political spectrum.

A typically American political process took over during the consideration of whether to build the Space Shuttle. In part to reduce tensions, as well as to buy time and avoid making a firm decision that might be premature, Nixon appointed a Space Task Group to study post-Apollo plans and to make recommendations. Chartered on February 13, 1969, under the chairmanship of Vice President Spiro T. Agnew, this group met throughout the spring and summer to plot a course for the space program. The politics of this effort were intense. NASA lobbied hard with the group—and especially its chair—for a far-reaching post-Apollo space program that included a mission to Mars, a space station, and a reusable shuttle. The NASA position was well reflected in the group's September 15, 1969, report, but those conclusions were included over the objections of senior White House officials [10]. Perhaps because of this disagreement over the reach of the space effort, Nixon did not act on the group's report immediately.

Without clear presidential leadership, NASA began laying plans for what eventually became the Space Shuttle [11]. It unveiled them publicly at the August 1968 annual meeting of the British Interplanetary Society, when George E. Mueller, NASA's Associate Administrator for Manned Space Flight, described what he called the Space Transportation System [12]. He made clear that the Space Shuttle program was never an entity unto itself. It was always viewed as part of a much larger program to build a space station and later to undertake a human mission to Mars. "In order to support the station," NASA officials declared, "efficient transportation to and from the earth is required." This could be best provided with "low-cost access by reusable chemical and nuclear rocket transportation systems" [13]. It was realized early that NASA wanted the shuttle because of the economies it could introduce into space flight. George M. Low, NASA's Deputy Administrator, said as much in a memorandum to the NASA leadership on January 27, 1970: "I think there is really only one objective for the Space Shuttle program, and that is 'to provide a low-cost, economical space transportation system.' To meet this objective, one has to concentrate both on low development costs and on low operational costs" [14]. From the outset, the economics of the shuttle outweighed any other features. This was a striking difference between the Shuttle and Apollo programs, and NASA had adopted the stance before any external debate took place. "Low-cost, economical" space transportation became NASA's criteria even before the proposal went to the political marketplace. Of course, this was a logical decision at the time. To complete the missions that NASA officials envisioned would require a well-designed reusable Space Shuttle that could be launched for about $10 million per flight (in 1970 dollars). This estimate, according to NASA was "based on a careful assessment of NASA and contractor studies" [15].

By making low-cost transportation to and from orbit the principal objective for the Space Shuttle, NASA leaders made a critical long-term error. Had they been more experienced in the normal policy-making process, they might have been able to avoid this pitfall. As a result, every briefing, every testimony, every action on the part of the space agency about the Space Shuttle had to rise or fall based on economics. At the same time, this gave critics a big target. This was despite what many supporters thought about the program, for officials repeatedly stated that the shuttle's real virtue was not in providing low-cost access to space but in furthering technological and scientific knowledge and in increasing the potential for practical applications. Dr. James C. Fletcher, a scientist and former president of the University of Utah who became NASA Administrator in the spring of 1971, expressed surprise at the economic arguments his agency was making. He "never did like the highlighting of the cost-benefit argument" and suggested, "The most important justification for the Shuttle is that it gives the nation an entirely new capability for working routinely and quickly in space" [16]. The economic rationale, although it eventually proved an embarrassment to the space program, provided a solid rationale for supporters. Even supporters were hesitant to fund an expensive new NASA project when other pressing issues required resolution.

During the summer of 1970, the outlook was bleak around NASA headquarters and in its centers where the human spaceflight program represented a major part of the work. Emerging difficulties between Paine and White House staffers worked against program approval. Paine made a nuisance of himself by continually agitating for additional funding for his agency and by publicly criticizing the administration for cutting NASA's budget [17]. His strident appeals became an embarrassment, and he soon lost credibility with the administration.

In a memo to Assistant to the President Peter M. Flanigan, analyst Clay T. Whitehead commented that "NASA is—or should be—making a transition from rapid razzle-dazzle growth and glamour to organizational maturity and more stable operations for the long term." Paine, unfortunately, had been unwilling to give that leadership. The analyst noted [18]:

> We need a new Administrator who will turn down NASA's empire-building fervor and turn his attention to (1) sensible straightening away of internal management and (2) working with [the] OMB and White House to show us what broad but concrete alternatives the President has that meet all his various objectives. In short, we need someone who will work with us rather than against us, [who] will seek progress toward the President's stated goals, and [who] will shape the program to reflect credit on the President rather than embarrassment.

When Paine left NASA on September 15, 1970, Deputy Administrator Low tried to heal the breach between NASA and other agencies in the Executive Branch. His good will was demonstrated in a friendly—even chatty—note in February 1971 from OMB Deputy Director Caspar (Cap) W. Weinberger to Low. He told Low he was happy with the turnabout in NASA's response to the budget exercise. "I thought your statement to the press in support of the NASA Budget for 1972 was particularly good; it was both accurate and loyal and these are traits I greatly admire," he wrote, leaving unsaid but certainly clear that his predecessor had not been loyal [19]. After he became NASA administrator in February 1971, Fletcher made similar efforts to smooth over these problems and to get on with the definition of a viable post-Apollo human spaceflight program. For instance, in a memorandum to Low, NASA Deputy Administrator, Fletcher remarked that Thomas Whitehead and William Anders of Nixon's staff "seem to feel that NASA and the various constituents of the White House Staff had Jim Webb and Paine both in an adversary position; that now that NASA has leveled off to a reasonable level, perhaps this is the time to all be working together towards common objectives, and particularly towards objectives which the President himself wants" [20].

The external pressure on the space agency over a period of several months in 1970 and 1971, coupled with internal studies pointing in the same direction, irritated NASA leaders but forced them to review a variety of options for the shuttle that would allow them to develop it for under $1 billion per year [21]. These events also stimulated NASA to alter the focus of its research, and to contract with three firms—Lockheed, the Aerospace

Corporation, and Mathematica, Inc.—to analyze the possible uses and potential savings that could be incurred by developing a reusable spacecraft. Mathematica made the primary effort, whereas the other firms provided Mathematica with estimates of future space traffic and payload characteristics [22]. NASA's earlier efforts to demonstrate the shuttle's economic feasibility had met with mixed success [23]. However, these sophisticated studies conducted by outside organizations were convincing to most Washington executives. A watershed came in March 1971 with a study by Mathematica, "An Interim Report on Shuttle Economics," which indicated that a two-stage, fully reusable Space Shuttle was not economically justified, whereas a one-and-a-half-stage vehicle was "cost effective" [24]. This was a breakthrough, for it suggested that a partially reusable shuttle could significantly reduce both developmental and operational costs, making the system palatable to administration officials concerned with the long-term investment. The study helped break an impasse that had stalled the decision process for months. The complete articulation of Mathematica's conclusions was in its final report. This was apparently adopted by Fletcher soon afterward; there was a complaint about Fletcher's apparent about-face of the fully reusable system in June 1971.

The interim report was followed in May 1971 with a formal two-volume work that suggested, "Present economic data so far shows the Shuttle to be cost effective" [25]. Another Mathematica study reviewed several possible configurations for the shuttle and recommended that a thrust-assisted orbiter Shuttle (TAOS) that was partially reusable would fit the funding criteria established by the OMB and would yield desired operational economies. In a memorandum from Mathematica to Fletcher on October 28, 1971, investigators Klaus P. Heiss and Oskar Morgenstern commented, "Our calculations show the emergence of an *economical and acceptable solution to the question of the best strategy for NASA to achieve a reusable space transportation system for the 1980's at acceptable costs*" [emphasis in original]. The TAOS concept, Heiss and Morgenstern added, "practically assures NASA of a reusable space transportation system with *major objectives achieved*" [26]. These were tremendously important assurances. The Mathematica report confirmed the direction NASA was heading with other studies from primary contractors and convinced Fletcher that the TAOS Shuttle would be palatable to the OMB, the White House, and Congress [27].

With NASA working both to constrain its budget and to find a viable follow-on space vehicle, support from the White House emerged. While Fletcher worked without any apparent success for approval of the shuttle, one key individual in the Nixon Administration decided that cuts in the NASA budget had gone far enough and that it was time to move forward with a post-Apollo effort. Weinberger, Deputy Director of the OMB, argued in an August 12, 1971, memorandum to the President, "There is real merit to the future of NASA, and to its proposed programs." As a means of further cutting the NASA budget, the OMB was considering not approving the start of Space Shuttle development and cancelling the last two Apollo missions, Apollo 16 and 17. Weinberger suggested that such cuts "would be confirming, in some respects, a belief that I fear is gaining credence at home and abroad: that our best years are behind us, that we are turning inward, reducing our defense commitments, and voluntarily starting to give up our superpower status, and our desire to maintain world superiority." He added that "America should be able to afford something besides increased welfare, programs to repair our cities, or Appalachian relief and the like." In a handwritten scrawl on Weinberger's memo, Richard Nixon indicated, "I agree with Cap" [28].

Fletcher did not know of this exchange, and in the summer and fall of 1971, he led an often heated debate with administration bureaucrats over the propriety of the Shuttle program. Fletcher played every card he had to win approval of the effort, including the promise of European participation, but it was certainly not his primary rationale. Instead, he based the justification of the shuttle on these reasons:

1. The United States cannot forego manned space flight.
2. The Space Shuttle is the only meaningful new manned space flight program that can be accomplished on a modest budget.
3. The Space Shuttle is a necessary next step for the practical use of space.
4. The cost and complexity of today's Shuttle are *one-half* of what they were six months ago.
5. Starting the shuttle now will have a significant positive effect on aerospace employment. Not starting would be a serious blow to both the morale and health of the [U.S.] aerospace industry [29].

In the fall of 1971, the debate over the shuttle shifted from whether or not to build it to what type it would be. No doubt, the politics of the presidency played a central role in the decision. John Erlichman referred to it as the "mother's milk" of Washington affairs. It was always present, and in this instance, the sorry state of the aerospace industry was a significant factor in building support for the shuttle. The supersonic transport had recently been cancelled by Congress, and this cancellation had thrown the industry both physically and psychologically into a downturn. Another major defeat in the government procurement arena could have meant especially negative consequences. In late 1971 and early 1972, Nixon was beginning a reelection campaign and was concerned about the battleground states. Erlichman recalled that there were few battleground states, but "when you look at employment numbers [for the aerospace industry], and you key them to the battleground states, the space program has an importance out of proportion to its budget." Consequently, although the shuttle created several thousand jobs that might not have been too important in the overall scheme of things, it helped in regions where Nixon's political fortunes could benefit. "So," Erlichman said, "you must not underemphasize that element, that employment element, in Nixon's decision on the whole manned space program" [30]. Other factors such as letters and personal meetings also brought this issue home to Nixon.

On December 11, 1971, Fletcher and Low met with presidential assistant Peter M. Flanigan; science advisor Edward E. David, Jr.; and representatives of the OMB to discuss the Shuttle program. They learned at that meeting that Nixon had decided in principle to go ahead with the project but that some additional decisions about size and cost had yet to be made [31]. On January 3, 1972, Fletcher and Low met with Weinberger and learned of the final decision to build, for $5.5 billion, a partially reusable TAOS configuration Shuttle with specific cargo capacity that would meet both DoD and NASA specifications. Fletcher and Low flew to San Clemente, California, for a meeting with the president on January 5, to announce the decision. The announcement brought relief both to the aerospace industry and space advocates. Both ballyhooed it as a great step forward in national capability. Critics derided it as an ill-timed, ill-considered, unnecessary expenditure of public funds. The positions taken on the decision, in too many instances, resulted from individual perspective and depended on whose ox was being gored or spared [32].

The Apollo Analogy

What does the Space Shuttle decision process tell us about governmental control of "big technology" programs? First, when the shuttle debate was taking place, NASA, an agency born of crisis in the immediate post-Sputnik world of the late 1950s, was suffering from the Apollo syndrome. Kennedy had mandated by fiat its rapid expansion to accomplish the technological marvel of landing Americans on the moon before the end of the 1960s. Its own success in this enterprise, a success partly attributable to enormous outlays of funds, sparked both a voracious appetite for money to support overhead and to continue technological progress, and a desire to press for even more daring missions. Very much like Martin Luther King, Jr., did after the successful campaign he led as a young man in the Montgomery Bus Boycott of the 1950s, NASA had to ask itself what it would do for an encore. Although NASA had the will, bravado, and expertise to carry forward with a path-breaking program, it did not have the political acumen to acquire the necessary funding. Its leaders asked for executive fiat, à la Kennedy, when the political and economic climate had changed and when it was apparent no mandate would be forthcoming.

Throughout this process, one of NASA's principal complaints was Nixon's reluctance to stand up as Kennedy had done a decade earlier and to endorse the Space Shuttle publicly. Indeed, the lack of strong presidential support has been blamed for the shuttle's failure to deliver on its promise of economical access to space, for it forced NASA to build the vehicle "on the cheap." Nixon was unwilling to take a strong position early in the shuttle debate for two principal reasons. First, although Nixon supported the space program, he had no intention of letting its spending escalate, and he was dedicated to holding it down. Nixon believed that NASA had to be reined in, and his wishes were reflected in the OMB's efforts to reduce NASA's funding. Economic conditions and other priorities also figured in this decision. Second, it was not Nixon's style to involve himself in the intricate debates between members of his staff. He let competing personalities and positions slog things out below him, allowing information to bubble up through the chain.

On issues of importance—with importance being defined as high dollar value and national interest—Nixon would make the final decision on a program, but he shut himself off from the debates among staff. He was only the final arbiter, not a participant. Those who brought difficulties to him regularly, especially if he considered the difficulties petty, were shut out even more by Nixon—and, in some instances, he would not even see them. Paine fell into this category just before he left as NASA Administrator during the fall of 1970. Agnew, who headed the Space Task Group's effort, also quickly fell into this category [33]. Once the staff had hashed out the shuttle decision below Nixon, it went to him for approval. Only at that point, after the policy-making process had worked, did Nixon make a formal announcement on the Space Shuttle, and in many respects, this represented the triumph of routine government policy-making over NASA's preference for presidential edict. This approach toward policy-making is a common one among presidents, and it can be exceptionally effective. Nixon's mentor, Dwight D. Eisenhower, was a master at it [34].

In seeking presidential directive and getting none, NASA made itself few friends within the Executive Branch. Paine, sometimes nicknamed the "swashbuckler," was every bit as zealous for his cause as his namesake had been, and Paine was both unwilling to compromise and publicly critical of the administration's lack of strong action. As he became more adamant, fewer opportunities to push the shuttle successfully through the process availed themselves. Fletcher and Low, both of whom understood what had happened with Paine and who probably had been warned about such behavior, were the ones who broke the Executive Branch impasse and moved the Space Shuttle program forward. They still were unable to obtain the kind of presidential decree that they wanted and had to compromise the program toward the center, but they preserved the program and believed they had secured a viable alternative system.

FINAL DESIGN

Although a variety of configurations were studied, some of them quite exotic, the Space Shuttle as NASA conceived it was, in 1969, a two-stage, fully reusable system with both stages piloted and capable of landing on a runway like conventional aircraft. Launched like a rocket, the two stages separated near the edge of the atmosphere, with the first stage, about the size of a Boeing 747, returning to Earth. The second stage—about the size of a Boeing 707—flew on under its own power into orbit, performed its mission, and then returned to Earth. Other officials on the president's staff pushed for a smaller vehicle, one more in the form of a space glider with a smaller payload bay. Pockets of resistance remained within NASA as well. Max Faget, from his post at the Houston space center, argued against adopting the delta-shaped wings that the DoD's cross-range requirements imposed. Faget preferred straighter, airline-type wings. Officials at NASA headquarters listened for a while, but then ignored him [35].

To build the system NASA wanted would cost an estimated $10 billion, although those figures were quite soft and were inflated by as much as another $5 billion. A $10 to $15 billion investment, however, was not very attractive to most government officials. As a result, Bureau of the Budget pressured NASA, and the agency's dream of a fully reusable Space Shuttle did not survive even a year in the political spotlight. NASA responded with study after study, trying to document long-term cost savings and searching for configurations that might meet the administration's funding constraints for a major space vehicle. NASA's work was in part predicated on a Bureau of the Budget intimation that the agency's budget would remain flat for the first five years of the 1970s, at the $3.2 billion mark [36]. This budget was, NASA officials believed, in part a ploy to abolish NASA as a major government agency by letting inflation eventually zero out the space program. That was what Fletcher concluded in 1977 [37].

People in every part of the United States worked on the Space Shuttle program. Thousands of workers at NASA and contractor sites around the country helped get the shuttle ready to fly again and again. Shuttle workers occupied 640 facilities nationwide and used over 900,000 pieces of equipment, totaling almost $20 billion in value. More than 1200 active parts suppliers and some 4,000 qualified suppliers supported the Space Shuttle program (Fig. 2.2).

The DoD requirements were critical to the Space Shuttle's final design. That military mission, as it came to coalesce around the new Space Shuttle in the 1970s, took as its raison d'être the deployment of reconnaissance and other national security payloads. During a return to Earth, the orbiter had a cross-range maneuvering capability of 1265 statute miles to meet requirements for liftoff and landing at the same location after only one orbit. This would enable great flexibility in deploying those space assets into orbit, while

Fig. 2.2 Map of Space Shuttle suppliers. (NASA)

masking their trajectories from the Soviet Union. Without those design modifications to support the military space program, the DoD would have probably withheld monetary and political support from the project. In essence, NASA embraced a military mission for the Space Shuttle program as a means of building a coalition in support of an approval that might not have been achieved otherwise. In return, military astronauts would fly on classified missions in low-Earth orbit. Most of those missions were for deploying reconnaissance satellites, but what else might have been accomplished on them is unknown in the nonclassified world [38].

In keeping with plans developed in the Carter Administration of the late 1970s, the Space Shuttle would thereafter carry all U.S. government payloads; military, scientific, and even commercial satellites could all be deployed from its payload bay [39]. To prepare for this, in 1979, Air Force Secretary Hans Mark created the Manned Spaceflight Engineer program to "develop expertise in manned space flight and apply it to Department of Defense space missions." In all, between 1979 and 1986, this organization trained 32 Navy and Air Force officers as military astronauts [40].

It might be easy to underestimate the national security implications of the Space Shuttle decision and the desire of some in the DoD to gain a military astronaut foothold that facilitated it. However, this goal seems to be critical to DoD support. Key Nixon advisor Weinberger believed the shuttle had obvious military uses and profound implications for national security. "I thought we could get substantial return" with the program, he said in a 1977 interview, "both from the point of view of national defense, and from the point of view

[of] scientific advancement, which would have a direct beneficial effect" [41]. He and others also impressed on the president the shuttle's potential for military missions. Erlichman even thought it might be useful to capture enemy satellites, a mission that would require military astronauts to, in effect, "lasso" those satellites during EVAs and to bring them into the shuttle payload bay for return to Earth [42]. The Soviet Union, which built the Buran in the 1980s and flew it without a crew, pursued a shuttle project as a counterbalance to the U.S. program, solely because it was convinced that the U.S. shuttle was developed for military purposes. As Russian space watcher James Oberg concluded, "They had actually studied the shuttle plans and figured it was designed for an out-of-plane bombing run over high-value Soviet targets. Brezhnev believed that and in 1976 ordered $10 billion of expenditures. They had the Buran flying within ten years and discovered they couldn't do anything with it" [43].

Building this entirely new type of space vehicle led to several challenges. Perhaps the most important design issue, after the orbiter's configuration, concerned the boosters to be developed and whether they should burn liquid or solid fuel. Also important was the development of the shuttle orbiter's unique reentry method. A question over how best to pass through the ionosphere arose: by using a high angle of attack that would bring the orbiter through it quickly and heat the outer skin to extremely high temperatures (but only for a short period of time), or by using a blunt-body approach like that of earlier capsules. NASA eventually decided on an approach that required developing a special ceramic tile to be placed on the underside and nose of the orbiter to withstand the re-entry heat. Because of these issues, as well as political and management questions, shuttle development slowed down considerably in the mid-1970s, prompting the project to be redefined and refinanced and causing a delay of its first operational flight from 1978 to 1981.

The shuttle's godfathers and designers intended it as a vehicle that was essentially "one-size-fits-all." It would carry all government payloads—military, scientific, and even commercial satellites could all be deployed from its payload bay. It could also retrieve satellites and either repair on orbit, as in the case of the Hubble Space Telescope, or return them to Earth. It could be used for all government launches, and NASA could even sell rides into orbit for commercial satellites. This offered considerable excitement, largely because NASA offered users a launch price at well below cost. NASA officials predicted a quick turnaround between missions. It appeared that deploying satellites would be a lucrative market, and in this environment, NASA leaders planned to launch a couple of orbiters each month [44].

Throughout the 1970s, NASA officials anticipated that they could launch the Space Shuttle as often as 24 times per year—if they could find enough missions to fly. Twenty-four flights divided into the mid-point between $1.3 and $1.4 billion of annual operational costs numbered about $57 million per flight. NASA officials believed that this compared fairly well to alternative rockets with similar payloads. Indeed, using the Space Shuttle compared favorably to using the Saturn IB (no longer in production) and the Titan IV (not yet in production). This logic pointed to using the Space Shuttle as the nation's primary launch vehicle. NASA officials relied heavily on emerging airline models as their means to this end. Airlines used robust engines, computerized checkout, and components that could be replaced without taking the aircraft apart to keep their aircraft in the air and earning revenue [45]. NASA officials planned to use similar techniques to cut the cost of space flight. They never achieved that end.

The Wages of Overpromising

Although the goals of "low cost, economical" access to space were appropriate for NASA, it eventually proved an embarrassment to the space program. Despite high hopes, the shuttle never provided inexpensive or routine access to space. The Space Shuttle—second to the Saturn V in both capability and cost—launched some 53,000 pounds of payload into orbit on each mission at a cost per launch of about $450 million. It was a high-end user, and the cost per flight was so astronomical that only the government could afford it. In addition, by January 1986, there had been only 24 Shuttle flights, although in the 1970s, NASA had projected that number of flights for every year. Although the system was reusable, its complexity, coupled with the ever-present rigors of flying in an aerospace environment, meant that the turnaround time between flights required several months instead of several days.

As only one measure of Shuttle program performance, the cost of the development effort came under heavy scrutiny, and overruns of the budget received considerable attention from such organizations as the General Accounting Office (GAO) [46]. However, the cost from when the program was approved through the first flight was $5.974 billion (in 1972 adjusted dollars), a 17 percent overrun above the $5.15 billion budget originally approved. For the development effort, NASA did not do too badly in estimating costs in an era of rampant inflation. On the other hand, the Apollo program, which has a reputation as a highly successful, well-managed program, spent $13.45 billion (in 1961 adjusted dollars; $21.4 billion in nonadjusted funds) from the start of the project to the first moon landing. NASA engineers had told NASA Administrator James E. Webb that Apollo would cost between $8 and $12 billion to complete the first mission. "Because no one could anticipate all contingencies, he [Webb] enlarged the figure NASA sent Kennedy to $20 billion for the first lunar journey...using administrative realism to counter technical optimism in setting Apollo's deadline and price" [47]. If Webb had accepted the lower numbers—and instinct was the only reason he did not—clearly NASA would have seriously underestimated the cost of the lunar landing program. Other factors beyond the management of the R&D effort for the shuttle must account for its poor reputation. Likewise, in terms of operational costs, individual shuttle missions were comparable to individual Apollo missions when adjusted for inflation, although, of course, there was a fundamental difference between an orbital mission on the Space Shuttle and the lunar landings [48].

Rather, it seems that NASA officials created seriously false expectations about what the shuttle would be able to do in terms of costs. At a fundamental level, from the perspective of NASA's operational era, it spent heavily on the Shuttle program. All NASA spending on the shuttle through the end of the program in 2011 totaled approximately $192 billion (in current dollars), which equated to about $1.5 billion per launch over the 135-mission life of the program. Accordingly, it has been widely acknowledged that the cost of operating the Space Shuttle did not meet its original expectations. However, surprisingly, if NASA could have launched 24 missions per year as it envisioned, it would not have been far off the original estimates. This has been a difficult reality for the Shuttle program [49].

Couple that with a persistent drumbeat for the shuttle as the end-all, one-size-fits-all space truck, and expectations could never be realized. Indicative of these broad expectations, NASA published in 1983 a marketing brochure titled *We Deliver* that touted the vehicle as "the most reliable, flexible, and cost-effective launch system in the world" [50]. It suggested that the Space Shuttle could satisfy every requirement in the United States (Fig. 2.3). NASA and the Space Shuttle program may have fallen victim to their own highly successful marketing. Public opinion polls—Harris, Media General, NBC/Associated Press, NBC, Gallup, CBS/New York Times polls, and ABC/WP polls conducted between the 1980s and the present—have consistently shown a perception of the Space Shuttle as a good investment [51]. Over time, as two shuttle accidents and other difficulties with the program became apparent, the Space Shuttle became perceived as a failure rather than as an unadulterated success like Apollo. The shuttle probably did not deserve that characterization, but perceptions and myths are almost as significant as facts and reality in public discourse, and the result has been a legacy of the entire Shuttle program as a mistake persisted in for more than 30 years [52].

So What?

In reality, the Space Shuttle was the first attempt at a reusable, experimental vehicle. The Shuttle program probably achieved about 90 percent of its technical objectives but failed miserably in meeting its economic objectives. There are many reasons for this, including the fact that there ended up being no reason to fly 24 missions a year (because satellites became more capable and longer lived). There was also a huge misstep in how NASA chose to

Fig. 2.3 The April 12, 1981, launch at Pad 39A of STS-1, just seconds past 7 a.m., carried astronauts John Young and Robert Crippen into an Earth orbital mission that lasted for 54 hours, ending with an unpowered landing at Edwards Air Force Base in California on April 14. It signaled a new era in human spaceflight. (NASA)

operate the vehicle: The standing army necessary to develop it was maintained into the "operational era" since 1982 because NASA managers (and their contractors) did not want to diminish capability by reducing personnel and budgets. Probably, NASA could have operated the vehicle with significantly fewer personnel—and about two-thirds of the budget—ultimately used. Beyond that, NASA's principal failing may have been in staying with the shuttle for so many years instead of treating it as an experimental program that would lead to future reusable vehicles.

In essence, NASA launched two major human spaceflight programs in two very different political contexts: Apollo was driven by Cold War rivalry with the Soviet Union and was a signature program for U.S. technological and political supremacy. There was no need to justify its costs, even though in the mid-1960s, Johnson began to worry about them. The Space Shuttle, by contrast, was embarked upon by an administration that was seeking detente with the Soviet Union. There was no question of competition with the Soviet Union concerning this program, and its relationship to waging the Cold War was less well defined. An entirely new kind of justification was needed. NASA had to "learn" that, and this took time. Flush with the success of Apollo, the space agency did not at first realize that it would have to find quite new kinds of justifications for human spaceflight. In the context of the day, economic arguments emerged as the solution to this problem, at least for Low and then Nixon and Weinberger (and that was not dumb at all). In short, this is a fundamental story about the way a changing geopolitical context forced NASA to rethink its rationale for human spaceflight and to make the political and technological adjustments to build support for a cheaper shuttle that would satisfy other stakeholders like the DoD (all irrelevant in 1961).

REFERENCES

[1] Hallion, R. P., *The Path to the Space Shuttle: The Evolution of Lifting Reentry Technology*, Air Force Flight Test Center History Office, Edwards AFB, CA, 1983; Peebles, C., "The Origins of the U.S. Space Shuttle – parts 1 and 2," *Spaceflight,* Vol. 21, Nos. 11 and 12, Nov. and Dec. 1979, pp. 435–442, 487–492; Sänger-Bredt, I., "the Silver Bird Story," *Spaceflight*, Vol. 15, No. 5, May 1973, pp. 166–181; Smith, E. P., Space Shuttle in Perspective—History in the Making," AIAA Paper 75-336, Feb. 1975.

[2] Logsdon, J. W., *John F. Kennedy and the Race to the Moon,* Palgrave Macmillan, New York, 2011; see also Brooks, C. G., Grimwood, J. M., and Swenson, Jr., L. S., *Chariots for Apollo: A History of Manned Lunar Spacecraft*, NASA, Washington, DC, 1979; Murray, C., and Bly Cox, C., *Apollo: The Race to the Moon,* Simon and Schuster, New, York, 1989; van Dyke, V., *Pride and Power: The Rationale of the Space Program*, University of Illinois Press, Urbana, IL, 1964.

[3] Divine, R. A., "Lyndon B. Johnson and the Politics of Space," in *The Johnson Years: Vietnam, The Environment, and Science,* University of Kansas Press, Lawrence, KS, 1987, pp. 217–253.

[4] Johnson, L. B., to Webb, J. E., White House Central Files, Executive Office Staff, Box 1, Lyndon Baines Johnson Presidential Library, Austin, TX, 30 Jan. 1964.

[5] Webb, J. E., to Johnson, L. B., White House Central Files, Ex. OS, Box 1, Johnson Library, 16 Feb. 1965.

[6] Redford, E. S., and White, O. F., *What Manned Space Program After Reaching the Moon? Government Attempts to Decide, 1962–1968*, Inter-university Case Program, Syracuse, NY, 1971, p. 207.

[7] Viorst, M., *Fire in the Streets: America in the 1960s,* Simon and Schuster, New York, 1979; Matusow, A. J., *The Unraveling of America: A History of Liberalism in the 1960s*, Harper and Row, New York, 1984; O'Neill, W. L., *Coming Apart*, Quadrangle Books, Chicago, 1971; Hodgen, G., *America in Our Time: From World War II to Nixon, What Happened and Why*, Doubleday, Garden City, NY, 1976; Dickstein, M., *Gates of Eden: American Culture in the Sixties*, Basic Books, New York, 1977.

[8] Paine, T. O. and Nixon, R. M., *NASA Historical Reference Collection*, NASA History Office, Washington, DC, 26 Feb. 1969.

[9] Memorandum for the Record of Paine, T. O., "Meeting with the President, January 22, 1970," NASA Historical Reference Collection, 22 Jan. 1970.

[10] President's Space Task Group, *The Post-Apollo Space Program: Directions for the Future*, Executive Office of the President, Washington, DC, Sept. 1969; *New York Times*, 16 Sept. 1969, p. 1; Crabill, D. E., Bureau of the Budget, to Director, Bureau of the Budget, "President's Task Group on Space-Meeting No. 2," Record Group 51, Series 69.1, Box 51-78-31, National Archives, 14 March 1969; Whitehead, C. T., Staff Assistant White House, to Flanigan, P. M., White House, Record Group 51, Series 69.1,

Box 51-78-31, National Archives, 25 June 1969; Mayo, R. P., Director, Bureau of the Budget, to Nixon, R. M., "Space Task Group Report," Record Group 51, Series 69.1, Box 51-78-31, National Archives, 25 Sept. 1969; Erlichman, J., *Witness to Power: The Nixon Years*, Pocket Books, New York, 1982, pp. 123–124.

[11] Lockheed-California Co., "Design Studies of a Reusable Orbital Transport, First Stage: Final Report—Research and Technology Implications," 21 May 1965, Boeing Co., "A Lifting Re-entry Horizontal-Landing Type Logistic Spacecraft," 3 Feb. 1964, and General Dynamics/Convair, "Reusable Orbital Transport, Second Stage Research and Technology Implications," April 1965, all in NASA Historical Reference Collection; see also Root, M. W., and Fuller, G. M., "The Astro Concept," *Astronautics & Aeronautics*, Vol. 2, No. 1, Jan. 1964, pp. 42–51; Bailey, R. A. and Kelley, D. L., "Potential of Recoverable Booster Systems for Orbital Logistics," *Astronautics & Aeronautics,* Vol. 2, No. 1, Jan. 1964, pp. 54–58; Moise, J. C., Henry, C. S., and Swanson, R. S., "The Astroplane Concept," *Astronautics & Aeronautics*, Vol. 2, No. 1, Jan. 1964, pp. 35–41; Chambers, A. N., "On the Role of Man in Space," Bellcomm, Inc., NASA Historical Reference Collection, 6 May 1968.

[12] Mueller, G. E., *Address on the Space Shuttle before the British Interplanetary Society*, University College, London, England, NASA Historical Reference Collection, 10 Aug. 1968, published as Mueller, G. E., "The New Future for Manned Spacecraft Developments," *Astronautics & Aeronautics*, 7, March 1969, pp. 24–32; see also Bono, P., and Gatland, K., *Frontiers of Space*, Macmillan, New York, 1969.

[13] *The Post-Apollo Space Program: A Report for the Space Task Group*, NASA, Sept. 1969, p. 1, 6.

[14] Low, G. M. and Myers, D. D., "Space Shuttle Objectives," NASA Historical Reference Collection, 27 Jan 1970.

[15] Fletcher, J. C., Administrator, to Mondale, W. F., NASA History Office, 25 April 1972.

[16] Barfield, C. E., "Technology Report/NASA Broadens Defense of Space Shuttle to Counter Critic's Attacks," *National Journal*, Vol. 4, No. 33, 19 Aug. 1972, pp. 1323–1332, quote on pp. 1323–1324; see also David Packard's remarks in Fletcher, J. C., NASA Administrator to Low, G. M., NASA Deputy Adminsistrator, "Luncheon Conversation with Dave Packard," NASA Historical Reference Collection, 20 Oct. 1971.

[17] See examples in Paine, T. O., NASA Administrator, to Agnew, S. T., Vice President, 12 Sept. 1969; Stoer, N. S., Economics, Science, and Technology Division, Bureau of the Budget, to Mayo, R. P., Director, Bureau of the Budget, "Analysis of NASA Report to Vice President on Recent Interest/Reaction to the Space Program," 2 Oct. 1969; Derman, D. A., Economics, Science, and Technology Division, Bureau of the Budget, to Mayo, R. P., Director, Bureau of the Budget, "Budget Appeals Session for NASA," 19 Nov. 1969; Paine, T. O., NASA Administrator, to Mayo, R. P., 19 Jan. 1970, all in Record Group 51, Series 69.1, Box 51-78-32, National Archives.

[18] Whitehead, C. T., White House Staff Assistant, to Flanigan, P. M., Assistant to the President, Record Group 51, Series 69.1, Box 51-78-32, National Archives, 8 Feb. 1971.

[19] Weinberger, C. W., Deputy Director, Office of Management and Budget, to Low, G. M., Acting Administrator, NASA, Record Group 51, Series 69.1, Box 51-78-32, National Archives, 18 Feb. 1971.

[20] Fletcher, J. C. and Low, G. M., "Luncheon with Tom Whitehead and Bill Anders," NASA Historical Reference Collection, 5 Nov. 1971; see also Rice, D. B., OMB Assistant Director, to Fletcher, J. C., NASA Administrator, 20 July 1971, Record Group 51, Series 69.1, Box 51-78-32, National Archives.

[21] Fletcher, J. C. interview by Logsdon, J. M., 21 Sept. 1977, NASA Historical Reference Collection; Williams, F., NASA, to von Braun, W., NASA, NASA Historical Reference Collection, 23 Aug. 1971.

[22] Cohen, N. B., Special Assistant to NASA Assistant Administrator for Policy, to Deputy Director, NASA Headquarters Contracts Division, "Consultant Activities of Oskar Morgenstern," NASA Historical Reference Collection, 15 May 1970; Eggers, A. J., NASA Assistant Administrator for Policy, to Morgenstern, O., NASA Historical Reference Collection, 8 June 1970.

[23] Wyatt, D. D., "Cost Effectiveness of the Shuttle," NASA Historical Reference Collection, 12 Feb. 1970; see also Fletcher, J. C. interview by Logsdon, J. M., 21 Sept. 1977, NASA Historical Reference Collection; Mark, H., *The Space Station: A Personal Journey*, Duke University Press, Durham, NC, 1987, pp. 40–42; Galloway, A., "Does the Space Shuttle Need Military Backing?" *Interavia*, Vol. 27, Dec. 1972, pp. 1327–1331.

[24] Lindley, R. N. to Gilruth, R. R., Johnson Space Center Archives, 29 March 1971; Mathematica Inc., "Economic Analysis of New Space Transportation Systems," NASA Historical Reference Collection, 31 May 1971; Mark, H., Director, Ames Research Center, to Faget, M., "Manned Spacecraft Center," NASA Historical Reference Collection, 22 June 1971.

[25] *Economic Analysis of New Space Transportation Systems: Executive Summary*, Mathematica, Inc., Princeton, NJ, 1971, pp. 3–27.

[26] Heiss, K. P., and Morgenstern, O., "Factors for a Decision on a New Reusable Space Transportation System," NASA Historical Reference Collection, 28 Oct. 1971, pp. 2, 10, followed by three-volume formal report: Heiss, K. P., and Morgenstern, O., *Mathematica Economic Analysis of the Space Shuttle System,* Mathematica, Inc., Princeton, NJ, 1972; see also Heiss, K. P., "Our R and D Economics and the Space Shuttle," *Astronautics & Aeronautics*, Vol. 9, Oct. 1972, pp. 50–62, and Heiss, K. P., "Space Shuttle Economics and U.S. Defence Potentialities," *Interavia*, Vol. 31, Nov. 1976, pp. 1071–1073.

[27] Fletcher, J. C. interview by Logsdon, J. M., 21 Sept. 1977, NASA Historical Reference Collection; Fletcher, J. C., NASA Administrator, to Weinberger, C. W., Deputy Director, Office of Management and Budget, NASA Historical Reference Collection, 29 Dec. 1971, assessment confirmed by Rice, D. interview by Logsdon, J. M., 13 Nov. 1975.

[28] Weinberger, C. W. Memorandum to the president, "Future of NASA," via Shultz, G., White House, Richard M. Nixon, President, 1968–1971 File, NASA History Reference Collection, 12 Aug. 1971; Huntsman, J. M. and Shultz, G. P., "The Future of NASA," Nixon Project, National Archives and Records Administration, Arlington, VA, 13 Sept. 1971.

[29] Fletcher, J. C. memorandum to Rose, J., Special Assistant to the President, Nixon Project, National Archives, 22 Nov. 1971.

[30] Erlichman, J. interview by Logsdon, J. M., May 6, 1983, NASA Historical Reference Collection; Kizis, F. to Nixon, R. M., Record Ground 51, Series 69.1, Box 51-78-31, National Archives, 12 March 1971; Melencamp, N. M., White House, to Kizis, F., Record Ground 51, Series 69.1, Box 51-78-31, National Archives, 19 April 1971.

[31] Logsdon, "The Space Shuttle Program," p. 1103; Flanigan, P. M., Assistant to the President, to Fletcher, J. C., NASA Administrator, Fletcher Papers.

[32] Fletcher, J. C., NASA Administrator, to Weinberger, C. W., Deputy Director, OMB, NASA Historical Reference Collection, 4 Jan. 1972; Low, G. M., NASA Deputy Administrator, "Meeting with the President on January 5, 1972," NASA Historical Reference Collection, 12 Jan. 1972; Barfield, C. E., "Technology Report/Intense Debate, Cost Cutting Precede White House Decision to Back Shuttle," *National Journal*, 4, 12 Aug. 1972, pp. 1289–1299; Heckler, *Toward the Endless Frontier*, pp. 284–295; Gillette, R., "Space Shuttle: Compromise Version Still Faces Opposition," *Science*, 175, 28 Jan. 1972, pp. 392–396; Hotz, R., "The Shuttle Decision," *Aviation Week & Space Technology*, 31 July 1972, p. 7; "Space Shuttle: NASA Versus Domestic Priorities," *Congressional Quarterly*, 26 Feb. 1972, pp. 435–439; Aspin, L., "The Space Shuttle: Who Needs It?" *Washington Monthly*,

Sept. 1972, pp. 18–22; "NASA In Trouble with Congress, Executive, Scientists," *Nature*, Vol. 231, No. 5302, 11 June 1971, pp. 346–348.

[33] Weinberger, C. W. interview by Logsdon, J. M., 23 Aug. 1977; Erlichman, J. interview by Logsdon, J. M., 6 May 1983; Erlichman, J., *Witness to Power*, Simon and Schuster, New York, 1982, pp. 120–127.

[34] Greenstein, F. I., *The Hidden-Hand Presidency: Eisenhower as Leader*, Basic Books, New York, 1982.

[35] Logsdon, J. M., "The Space Shuttle Decision: Technological and Political Choice," *Journal of Contemporary Business*, Vol. 7, No. 3, 1978, pp. 13–30; Logsdon, J. M., "The Space Shuttle Program: A Policy Failure?" *Science*, Vol. 232, No. 4754, 20 May 1986, pp. 1099–1105.

[36] Mayo, R. P., Director, Bureau of the Budget, to Paine, T. O., NASA Administrator, 7 Aug. 1969, Record Group 51, Series 69.1, Boxes 51-78-31 and 51-78-32, National Archives; Timmons, W. E., White House, to Gifford, B., White House, 30 Oct. 1970, Record Group 51, Series 69.1, Boxes 51-78-31 and 51-78-32, National Archives; Fletcher, J. C., NASA Administrator, "Dr. Fletcher's Notes from Meeting with Mr. Weinberger," 5 Aug. 1972, NASA Historical Reference Collection.

[37] Fletcher, J. C. interview by Logsdon, J. M., 21 Sept. 1977, NASA Historical Reference Collection.

[38] Day, D. A., "Invitation to Struggle: The History of Civilian-Military Relations in Space," *Exploring the Uniontown: Selected Documents in the History of the U.S. Civil Space Program, Volume II, External Relationships*, edited by Logsdon, J. M. (gen. ed.), NASA SP-4407, 1996, p. 264; Draper, A. C., Buck, M. L., and Goesch, W. H., "A Delta Shuttle Orbiter," *Astronautics & Aeronautics*, 9, Jan. 1971, pp. 26–35; Mathews, C. W., "The Space Shuttle and its Uses," *Aeronautical Journal,* 76, Jan. 1972, pp. 19–25; Logsdon, J. M., "The Space Shuttle Program: A Policy Failure," *Science*, 232, 30 May 1986, pp. 1099–1105; Pace, S., "Engineering Design and Political Choice: The Space Shuttle, 1969–1972," M.S. Thesis, MIT, Cambridge, MA, May 1982; Scott, H. A., "Space Shuttle: A Case Study in Design," *Astronautics & Aeronautics*, Vol. 17, No. 6, June 1979, pp. 54–58.

[39] Jenkins, D. R., *Space Shuttle: The History of the National Space Transportation System, The First 100 Missions,* 3rd ed., Cape Canaveral, FL, 2001.

[40] USAF Fact Sheet 86-107, "Manned Spaceflight Engineer Program," 1986; Cassutt, M., "The Manned Spaceflight Engineer Program," *Spaceflight*, Vol. 23, No. 1, Jan. 1989, p. 32.

[41] Weinberger, C. W., interview by Logsdon, J. M., 23 Aug. 1977, NASA Historical Reference Collection, NASA History Division, NASA Headquarters, Washington, DC.

[42] Smart, J. E., NASA Assistant Administrator for DoD and Interagency Affairs, to Fletcher, J. C., NASA Administrator, "*Security Implications in National Space Program,*" 1 Dec. 1971, with attachments, James C. Fletcher Papers, Special Collections, Marriott Library, University of Utah, Salt Lake City, UT; Fletcher, J. C., NASA Administrator, to Low, G. M., NASA Deputy Administrator, "Conversation with Al Haig," 2 Dec. 1971, NASA Historical Reference Collection, NASA History Division, NASA Headquarters, Washington, DC.

[43] Oberg, J., "Toward a Theory of Space Power: Defining Principles for U.S. Space Policy," 20 May 2003, p. 5, copy of paper in possession of author.

[44] Jenkins, *Space Shuttle.*

[45] Heppenheimer, T. A., *The Space Shuttle Decision: NASA's Search for a Reusable Space Vehicle*, NASA SP-4221, Washington, 1999, D.C., pp. 246–254.

[46] General Accounting Office, *Status and Issues Relating to the Space Transportation System*, General Accounting Office, Washington, DC, 21 April 1976; General Accounting Office, *Space Shuttle Facility Program: More Definitive Cost Information Needed*, General Accounting Office, Washington, DC, 9 May 1977; General Accounting Office, *Space Transportation System: Past, Present, Future*, General Accounting Office, Washington, DC, 27 May 1977.

[47] Lambright, W. H., *Powering Apollo: James E. Webb of NASA*, Johns Hopkins University Press, Baltimore, MD, 1995, p. 101; Ray, T. W., "Apollo's Antecedents: The Conceptualization, Planning, Resource Build-Up, and Decisions that Led to the Manned Lunar Landing Program," Ph.D. Diss., History Dept., University of Colorado, Boulder, CO, 1974, p. 192.

[48] Calculations based on budget data contained in *President's Report for Aeronautics and Space, Fiscal Year 2004*, *NASA Pocket Statistics*, 1997, and various project summaries, NASA Historical Reference Collection.

[49] Jenkins, D. R., *Space Shuttle: The History of the National Space Transportation System, the First 100 Missions*, 3rd ed., Voyageur Press, North Branch, MN, 2001, p. 256; Pielke, Jr. R., and Byerly, R., "Shuttle Programme Lifetime Cost," *Nature*, Vol. 472, No. 7341, 7 April 2011, p. 38.

[50] *We Deliver* brochure, p. 2, reproduced *Exploring the Unknown: Selected Documents in the History of the U.S. Civil Space Program, Volume IV, Accessing Space*, edited by Logsdon, J. M. (gen. ed.), NASA SP-4407, Vol. 4, No. 423, 1998.

[51] Polls available in NASA Historical Reference Collection; Launius, R. D., "Public Opinion Polls and Perceptions of U.S. Human Spaceflight," *Space Policy*, Vol. 19, No. 3, 2003, pp. 168–170.

[52] Robertson, J. O., *American Myth, American Reality*, Hill and Wang, New York, 1980; Kamman, M., *Mystic Chords of Memory: The Transformation of Tradition in American Culture*, Knopf, New York, 1991; Lowenthal, D., *Possessed by the Past: The Heritage Crusade and the Spoils of History*, Free Press, New York, 1996; Linenthal, E. T., *Sacred Ground: Americans and Their Battlefields*, University of Illinois Press, Urbana, IL, 1991.

Chapter 3

MANAGING THE SPACE SHUTTLE: LEADERSHIP, CHANGE, AND BIG TECHNOLOGY

W. HENRY LAMBRIGHT

INTRODUCTION

In early June 1971, James C. Fletcher, NASA administrator, decided that if President Richard Nixon gave NASA the go-ahead to build the Space Shuttle, it would depart from the Apollo management model and use a *lead-center* approach. This meant that there would not be a strong program manager at NASA's Washington, D. C., headquarters directing NASA's field centers; rather, a specific field center, namely the Johnson Space Center (JSC) in Houston, then known as the Manned Spacecraft Center (MSC), would steer decision-making for shuttle development and use. Sister centers would take marching orders from the JSC. The decision was extremely controversial at the time, and subsequently the organizational approach was changed—only to be shifted again, and then again. Indeed, there were various perturbations in the shuttle's management system over the years as NASA adapted to events and the course of the shuttle's technological evolution.

Why did NASA go from the Apollo model to a new design? Why did it alter that shuttle approach? What difference did it make for the shuttle that one part of the NASA organizational system, rather than another, was "in charge"? The case of the Space Shuttle provides evidence that one approach can work better than another. It suggests the way a program is structured can affect success in a large-scale federal technology program. The four-decade-long shuttle history provides many lessons learned, both positive and negative.

APPROACH

The Space Shuttle was a supremely complex technological system. It may have been the most complex machine built up to its time. It has also embodied an exceptionally complex organizational system. That is because the NASA shuttle organization—embracing headquarters, field centers, and contractors—is by no means a monolith. Indeed, every major organizational division (and subdivision) has unique interests. There is always a struggle between the parts and the whole. Centripetal and centrifugal forces are at play. A unified NASA and shuttle program have required more parochial interests to be checked and balanced in the interest of the whole.

Bringing the parts together to make a program succeed is not easy. Where leadership is assigned matters. The history of the shuttle reflects three leadership structures. The first is the *headquarters-control* model. In this approach, a strong program director in Washington, D. C., controls work, personnel, and budgets. The technical work is performed in the field centers and by contractors. However, power over policy (what) and management (how) are in Washington.

The second is the *lead-center* structure. Here, a field center is assigned the program management task. Other centers report to the lead official at the center. Headquarters' roles in this management approach are to get money and to defend the program to the president and Congress. Headquarters has that same policy/political role as in the headquarters-control model. It is expected to get resources and advocate. The difference is that policy and management are sharply differentiated in the lead-center approach, with management lodged in a center. Headquarters' staffing is minimal, especially in terms of high-powered technical oversight. The lead-center has power, and managerial "how" can stretch into policy "what" (that is, program direction).

The third structure exists only when leadership is highly dispersed. It is seen when management as well as technical work are outsourced from NASA to industry or quasi-autonomous governmental partners. International programs such as the International Space Station (ISS) best reveal this *pluralist* approach. The Space Shuttle had international partners, but their role was modest, and they performed more as contractors. The pluralist model can be managed, but the added complexity makes central direction more difficult. The shuttle was managed through this model in the mid-1990s, when NASA consolidated many contracts under a single company and sought to privatize Space Shuttle processing.

These three models of management—headquarters, center, and pluralist—are formal. They are set forth in organization charts and official divisions of labor. However, they tell only part of the story for the shuttle or any large-scale technology program. The *informal* relations among individuals involved in a program are extremely significant. These relationships are difficult to detect without intensely studying a program, its organization, and its dynamics of decision. The existing literature on the shuttle barely touches on such informal relations, although the Columbia Accident Investigation Board (CAIB) report discusses the importance of such informal aspects when it deals with "culture" and communication networks.

The three systems of NASA management are not closed. The political environment influences each and is, in turn, influenced by them. NASA deals constantly with the White House, Congress, the media, interest groups, and the public. The field centers relate to regional political officials and "their" representatives in Congress. Contractors have lobbyists walking the corridors of Washington. International partners assert their claims. More than any other individual, the NASA administrator copes with these diverse pressures, regardless of the management system.

THE EVOLUTION OF SPACE SHUTTLE MANAGEMENT SYSTEMS

Shuttle management has evolved through five major phases. The first was lead-center and extended from 1971 to 1974. In this era, the NASA administrator adopted the lead-center philosophy. The JSC was intended to be in charge.

A strong Director of NASA's Office of Space Flight in headquarters informally altered the lead-center approach. He complemented lead-center technical leadership with informal personal power to supervise and fund priorities during the shuttle's development period. When he left, and when the shuttle shifted from development to operations in the early 1980s, management controls weakened in both Washington and the lead-center. This phase lasted until 1986, when the *Challenger* shuttle accident took place.

The third era, *headquarters control*, began in the wake of *Challenger*. Ironically, Fletcher authorized this system just as he had the original lead-center approach. This headquarters control model lasted until 1996.

The fourth era began then. In that year, Daniel S. Goldin, NASA administrator, returned to the *lead-center approach* and extended it to a *pluralist model* because Space Shuttle processing was outsourced to industry. To the extent that the shuttle held together as a management system was because of the strong lead-center director at the JSC.

The fifth era, characterized by a return to *headquarters control*, began in 2002 with Administrator Sean O'Keefe and was amplified greatly by the 2003 *Columbia* shuttle disaster. O'Keefe's initial move to centralize power happened because of the ISS, not the shuttle. However, pressure to build the ISS put schedule pressure on the shuttle. The *Columbia* disaster redirected focus to the shuttle. The strong headquarters control approach continued under successive administrators until the end of the Shuttle program in 2011.

What one sees is an oscillation in approach. The pendulum of power within the management system has swung back and forth. A particular approach is used because of problems it is perceived to mitigate. That "solution" gives rise to a different set of problems, bringing about a change in approach. However, for one reason or another, that change does not last, and the pendulum swings in another direction. Overall, headquarters control seems to have worked better than lead-center, which is much less of a pluralist model. However, history shows that headquarters control, too, can have flaws.

Managing the shuttle has thus been a highly dynamic process that points up the limits of any organizational design for large-scale, complex, and risky technology. Theorists, historians, and practitioners of public administration would do well to remember an admonition issued years ago by Dean Acheson, who served as President Truman's Secretary of State: "Organization—or reorganization—in government, can often be a trap for the unwary. The relationships involved in the division of labor and responsibility are far more subtle and complex than the little boxes which the graph drawers put on paper with the perpendicular and horizontal connecting lines" [1].

There are people in those "little boxes," and lines of hierarchical relations do not necessarily work in practice. "Organization" is another word for "politics," with units of organization vying for power to control the agency's direction. Lead-center, headquarters control, and pluralism enhance certain interests while diminishing others. Somehow, agency leaders have to convert segmented wills into a larger will to have a chance to accomplish great goals. Not easy!

THE APOLLO ANALOGY

In the Apollo era, there were strong centers overseen by a stronger headquarters. James E. Webb, the NASA administrator throughout much of the 1960s, was not only an exceptional manager and canny bureaucratic politician, but he also was a management scholar. He read widely, and at an early point in his career, following a stint as Truman's budget director, he had an offer to be a dean of the University of North Carolina Business School. He thought deeply about management and wanted NASA to be—and to be perceived as—the best-run agency in Washington. Given the Apollo mission, he knew he needed strong and capable centers. He often spoke of the need to build and maintain technical "in-house competence." NASA needed real competence because the centers managed contractors, and industry performed 90 percent of Apollo's tasks. He wanted NASA to run industry and not vice versa.

However, strong centers, especially the human spaceflight centers, were constantly seeking autonomy and the expansion of their missions. The Marshall Space Flight Center (MSFC), headed by Wernher von Braun; the

Fig. 3.1 Apollo 11 mission officials relax in the Launch Control Center following the successful Apollo 11 liftoff on July 16, 1969. From left to right are: Charles W. Mathews, deputy associate administrator for Manned Space Flight; Wernher von Braun, director of the MSFC; George Mueller, associate administrator for the Office of Manned Space Flight (OMSF); and Lt. Gen. Samuel C. Phillips, director of the Apollo program. (NASA photo no. 108-KSC-69P-641)

MSC (later JSC); and what would be called the Kennedy Space Center (KSC) all jealously guarded their turf, even as they sought its enlargement (Fig. 3.1).

Webb created a very powerful headquarters directorate that counterbalanced the centers and coordinated their work. He believed in managing through checks and balances—among the centers, between centers and headquarters, and between NASA and industry. However, headquarters had to be in charge, and he in charge of headquarters [2]. To make sure NASA had the requisite expertise for Apollo, he brought in from the Air Force and industry managers who were steeped in "systems engineering." This culture complemented the Marshall rocket builders and Johnson spacecraft makers, many of whom brought different traditions from their predecessor organizations [3]. The NASA headquarters directorate for Apollo had plenty of help: Five hundred people with technical expertise were drawn from support contractors (Bell Com, General Electric, Boeing, and other firms) [4].

Air Force General Samuel C. Phillips, who was experienced at managing large-scale technological programs, served as Apollo director at the headquarters. The three field centers (MSC, Marshall, and Kennedy) reported to him. He had the power to hire his own cadre of contractors to enable him to have capacity independent of the centers. This "headquarters control" model came directly from the Air Force Ballistic Missile Program [5].

Webb genuinely tried to create the "perfect" management system. He came close. As management scholar Gary Brewer wrote: "Apollo successes created the powerful image of the space agency as a 'perfect place,' as the 'the best organization that human beings could create to accomplish selected goals'" [6].

NASA was never perfect, as the Apollo fire that took three astronauts' lives in 1967 revealed. Nevertheless, NASA was unusual in its ability to rebound quickly from the accident and achieve the lunar goal in 1969. Webb said that headquarters had to "penetrate" the centers and contractors to know what was really going on and to anticipate problems. Headquarters did an outstanding job at managing, but it missed the signals that led to the fire, because, as noted above, no matter how carefully you arrange organizational boxes, human beings still remain in them, and humans are imperfect. Moreover, where a complex technology is at issue, human error—commission or omission—reverberates widely.

Flaws notwithstanding, the Apollo approach worked! As NASA moved towards its lunar destiny in 1968, Dael Wolfe, in an editorial in *Science*, wrote [7]:

> In terms of number of dollars or of men, NASA has not been our largest undertaking, but in terms of complexity, rate of growth, and technological sophistication, it has been unique ... It may turn out that [the space program's] most valuable spin-off of all will be human rather than technological; better knowledge of how to plan, coordinate, and monitor the multitudinous and varied activities of the organizations required to accomplish great social undertakings.

The Lead-Center Decision (1971)

After the moon landing in 1969, NASA looked ahead and saw no approved post-Apollo program to maintain its organizational viability. President Richard Nixon had basked in the glory of the 1969 moon landing but showed no interest in aggressive post-Apollo proposals. Webb's successor, Tom Paine, made proposals that involved Mars, a space station, and a space shuttle to and from the station. Paine left in frustration. NASA's future seemed grim indeed, its budget falling, its personnel dispirited. Many of the Apollo managers, such as General Phillips, went back to the DoD and industry. The headquarters support team atrophied. NASA desperately needed a new large-scale technological program for continued viability, and the only one with a chance for Nixon acquiescence was the shuttle.

George Low, who had been Paine's deputy, served as acting administrator during this era. He activated the shuttle's planning, which went beyond the technical to the organizational. The key person involved in shuttle planning was Charles Donlan, deputy associate administrator for manned space flight at NASA headquarters. He headed a nascent program office for the shuttle, even though the president had not decided to have a program.

DONLAN'S DESIGN

Donlan, who had begun his career at NASA's Langley Research Center, reasoned that the shuttle would not need to, nor could, afford the large management force in headquarters that Apollo had enjoyed. He argued that the organization that had preceded NASA, the NACA, was decentralized and that Langley had enjoyed near autonomy. The NACA's headquarters had played a relatively modest role vis-à-vis the centers, and he wondered if the shuttle program could be managed in a similar way. Moreover, the human spaceflight field centers were getting into conflicts as they struggled for pieces of a shrinking NASA budget. There was a need to give them work and to clarify roles in hopes they could carry out a shuttle program, assuming one was approved. Donlan later described his reasoning at the time, 1971, for the lead-center approach.

"The shuttle is a system," he reasoned. "It is a system composed of these elements: the Orbiter, the [external] tank [carrying fuel], the boosters [rockets], and everything else, and ought to be designed not only as a system, but managed as a system." Of all the parts, the Orbiter [carrying astronauts] was most important, Donlan declared. It should be the driver. He said, "At all times, the status of the Orbiter should be the principal consideration… Anything that affects weight or its performance has got to be weighed very carefully. That means if, for whatever reason, the Orbiter's size is to change and requires a change in the tank, that would have to be done, and so on, right through the system." He went on [8]:

> So how do you manage something like that? My concept was to have the prime management where the Orbiter was the prime element; that the boosters and the solids [rockets] were subordinate elements. I structured a way in which Johnson [Space Center], as the Orbiter manager, would retain the prime field management for the system. And that Marshall would be delegated to carry out the management of the tank and solids, would indeed do that independently; but would report through the Orbiter office of the shuttle program office, and similarly for KSC, where all the requirements for launch facilities were being delegated.

MOVING TOWARD CHANGE

Donlan took his ideas to his boss, Dale Myers, the head of NASA's OMSF. He persuaded Myers that this organizational design was wise. They then won support for the plan from Low, the deputy administrator. After that, they discussed the arrangement with the senior managers of the three centers most involved.

The leadership of the MSC, the big winner in the plan, was naturally pleased. In response to a headquarters request in early April 1971, it sent a plan for how the lead-center approach would work. MSC provided a rationale for the change from the strong headquarters control model. It argued that centralization in the Apollo era was a function of the new and untested nature of the space field, the size of the program, its national priority, the challenge of developing a new launch vehicle and facilities from scratch, and the task of rapidly assembling the teams to accomplish the mission. Under those conditions, there had to be a strong headquarters directing overall systems engineering and integration. The technical support structure (such as from Bell Com and GE) were part of this capacity.

MSC officials commented, however, that "the basic situation has changed drastically." This was because both NASA and industry had greater "competence in manned spaceflight technology," "funding and cost effectiveness have become critical issues," and "both NASA and industry have developed specialized expert capabilities on manned spaceflight technology which are available for immediate application to the Space Shuttle Program." MSC listed the "very nature of the space shuttle concept" as requiring "a different management approach." This had to be "at the lowest possible level where continuous, direct technical communications can best be established and maintained" [9].

In a separate memo accompanying the planning document, Christopher C. Kraft, deputy director of MSC, emphasized that the lead-center design required a strong and clear statement of his center being in charge. He declared, "MSC should be given the responsibility for complete systems engineering, program management and control including financial management." The role of a headquarters shuttle director would be to "review the MSC decisions and concur in these decisions." If there were contentious program issues, their resolution should be "entrusted to the program management at MSC." He warned against any attempt to "water down this direct management responsibility for the lead-center concept." Such a move from "biting the bullet" of change from Apollo would only "drive the management [of the shuttle] back to the headquarters level." To allow each of the centers "equal" say in program management would require "a greatly expanded headquarters organization." Given the resource and personnel squeeze NASA faced, Kraft did not see headquarters getting the required added capability to oversee the shuttle program effectively [10].

Fletcher Decides

Once the centers heard about the proposed change, there was a "furor," particularly in the case of Marshall. Fletcher arrived as administrator on April 27, 1971. He had to deal not only with the turmoil of shuttle project management engulfing the agency but also the larger questions of shutting down the Apollo program and reorganizing for a smaller, less expensive space effort. He consulted and heard from many different sources. Among those against the reorganization of control was the senator from Alabama, John Sparkman, where the MSFC was located. Sparkman made it a point to let Fletcher know his views. Also among those who faulted the plan was von Braun, whom Paine had brought to Washington as a headquarters official to help sell a comprehensive post-Apollo program to Nixon in 1971. George Mueller, who had been the predecessor of Myers as OMSF director in the Apollo era, also made known his opposition to this management approach. He was "furious" with the idea of decentralizing management to a lead-center. The associate administrator with the general manager position (number three in the agency), Rocco Petrone, also registered opposition. The senior managers of Marshall particularly made known their dissent. Fletcher turned to General Phillips, who had been Apollo director at headquarters, and asked him what he thought. He said he was "neutral," adding that with enough effort "any organization could be made to work" [11].

Notwithstanding the opposition, Fletcher decided to go with the lead-center model, announcing on June 10, 1971, the choice of the MSC as lead-center for the Space Shuttle program. The announcement was clear as to which NASA entity was responsible for the lead-center management scheme, but it failed to convey the unequivocal power, especially financial control, that Kraft had sought. Headquarters (OMSF) would be responsible for providing a detailed assignment of responsibilities, delineating basic performance requirements, designating major milestones, and deciding what funding allocations would be made among the various field centers. Program management, though, would be under MSC. This role provided authority for the "total system" [12].

Under this plan, each of the three human space flight centers had its own work package. Marshall got all propulsion components, including external tank, solid rockets, and main engine. KSC would direct the launch and operate facilities associated with preparing, launching, and recovering the Space Shuttle during operations. JSC would have the orbiter and astronaut training. There would be project managers at each center for the work package at that center. An overall shuttle manager would be responsible for integration at JSC. The project directors would report to this program director.

NASA spoke about "levels" of program management, with headquarters in charge of policy (level 1), JSC in charge of management (level 2), the project directors at the three centers responsible for component work (level 3), and the industrial contractors accountable for hardware manufacture (level 4).

In theory, the division of labor was straightforward. In practice, it remained to be seen if this system would be effective. On January 5, 1972, President Nixon authorized NASA to proceed to develop the Space Shuttle. NASA projected it to cost $5.15 billion and to be completed within the decade, as early as 1978.

IMPLEMENTATION BEGINS

Implementation could now begin, with the selection of prime contractors. The Orbiter contract was critical. The choice was between North American Rockwell and Grumman. The decision went to North American Rockwell. A contract was issued in August 1972. By the end of 1973, all the key contracts were set or in process. These included—in addition to the Orbiter—the external tank to Martin Marietta; the main engine to Rocketdyne; and the two SRBs to Morton Thiokol.

The Morton Thiokol contract was the most controversial. Fletcher was from Utah, and the contractor was based in Utah. The chair of the Senate Space Committee, Frank E. Moss, was from Utah. Even the Mormon Church (of which both Fletcher and Moss were devout members) was rumored to have had some influence on the decision. At one point, Fletcher even wrote Moss a sharp letter on the awarding of the contract [13]:

> One of your staff—I think you probably know who I am referring to—went so far as to insinuate sometime ago that I had a moral, if not a spiritual, obligation to acquiese on [*sic*] some of [the] business issues previously raised by President [N. Eldon] Tanner [of the Mormon Church]. This person voiced an unthinkable opinion to the effect that my Church membership took precedent [*sic*] over my Government responsibilities.
>
> Knowing that you share similar sentiments with me in the clear separation of Church and State, I would like to request that you take this unpleasant matter under advisement with the individual in question and explain just how serious and unconscionable those references were.

The NASA administrator denied any political influence. (Later, after 1986 and the *Challenger* accident, the Fletcher–Morton Thiokol link came up again, but nothing was proven.) [14]

In any event, with the contractor decisions, the main questions about organization and division of labor were settled. Particular individuals in the centers were chosen to run the specific work packages along with the program manager at JSC, Robert F. Thompson, to integrate the projects into a program. Thompson's position was key under the lead-center model [15].

Myers, as OMSF director at headquarters, had overall responsibility, and Donlan was his subordinate. As advocates of the lead-center approach, they had high stakes in making it work. On March 15, 1973, Myers promulgated an extremely detailed "management instruction" for the lead-center model [16]. Reading the management instruction, one could see the centrality of the shuttle program manager in Houston. Although this individual had nowhere near the power Kraft had wanted, the authority he possessed was substantial and now better spelled out. For example, he would set requirements and make the "integrated funding request" that went to headquarters. NASA was making a distinction between policy and management, with management being a very "big M" at MSC. Much authority was delegated to MSC. The architects of this management instruction did not stay to monitor its execution, however; Donlan left his position in 1973, and Myers left in 1974.

A Reoriented Lead-Center Strategy (1974)

On May 9, 1974, Fletcher named John Yardley as associate administrator for the OMSF at NASA headquarters. Fletcher also appointed Myron Malkin to be Donlan's successor as deputy for the shuttle. Yardley, then 49, had been vice president and general manager at McDonnell Douglas. As a contractor, he had worked on human spaceflight for NASA in the 1960s and was eager to lead NASA in shuttle development. He took a large pay cut to come to NASA.

At OMSF, Yardley also had activities other than the shuttle with which to deal, but the shuttle was his top priority. It was also the overriding priority for Fletcher. The future of NASA and human spaceflight depended on the shuttle, and everyone associated with the program knew that. Yardley had to make the shuttle succeed. The shuttle was not only a great engineering and management challenge but also the greatest personal challenge of Yardley's career. The organizational design Yardley inherited implied that he had to cede some control of the program to Thompson at the JSC, the renamed MSC in 1973. However, Yardley was not about to give up that much.

Yardley had a tough and demanding administrative style. He possessed a technical acumen such that "no one could snow him." He "did not tolerate sloppy work," and he could berate and embarrass those who fell short of his standards. He was not one to delegate to another person who was working with the centers. In fact, he would get involved in the technical decisions on the shuttle to the extent he could. Fletcher called him "brilliant" [17]. Thus, the sharp line between policy and administration established by the lead-center approach did not suit Yardley. He made the rounds of the field centers and spoke to the principals concerned with building the Space Shuttle. He concluded the management system needed adjustment. Around Christmas 1974, he wrote a memo to Petrone. "Management surveillance and evaluation of sister center activities [by JSC] are very weak," he said. He reported that Thompson as program manager at JSC "feels that once the technical direction has been given, it's clearly the other center's job to produce, and he feels little, if any responsibility if problems arise." Yardley had also spoken to Kraft, by then the JSC director. He reported that Kraft had told him that JSC "could not direct MSFC as a practical matter." Complicating everything, said Yardley, was the fact that the management styles of the two centers—MSFC and JSC—were entirely different. MSFC had an informal decision-making style. That of JSC was more structured. The lead-center approach further exacerbated relations among the various players in the centers, including the program managers and center directors. He implied that there was distrust on the part of Marshall and KSC about the way work and money were distributed to them given JSC's role.

What was to be done? Ironically, Yardley did not recommend abandoning the lead-center design. Instead, he called for more headquarters activity in neglected areas of lead-center management. Headquarters should engage more in supplying "the missing management surveillance function." Also, headquarters needed to play a greater "role in the fiscal area." It was necessary for headquarters to "assure objectivity" in JSC management and budget decisions, he declared [18].

In other words, he would keep the model but add greater checks and balances to it via headquarters—that is, his own role. The informal would complement the formal. JSC's "management" role was limited to the more narrowly technical. That was still significant, but decisions about money and oversight across centers gravitated upward to Yardley as he put his imprint on the program. This was what Kraft had predicted would happen in his 1971 memo in which he had demanded unambiguity in MSC's leadership charge. There was enough ambiguity between policy and management for Yardley to take back a significant measure of command his predecessor had apparently been willing to relinquish.

Indeed, it appears that Fletcher himself thought the shuttle too central to NASA to allow the kind of delegation implied in the original lead-center vision. With Fletcher as administrator, the shuttle continued to have the highest priority at NASA. Thereafter, the 1974–1977 period was fairly stable. Shuttle development moved ahead under the reoriented management approach. The consensus view within and outside NASA was that the shuttle would make space transportation routine and less expensive and that it would launch the bulk of military, science, and commercial payloads.

A New Administrator Makes Management Changes (1977)

In 1977, President Jimmy Carter came to power. His appointee as NASA administrator was Robert Frosch. Yardley continued as the head of what was now called the Office of Space Flight. What Frosch inherited from Fletcher was a development program aiming eventually for a fleet of four operational shuttles. Frosch hoped to sell Carter on a fifth shuttle, but circumstances changed that calculus. Management and technical issues would force Carter to consider major policy changes, including the shuttle's termination.

Superficially, the shuttle program appeared in good shape, but several issues were festering. One was the erosion of organizational capacity at Marshall for the launch phase. The Germans under von Braun who had dominated the center since 1960, and others who had made the Saturn rocket so successful were getting older; many were retiring. This cohesive group, a unit that had poured its experience, skill, and total attention into the Saturn, was not easily replaced. As NASA concentrated on hardware development, Fletcher looked ahead to the launching of Space Shuttles.

On July 7, 1977, Fletcher wrote an "eyes only" memo to Alan Lovelace, the senior associate administrator. He warned that Marshall, "from Bill Lucas [the director] on down, does not feel it has the same responsibility for the shuttle (during the launch phase) that it had for the Saturn. In fact, Bob Thompson and Chris Kraft feel this is Houston's responsibility, just as the orbit phase and the re-entry phase are their responsibility" [11]. The problem, Fletcher said, with this division of responsibility was that JSC did not really understand rockets. That was not their expertise. Nor was it the expertise of Yardley and other headquarters managers. Launching a shuttle was complicated and new. Fletcher saw a mismatch between shuttle requirements and organizational capacity. He pointed out that Lovelace's predecessor, Petrone, had brought up this problem of expertise and responsibility "several times, and our unresponsiveness may have contributed to his leaving" [19].

Fletcher did not attribute this organizational problem to the lead-center arrangement. Nevertheless, the division of labor created by Fletcher's 1971 decision may have been part of the issue, and he recognized that it needed addressing before any attempt at flying the new vehicle could be made. Marshall was the rocket/propulsion center. Its officials saw the center's job as done once the shuttle was ready for launch. That had not been the case where Marshall was concerned when the Saturn rocket was at the launch phase. Marshall's expertise carried over to participation in the launch phase. Now, under the division of labor set in motion by the lead-center decision, there was greater compartmentalization. Marshall was handing off the shuttle to JSC, and Fletcher questioned JSC's expertise at the launch stage [20].

It is not clear whether Lovelace did anything in response to Fletcher's expression of concern. The launch of a shuttle was not imminent. Technical and budget issues had set test launches back from 1978 to 1979. Other problems than this potential "mismatch" were coming to the fore in 1977.

Management Troubles

At the beginning of 1977, Yardley had told Congress that all was well with the shuttle, but by October 1977, he went back to ask Congress for more money to cover additional costs. The main engine was causing problems, Yardley stated. Congress did not care for dealing with a supplemental budget request that came as a surprise. It asked the National Research Council (NRC) to investigate the shuttle program independently of NASA. In December, before NRC reported, a fire destroyed a shuttle engine during testing. In February 1978, the NRC issued its report. It reviewed the various technical problems it had found and declared it was "unlikely that the first manned orbital flight will occur before April or May, 1980" [21].

NASA said it was making progress and dealing with its technical problems. The Senate Space Committee in April 1979 reported favorably on the NASA FY 1980 Authorization Bill, incorporating in it the 1979 supplemental funding request. Soon after that action, however, Frosch had to inform Congress he would need even more money, meaning there would be another supplemental request. Three senior senators on the committee wrote Frosch demanding to know what was going on [22].

A new problem was happening—or a new awareness of an old problem. This was that tiles used on the Orbiter to protect it from the intense heat of reentering the atmosphere were not adhering. This issue was adding to development costs and delays. What seemed to really bother the senators, though, was that the request for supplementals made them wonder if NASA headquarters leaders knew what was taking place in NASA centers and with contractors. In short, was there a management problem?

In May 1979, Frosch, chagrined, ordered Lovelace (now his deputy) to conduct an internal study of NASA's shuttle management. "It is clear," Frosch said, "at a minimum that we have some difficulties in vertical and horizontal communications." He directed Lovelace to make this management evaluation his top priority [23].

In October 1979, the result of the study became known. The lead-center organizational design remained acceptable, but people in headquarters positions were doing too much. The informal reorientation upward that Yardley had adopted in 1974 was no longer working well. Yardley, in particular, was not only supervising development efforts, he was becoming so involved in day-to-day activities that he was even spending time negotiating operational agreements with the user communities as development progressed and as the flight era of shuttle beckoned. Frosch decided to divide Yardley's shuttle development and utilization tasks in half. Hereafter, Yardley would be in charge only of an Office of Space Transportation System Acquisition. Frosch said another associate administrator would be named to head a new Office of Space Transportation Systems Operations. The Office of Space Flight was thus bifurcated [24].

The aim was to allow Yardley to regain control of technical issues at the center and contractor level. In the process, Yardley went deeper into technical management, and the shuttle's policy dimension moved upward to the White House.

President Carter was aware of the shuttle's technical and financial problems, and especially its delays. He also knew more requests for funding were coming. Accordingly, he was not interested in considering a fifth operational shuttle. He was instead more concerned with whether or not to terminate the Space Shuttle in its development stage. While NASA conducted its internal assessment, Carter awaited results from a study of his own. Carter had not given the shuttle or space policy generally much attention up to this point. However, he was quite concerned about being able to monitor from space any adversaries' weapons development and nonproliferation agreements. DoD needed to launch surveillance satellites and was likely to be dependent, as a user, on the shuttle for some of its heavier and most sophisticated satellites. That particular matter was especially important to Carter as he weighed decisions about the shuttle.

On November 14, 1979, Carter met privately in the Oval Office with Frosch. The two then walked to the White House's Cabinet Room, where several others joined them. From NASA were Lovelace, Yardley, and Bill Lilly, NASA's budget officer. From DoD was Hans Mark, the deputy secretary. From the White House staff were the president's science adviser, Frank Press, and director of the OMB, James McIntyre. A few others in support of these principals also attended the meeting.

Carter had Frosch present to the group a status report on the shuttle. Carter then ordered Frosch to write a strong statement of support in the president's name. He designated McIntyre as Frosch's point of contact in the White House. The group discussed the money issue—$1 billion beyond what OMB had contemplated to finish building the shuttle. Carter turned to McIntyre and directed him: Find the money [25]!

Additional budget discussions had ups and downs for Frosch. He pursued the argument that seemed to have helped most with Carter—the national security connection. In the end, NASA's total request for FY 1981, plus a supplemental for FY 1980, came to $1.14 billion more than OMB had projected before the meeting with Carter. Most of this increase went to the shuttle, and it came from the DoD. Congress that year largely concurred [26].

TILE ISSUES (1980–1981)

By 1980, many challenges remained, but with additional funding, NASA was projecting that the first shuttle would be launched in 1981, three years later than promised in the original schedule. No longer was the main engine the problem delaying development. The pressing item involved the thermal protection system, especially the tiles. NASA had expected the tiles-adhesion problem to have been resolved by now. In 1979, it had moved its first finished shuttle, *Columbia*, from North American Rockwell's manufacturing plant in Palmdale, California, to KSC. Only 24,000 of the 30,000 thermal protection tiles were installed at that point (Fig. 3.2). Progress had since been made, but in early 1980, there were still 9,500 to 12,500 tile repairs to go [27].

In February, Yardley stated that NASA had worried so much about whether the tiles would be capable of thermal protection, it had not really contemplated how hard it would be to keep them on. They seemed surprisingly fragile, especially during take-off and landing. Also, the tiles were going to cost three or four times more than what was originally estimated. Blame was shared, said Yardley, between NASA and contractor [28]. Efforts to get them

Fig. 3.2 NASA had serious difficulty making the Space Shuttle's thermal protection system adhere to the vehicle. Here, some of *Columbia*'s thermal tiles are missing from the vehicle. (NASA photo)

to adhere properly continued until near the time of the first launch, in April 1981.

NASA's Priorities Change

In January 1981, Ronald Reagan became president. His appointee as NASA administrator was James Beggs. Beggs had two priorities: 1) make the Space Shuttle operational, and 2) get Reagan to authorize the Space Station. With the first launch of the Space Shuttle, on April 12, 1981, this first priority seemed on the verge of realization. The mission was a great success. One month later, on May 18, NASA announced Yardley was leaving to return to industry to become president of McDonnell Douglas Corporation. He felt he had done his job bringing the shuttle to this point. Development was perceived as largely over.

The following October, Major General James Abramson was named Yardley's successor. Beggs did not judge Frosch's notion of dividing leadership of the program in half necessary, and both the developmental and operational elements coalesced under Abramson. Meanwhile, *Columbia* continued its test flights while other shuttles were being constructed. There were a total of four tests, with the final flight completed on July 4, 1982, when *Columbia* landed at Edwards Air Force Base. President Reagan was there to declare, "Beginning with the next flight, the *Columbia* and her sister ships will be fully operational, ready to provide economical and routine access to space for scientific exploration, [for] commercial ventures, and for tasks related to the national security" [29].

The appointment of an able manager with strong ties to the military users, coupled with the successful shuttle tests, freed Beggs to concentrate on his space station quest. In late 1983, Beggs persuaded Reagan to go ahead with what became the ISS, and the president announced his decision in his State of the Union address at the beginning of 1984. This decision made a big difference for the Space Shuttle; it was now not the primary focus of the agency's top executives. The fifth Orbiter receded as a proposal by the agency. The emphasis in headquarters and at centers was the space station.

Abramson left NASA in April 1984, and Hans Mark, Beggs' deputy, departed soon after. Phil Culbertson, who had been Abramson's deputy, was involved in the management of the shuttle, even as he moved upward in the organization. He increasingly had many other tasks to pursue, including helping Beggs forge partnerships with other countries in building the space station. Moreover, Culbertson's management style was not to intervene closely in center affairs [30]. The lead-center approach remained, but as NASA entered an era of operations, communication among centers and between centers and headquarters was less intense than when the shuttle had been in development. Also, Beggs was personally not particularly approachable, as many in NASA found. According to Culbertson, center directors found it difficult to bring problems to Beggs. But, Culbertson noted, there was also a "long tradition of [center] independence from Washington in solving problems, not passing them on" [31].

Culbertson recalled that in the early 1980s, "everybody was caught up with the spirit of yes, we must become operational." Being operational meant being "competitive," being "responsive" to users. "It was clear the greater the launch rate, the more economical the system would be to operate. And the more effectively we drew missions to the Shuttle, the less it was gonna cost NASA for shuttle missions" [32].

Through all of this, the shuttle seemed to be performing well. Between 1982 and 1986, the shuttle retrieved communications satellites that had suffered upper-stage misfires after launch, repaired another communications satellite on orbit, and flew science missions with the pressurized spacelab module in its payload bay that European partners had built. The shuttle launched not only U.S. astronauts, but also citizens of Germany, Mexico, Canada, Saudi Arabia, France, and the Netherlands; two payload specialists from commercial enterprises; and two U.S. members of Congress, Senator Jack Garn (R-Utah) and Representative Bill Nelson (D-Florida). In 1985, when four orbiters were in operation, the vehicles flew nine missions. This number turned out to be the most shuttles NASA would ever launch in a single year [33].

Although the shuttle operated well, it did have difficulties. Rather than needing 10 working days to process the shuttle between flights, as originally planned, an average of 67 days elapsed before a shuttle was ready to launch [34]. This additional work caused NASA to issue a shuttle processing contract to Lockheed.

There were technical problems during this period, and some made their way to headquarters. However, none registered alarms, especially given turnover at headquarters—and because of centers' reluctance to raise issues to Washington. This was especially true for officials at the MSFC. Under the lead-center approach, it could have reported problems to JSC, but it seldom did so. The NASA management system that had been established at the outset of the shuttle program was not working effectively in early 1986. The management breakdown was exacerbated in late 1985, when Beggs had to take a leave of absence (and eventually resign) to fight what turned out to be a false charge of malfeasance when he was in industry. The White House had forced acting NASA administrator, Bill Graham, on Beggs. Graham did not have the respect of Beggs nor the cooperation of some of his key NASA associates. There was a vacuum of leadership at headquarters when disaster struck [35].

BACK TO HEADQUARTERS CONTROL, 1986

On January 28, 1986, the *Challenger* shuttle failed 73 seconds after launch, and its 7 crewmembers perished, including the first "teacher in space." The explosion was seen on television by millions. NASA was devastated. The president established an independent commission, the Rogers Commission, which came to be known after its chair, former Secretary of State William P. Rogers. The commission determined the fault was both technical and managerial, with NASA's institutional culture and communications system being partly to blame. Without question, NASA's lead-center approach to shuttle management had been a significant part of the problem.

A relatively small group of engineers from Marshall, Kennedy, and Morton Thiokol, the solid rocket motor contractor in Utah, had conversed by telephone the evening before the launch. The issue was whether to launch or delay because of the deleterious effects of prevailing cold temperatures on the rubber O-rings that sealed segments of the solid rockets together. Marshall's Larry Mulloy, SRB project manager, was impatient with those who argued for delay: "My God, Thiokol. When do you want me to launch, next April?" he asked. Thiokol supervisors, apparently anxious to please the customer, relented [36].

What especially bothered the Rogers Commission was that there was no communication about this debate to JSC (the supposed "lead") or headquarters.

Schedule pressure trumped prudent risk management, and evidence showed a NASA proclivity to turn deviations from sound operating procedures into "normal" behaviors. When the Rogers Commission issued its report on June 6, 1986, it addressed both technical and management concerns. It criticized the lack of accountability of project managers at centers to headquarters and the isolation of Marshall in particular. It called for structural changes in program management, charging the existing system had failed [37].

The NASA administrator who received the Rogers Commission report was Fletcher. He had returned to NASA's helm in May 1986 at President Reagan's behest to help NASA respond to the withering criticism it had received and to direct the recovery. Richard Truly, a rear admiral in the Navy and former astronaut, had become associate administrator for space flight on February 20, 1986. There had been many changes in shuttle leadership in Washington and at the centers, with more to come in the aftermath of the *Challenger* accident [38].

One of Fletcher's first moves was to ask the former Apollo manager (headquarters), General Phillips, to conduct an overall review of NASA organization and management. Truly, in turn, charged astronaut Robert Crippen with assessing the shuttle management structure. The findings of Phillips and Crippen were consistent that fundamental changes were required. Fletcher gave the go-ahead for a new era in shuttle management.

On November 5, 1986, Truly announced the change. The new structure centralized management at headquarters and looked a lot like the management system of Apollo, although it would not have the huge technical support structure Apollo had employed. Truly brought Arnold Aldrich, who had been shuttle director at JSC, to Washington as shuttle program director. The JSC, Marshall, and Kennedy project managers would report directly to him. Aldrich was given broad authority over requirements, hardware decisions, and budget matters.

In addition, Truly restored the moribund OSFM Council. Truly chaired this body, which included the JSC, Marshall, and KSC directors, along with Aldrich and a few others. Under Truly, it met monthly and brought to the surface issues that were typically buried. The aim was to maximize communication about problems, whether technical or managerial.

Fletcher and Truly addressed other management issues that the Rogers Commission had recommended, such as raising the authority of the safety office and putting more astronauts in management positions. It took 32 months for NASA to return to flight, this flight taking place on September 29, 1988. Along the way, President Reagan made important decisions affecting the program. A replacement shuttle would be built. Schedule pressure would be reduced by eliminating shuttle service for commercial and national security payloads. The primary task now was the launch of science payloads too big for expendables. The leading example of this kind of mission was the Hubble Space Telescope, launched in the spring of 1990.

THE AUGUSTINE REPORT (1990)

When George H. W. Bush became president in 1989, he promoted Truly to NASA administrator. The headquarters-control system remained in place. However, another management problem arose. In 1989, the president called for a return to the moon, this time to stay, and then on to Mars. This new mission came as the shuttle was having various technical problems and as it was becoming very difficult to get the space station program built. When the Hubble Space Telescope was found to have blurred vision, the White House grew alarmed and empanelled a commission headed by Norman Augustine, a retired aerospace industrialist, to analyze the problem and recommend a solution. The problem, said Augustine, was that NASA was trying to do too much with too little money. The Augustine commission, in its 1990 report, called for a 10 percent per year funding increase through 2000 for NASA. Without the extra money, the Augustine report said, NASA should do less [39].

Truly and the White House, especially Vice President Dan Quayle, did not get along. Truly's main concern was the shuttle, with the space station being the second priority. Moreover, Truly pressed for a fifth shuttle. Out of sync with the White House, which wanted Truly's attention on Bush's Moon-Mars program, Truly was forced out of office. In April 1992, Daniel S. Goldin replaced him (Fig. 3.3) [40].

Fig. 3.3 NASA Administrator Daniel S. Goldin lectures U.S. Secretary of State Madeleine Albright at the KSC's Apollo/Saturn V Facility while awaiting the launch of STS-88, the first U.S. launch for the ISS, in 1998. (NASA photo no. KSC-98PC-1769)

Return to Lead-Center (1996)

Goldin thought that NASA could do more with less, as long as it embraced management efficiencies. Bush's Moon-Mars mission ended once Bill Clinton became president in 1993, but everything else continued. So did Goldin, who called his (Goldin's) reforms "faster, better, cheaper." Initially, the shuttle remained under the centralized headquarters approach. However, the emphasis on the Space Station (now known as ISS) increased. Seeking money for ISS, within a human space flight budget that combined the Space Station and shuttle programs, Goldin favored ISS. The shuttle was increasingly constrained in money and people. Goldin kept trying to do more within a budget that actually decreased for most of the Clinton years [41].

In January 1995, chastened by Republican control of Congress after the November 1994 election, Clinton proclaimed "the end of big government." With budget cuts across most agencies, Clinton told NASA to find $5 billion in savings in its already austere five-year spending plan. Goldin again sought to maintain programs with less money than expected. With ISS essentially "fenced off" from cuts, and Goldin and the White House favoring science missions, cuts fell heavily on the shuttle.

To help save money, Goldin adopted restructuring as a management strategy. What restructuring meant for the shuttle was a return to lead-center management. It also meant "privatization." According to a report chaired by former JSC Director Kraft, the shuttle was mature. Thus, much shuttle work could be "privatized"; that is, done by the private sector. Lockheed Martin and Rockwell formed United Space Alliance (USA) to consolidate multiple contracts into one major contract for shuttle processing. Goldin also made JSC once again the lead-center. This time, JSC had both the shuttle and ISS to manage. The return to lead-center management, which took effect in February 1996, caused Bryan O'Connor, a former astronaut, to resign his position of shuttle director, headquarters. He charged that NASA was resuming the organizational design that helped cause the *Challenger* disaster. Goldin countered that too many civil servants obfuscated responsibility and hurt safety.

Privatization and decentralized control would allow NASA to downsize personnel, especially at headquarters. Goldin put his close associate, George Abbey, in charge of the Space Station and shuttle in the role of JSC director. Given the USA privatization strategy, NASA was extending the lead-center model to one of pluralistic management. Abbey had great power—formal and informal—and held the sprawling human space flight system together. However, the limits of this approach gradually became clear.

Back to Headquarters Control (2002)

By 2001, a host of problems had arisen with Goldin's many reforms. Where the shuttle was concerned, it was clear that there were safety issues.

Moreover, the lead-center approach once again was proving highly problematic. Goldin himself relieved the JSC director, Abbey, of command of JSC, the shuttle, and ISS.

In November 2001, Goldin left NASA, and President George W. Bush replaced him with Sean O'Keefe, the deputy director of OMB. O'Keefe was appointed (and Abbey replaced at JSC) because a $4.8 billion cost overrun on the ISS seemed to have appeared out of the blue [42]. In that context, the shuttle was not seen as "a problem" demanding topside attention.

As a result of the space station overrun, O'Keefe pulled power over ISS up to headquarters. The shuttle came along with it. O'Keefe made Bill Readdy associate administrator for space flight. O'Keefe's priority was ISS, but he was cognizant of the shuttle and its risks. On O'Keefe's first day on the job, he asked to see NASA's plan for dealing with a shuttle disaster. In 2002, he proposed a shuttle-upgrade program that could extend the shuttle to 2020. He also proposed an Orbital Space Plane to take some of the load off the shuttle. Catastrophic events put an end to these proposals, however.

On February 1, 2003, the *Columbia* orbiter broke up during reentry, and in the process, seven astronauts were lost. This accident returned the shuttle to center stage for headquarters—and all of NASA. Once again, a panel, the Columbia Accident Investigation Board (CAIB), was assembled to investigate what went wrong. As the Rogers Commission had before, CAIB found both technical and managerial error. The technical problem was that a piece of foam fell off the external tank and caused a hole in the shuttle's thermal protection system, at the leading edge of a wing. When the shuttle entered Earth's atmosphere, the intense heat destroyed it.

The managerial problem was that warnings about foam debris had been made along the way, and these were unheeded. After the *Challenger* launch, various engineers detected foam that had fallen off, and they made mission managers at JSC aware of that fact. Apparently, although the information was widely circulated around NASA on the internet, JSC mission management did not take this information seriously at the time, nor was it successfully elevated to NASA headquarters. The Rogers Commission had found serious communication problems, and so did CAIB. Although individuals made choices, organizational structures and culture mattered too. CAIB chastised the lead-center approach, privatization, management turbulence (too much change), the belief that the shuttle was "operational" in the routine sense, and the fact that official safety programs lacked clout with shuttle program officials.

Ironically, although lead-center approaches may have been part of the historical forces that over time led to the *Columbia* disaster, according to CAIB, centralization under O'Keefe had also played an indirect role. While at OMB, O'Keefe had put ISS on probation for being late and overly expensive. At NASA, he strove to get ISS off probation by getting ISS built faster via the shuttle. He thus added to schedule pressure.

Thus, the headquarters control model did not necessarily alleviate communication problems. In *Columbia*'s case, communication proved inadequate vertically and horizontally. CAIB had a number of important points to make about principles that would guide reforms in NASA's organizational system. These would go beyond formal structure to emphasize informal culture. First, leaders—not just NASA leaders, but also national political leaders —create culture by putting financial stress on the organization, thereby compromising sound technical principles. Leaders create culture, but they can also change it. Second, although change might be desirable, attention must be paid to impacts and unintended consequences of management reforms.

O'Keefe took the CAIB report, which covered many of the shuttle issues under both Goldin and O'Keefe himself, seriously. He made a number of shuttle-related changes. These included creating a technical safety organization independent of the program hierarchy. He searched more for "cultural" improvements in communication, and he spoke of "one NASA." O'Keefe personally participated in meetings about the shuttle's return to flight. He imposed stronger control of field centers via headquarters. He changed personnel. O'Keefe, however, retained Readdy as head of the shuttle and ISS. Also, O'Keefe used the window of opportunity *Columbia* had created to persuade President Bush and Congress to give NASA what CAIB argued it needed: a new, overarching goal that would pull the agency together and engender public support. In 2004, Bush announced a decision for NASA to go back to the moon and then to Mars and beyond.

Control by the NASA Administrator (2005)

O'Keefe left NASA in early 2005, and Michael D. Griffin succeeded him. Griffin kept the structure he inherited, but he placed a host of his appointees in the top positions at headquarters and in the centers. His priorities were to return the shuttle to flight, complete building the Space Station, and get the Moon-Mars exploration system under way. Above all, he wanted to narrow the gap between the retirement of the shuttle, which Bush had set for 2010, and the initiation of a successor in 2014 [43].

Griffin was an extremely well trained and confident technical manager. He personally understood and intervened in technical decisions. He said that after spending $1 billion to refurbish the shuttle since the *Columbia* disaster, it was time to launch, and on July 13, 2005, after two and a half years of repairs, the shuttle flew. However, there were still foam problems, so Griffin grounded the shuttle once more.

In June 2006, shuttle program managers signaled that the spacecraft was "ready." However, the agency's top safety official as well as its chief engineer disagreed. The decision was left to Griffin. He said the risk was acceptable. On July 4, the shuttle roared into space. When it landed safely on July 16,

Griffin was vindicated. He said it was now time to resume shuttle flights to finish the Space Station. Asked by media about his "feelings" at this time of triumph, Griffin said, "I'll have time for feelings after I'm dead." There was no question that under Griffin, the formal and informal structure was headquarters dominated: Griffin and the man he chose to run space operations, William Gerstenmaier, were in clear control. Gerstenmaier was a veteran technical manager who was highly regarded within and outside NASA. In the wake of *Columbia*, communication was not a serious issue.

Under Griffin, all went well subsequently with the shuttle, except that President Bush did not provide NASA with adequate funds for the "old" program (that is, the shuttle and ISS) and for a new exploration mission (called Constellation). One result was that the gap between ending the shuttle and launching its successor grew.

MANAGEMENT AND THE END OF THE SPACE SHUTTLE PROGRAM

In January 2009, Barack Obama became president and chose Charles Bolden as NASA administrator. In February 2010, Obama ended the Constellation program and called for commercial firms to replace the shuttle. After many months of White House/Congressional conflict, there was a compromise on building a new rocket and spacecraft to take NASA beyond low-Earth orbit, while nurturing a private enterprise to perform the shuttle's task to service ISS. Gerstenmaier provided continuity during this period. He guided the last 21 shuttle missions that enabled completion of ISS [44]. The last shuttle flight landed on July 21, 2011.

CONCLUSIONS AND LESSONS LEARNED

A longstanding truism was the case for the Space Shuttle program: "Organizational arrangements are not neutral" [45]. They favor certain interests and hurt others. NASA is not a monolith. It has a headquarters and 10 field centers. Contractors perform most of its work. It has many programs. Industrial and university contractors have interests of their own. Bringing all these different entities into a coherent whole is a supreme test for management. The closest NASA ever came to having one overarching interest that brought it and those who worked for NASA together was in the 1960s, with Apollo and the moon goal. Since then, getting cohesion around a program, much less NASA as an agency, has seemed impossible.

Looking at the Space Shuttle program, we can see that over a four-decade history, there have been two primary structural approaches, with a third being in nascent form. One is the headquarters-control model. A second has been the lead-center approach. A third was the pluralist model, which emerged to a modest extent with the role played by USA, the private firm created as a partnership by Boeing and Lockheed Martin to manage shuttle operations in the late 1990s. There were variations among these approaches because formal structures were complemented by informal relationships. Some models were more tightly controlled than others, owing to the personalities and skills of the individuals involved in authoritative positions. Organizational arrangements are inevitably about power—who is in charge, how authority is exercised, whose interests prevail, and whether the values of the organization as a whole, or those of subordinate entities, are maximized. Nothing is more important than leadership in turning organizational relationships into achievements. The key questions are these: Who leads? How? With what consequences? The following lessons may be derived from the shuttle experience:

1. The "pure" lead-center approach, as formulated at the outset of the shuttle program, might have made sense from a rationalistic standpoint. However, this approach failed to take account of discrete field center interests and rivalry among JSC, KSC, and Marshall. Sister centers were not about to take orders from JSC, the designated lead, and JSC's key decision makers for the shuttle were neither willing nor able to enforce their will.
2. Particular leaders matter. Yardley recognized the weakness of the lead-center model in the development stage of the shuttle. He kept the model as a formal system, but he complemented it informally through personal interactions. The lead-center approach was maintained for technical decisions, but supervisory and financial decisions affecting the shuttle were made or strongly influenced by headquarters. The development of the shuttle was most influenced by the personality and skills of Yardley. However, because Yardley was distracted in the late 1970s, the informal headquarters supervision weakened, and technical issues worsened.
3. Priority matters. Fletcher and Frosch supported Yardley. Ultimately, President Carter supported NASA management with the infusion of needed funds. The shuttle had top priority among space programs through its development period, and this priority was critical to moving the shuttle forward.
4. Once the shuttle transitioned to the operational stage in the first half of the 1980s, the formal structure of lead-center continued. However, the informal headquarters' oversight role weakened because the priority of the administrator (Beggs) and his management team shifted to getting and then implementing the Space Station. As the priority of the shuttle lowered, so did informal headquarters' vigilance. Organizational culture adapted to a "routine" program. Lead-center controls did not work. Marshall, in particular, acted as an independent entity from a communication standpoint.

5. Disasters change priorities. After *Challenger*, NASA reverted to a strong headquarters-control approach. As associate administrator for space flight, Truly took charge, backed by Fletcher. For both men, return-to-flight was "the" priority. Thanks to the Rogers Commission report, the lead-center model fell into disfavor.
6. Too much complexity in management creates problems. In 1996, NASA went back to the lead-center approach. The decision was driven by external forces (budget cuts) and by Goldin's own philosophy, which favored a lean headquarters. The lead center was JSC. In addition, a measure of the pluralist model was at play because NASA consolidated contracts and outsourced most shuttle processing between flights to just one company (USA). Also, Goldin delegated leadership over both the shuttle and ISS to the JSC director, Abbey. Abbey was by all accounts a powerful director and probably did as much as anyone could have done to manage the extremely complex situation with which he had to deal: The shuttle schedule was linked to the building of the Space Station. Abbey had to manage both the shuttle and ISS—and ISS entailed difficult relations with a relatively autonomous partner, Russia. Money was in short supply, with trade-offs between the shuttle and the Space Station favoring the Space Station. The close coupling of the shuttle to ISS meant that the lead-center model further merged with pluralism in structure.
7. In the wake of huge ISS overruns, strong and visible leadership by headquarters became essential for public relations as well as organizational purposes. In 2002, O'Keefe elevated shuttle decision-making along with ISS decision-making to headquarters and to his associate administrator for space flight, Bill Readdy. The return to headquarters control did not translate to strong topside vigilance for the shuttle. NASA's priority remained ISS. The shuttle's schedule was geared to Space Station building, and there was "schedule pressure" on the shuttle from headquarters. Information about potential foam damage on *Columbia* did not reach decision-makers at headquarters or affect decisions at JSC. The lesson is that elevating control to headquarters does not necessarily lead to "better" decisions if informal relations stay the same. Vertical and horizontal communications will remain flawed.
8. Disaster—and the top leader—can concentrate organizational attention. After *Challenger*, and *Columbia*, return-to-flight became the overriding agency goal. The NASA administrator and his associate administrator for space flight were personally engaged. Other managers were replaced or got the message that failure was not an option. When Griffin became administrator, he himself made critical shuttle decisions. He moved decisions that once had been made at the center level to the very top of NASA's hierarchy. This was because of the external

pressure on NASA for safety in the wake of Columbia and because Griffin had confidence in his own ability to make technical decisions.

9. The headquarters-control model is more suitable to large-scale, complex technologies like the shuttle. However, informal power and communications structures are just as important as formal structure. Responsibility must be clear. Those with responsibility should also have the resources they need to carry out their tasks. The attention of the agency administrator is essential if the organization is to give a program keen vigilance. Because of competing institutional interests and individual incapacities, checks and balances throughout the organization are essential to uncover problems. Leaders can be wrong, and they need to be informed of their errors, including the deleterious impacts of their decisions. The more communication and transparency there are in decision-making, the better.
10. There are limits to all organizational designs. Where technologies are especially large and complex, those human beings charged with their care may well err, sooner or later. Individuals and organizations must nevertheless always strive for perfection and try to come as close as possible to that ideal through learning and continual watchfulness.

Perhaps we should remember one result from the management of the Space Shuttle program above all others: Leaders should always be humble, especially when dealing with the space frontier.

References

[1] Acheson, D., "Thoughts About Thoughts in High Places," *New York Times Magazine*, October 11, 1959, as quoted by Harold Seidman, *Politics, Position, and Power: The Dynamics of Federal Organization*, Oxford University Press, New York, 1975, p. 14.

[2] Lambright, W. H., *Powering Apollo: James E. Webb of NASA*, Johns Hopkins University Press, Baltimore, MD, 1995.

[3] McCurdy, H. E. *Inside NASA: High Technology & Organizational Change in the U.S. Space Program*, John Hopkins University Press, Baltimore, MD, 1993.

[4] Day, L., in Heppenheimer, T.A *Development of the Space Shuttle, 1972–1981,* Smithsonian Institution Press, Washington, DC, 2002, p. 35.

[5] McCurdy, *Inside NASA*, p. 97. For systems management as a NASA technique, see Johnson, S. B., *The Secret of Apollo: Systems Management in American and European Space Programs*, Johns Hopkins University Press, Baltimore, MD, 2002.

[6] Brewer, G. D. "*Prefect Places: NASA as an Idealized Institution*," *Space Policy Reconsidered*, edited by R. Byerly, Jr., Westview Press, Boulder, CO, 1989, p. 155.

[7] Wolfe, D. "The Administration of NASA," *Science*, Vol. 163, 15 November 1968, Vol. 162 (3855), 753.

[8] Donlan, C., in Heppenheimer, *Development of the Shuttle*, 1972–1981. Smithsonian Institution Press, Washington, DC, 2002, p. 36.

[9] NASA Space Shuttle Program Management, "*Manned Spacecraft Center's Recommended Plan*," "Lead Center" files, NASA Historical Reference Collection, NASA History Office, NASA Headquarters, Washington, DC, 21 April 1971.

[10] Kraft, C., "Memorandum for Record," 8 April 1971, "Lead Center" files, NASA Historical Reference Collection.
[11] Donlan, in Heppenheimer, *Development of the Space Shuttle*, p. 36; Fletcher, J. C., Memorandum, to Al Lovelace, "Personal Concern About the Launch Phase of Space Shuttle," *Exploring the Unknown, Vol. IV: Accessing Space*, 7 July 1977, edited by J. M. Logsdon, NASA SP-4407, Washington, DC, 1999, pp. 352–54.
[12] NASA News Release, "Lead Center" files, 10 June 1971, NASA Historical Reference Collection.
[13] Fletcher, J. C., and Moss, F. E., *James C. Fletcher Papers,* Special Collection, Marriott Library, University of Utah, Salt Lake City, 23 Feb. 1973.
[14] Heppenheimer, *Development of the Space Shuttle*, pp. 73–75; Launius, R. D. "A Western Mormon in Washington, DC: James C. Fletcher, NASA, and the Final Frontier," *Pacific Historical Review,* Vol. 64, May 1995, pp. 217–241.
[15] Thompson, R. F., T*he Space Shuttle: A Future Space Transportation System.* Johnson Space Center, Houston, TX, 1974.
[16] NASA, "Management Instruction," 15 March 1973, "Lead Center" files, NASA Historical Reference Collection.
[17] Trento, J. C., *Prescription for Disaster*, Crown, New York, 1987, pp. 140–41.
[18] Heppenheimer, *Development of the Space Shuttle*, pp. 73–75; Launius, R. D. "A Western Mormon in Washington, DC: James C. Fletcher, NASA, and the Final Frontier," *Pacific Historical Review,* Vol. 64, May 1995, p. 40.
[19] Neal, V., "Space Policy and the Size of the Space Shuttle Fleet," *Space Policy,* Vol. 20, Aug. 2004, pp. 157–169.
[20] Fletcher, Memorandum, to Lovelace, 7 Jul. 1977.
[21] Williamson, R. A., "Developing the Space Shuttle," in Logsdon, J. et al. (eds.), *Exploring the Unknown*, NASA, Washington, DC, 1999, Vol. IV, p. 175.
[22] "NASA Ordered to Further Explain Shuttle Cost Problems, Management Remedies," *Aerospace Daily*, Vol. 7, May 1979, p. 31.
[23] Baumbach, D. "Frosch: Shuttle III-Run," *Today*, 17 May 1979, p. 1A.
[24] "NASA Creates Space Transportation System Operations Office," *NASA News Release* 19–197, 9 October 1979.
[25] Heppenheimer, *Development of the Space Shuttle*, pp. 73–75; Launius, R. D. "A Western Mormon in Washington, DC: James C. Fletcher, NASA, and the Final Frontier," *Pacific Historical Review,* Vol. 64, May 1995, p. 354.
[26] Heppenheimer, *Development of the Space Shuttle*, pp. 73–75; Launius, R. D. "A Western Mormon in Washington, DC: James C. Fletcher, NASA, and the Final Frontier," *Pacific Historical Review,* Vol. 64, May 1995, pp. 354–355.
[27] Launius, R. D., and Dennis, R. Jenkins, *Coming Home: Reentry and Recovery from Space,* NASA SP-2011-593, Washington, DC, 2012, pp. 185–219.
[28] Bailey, D. "Tiles Are Still Sticking Point," *Today*, 22 Feb. 1980, p. 18A.
[29] Reagan, R., President "Remarks on the Completion of the Fourth Mission of the Space Shuttle Columbia," July 4, 1982, p. 870, in *Public Papers of the Presidents of the United States: Ronald Reagan*, Government Printing Office, Washington, DC, 1982–1991.
[30] Oral History, Interview of Phil Culbertson, April 1, 2004, by Orville Butler, Kennedy Space Center, FL.
[31] Trento, *Prescription for Disaster*, pp. 260–61.
[32] Trento, *Prescription for Disaster*, pp. 260–61.
[33] *Columbia Accident Investigation Board (CAIB): Report,* Vol. 1, Government Printing Office, Washington, DC, 2003, p. 24.
[34] *Columbia Accident Investigation Board (CAIB): Report,* Vol. 1, p. 24.

[35] Trento, *Prescription for Disaster*, pp. 257–70.

[36] Mahler, J., with Casamayou, M. H., *Organizational Learning at NASA: The Challenger and Columbia Accidents*, Georgetown University Press, Washington, DC, 2009, pp. 52–53.

[37] Chapter 5, discusses the Rogers Commission findings. Mahler with Casamayou discusses Rogers and CAIB reports along with factors leading to the disasters, CAIB, Report, Vol. 1.

[38] Logsdon, J. M., "Return to Flight: Richard H. Truly and the Recovery from the Challenger Accident," *From Engineering Science to Big Science,* edited by P. E. Mack, NASA SP-2210, Washington, DC, 1998, Chap. 15.

[39] NASA, Report of the Advisory Committee on the Future of the US Space Program Augustine Report, NASA, Washington, DC, 1990.

[40] Albrecht, M., *Falling Back to Earth*, New Media Books, Lexington, KY, 2012.

[41] Lambright, W. H., *Transforming Government: Dan Goldin and the Remaking of NASA,* IBM, Washington, DC, 2001.

[42] Lambright, W. H., *Executive Response to Changing Fortune: Sean O'Keefe as NASA Administrator,* IBM, Washington, DC, 2005.

[43] Lambright, W. H., *Launching a New Mission: Michael Griffin and NASA's Return to the Moon,* IBM, Washington, DC, 2009.

[44] Gerstenmaier, W. H. Assoc. Admin., Human Exploration and Operations, NASA Home Page, http://www.nasa.gov/about/highlights/gerstenmaier.bio.html.

[45] Seidman, *Politics, Position, and Power*, p. 14.

Chapter 4

Engineering the Engine: The Space Shuttle Main Engine

Robert E. Biggs

Introduction

In July 1971, the Rocketdyne Division of Rockwell International won the competitive bid to design, develop, and produce the Space Shuttle Main Engine (SSME). After 10 months of delay because of a protest lodged by Pratt & Whitney, the work was started in April 1972. The engine, to be developed under contract with the NASA Marshall Space Flight Center (MSFC), was a significant departure from the Apollo human-rated rocket engines of the 1960s. A liquid oxygen (LOX)/liquid hydrogen (LH_2) engine, it was rated at approximately half a million pounds of thrust, with the capability to throttle from 50 percent to 109 percent of rated power. It was to be computer controlled with a fully redundant, fail-operate, fail-safe control system and reusable for up to 100 flights [1].

The development of the SSME was an arduous task beset with many technical, logistical, and managerial difficulties. The team that solved these problems pushed and advanced the state of the art in many different fields. Thousands of specialists at Rocketdyne and the MSFC worked to produce a device that ranks high on the list of humankind's greatest engineering achievements. For many, personal sacrifices took the form of long nights, long weekends, holidays lost, and punishing cross-country air travel with days and weeks at a time away from their families and loved ones.

THE ENGINE

The SSME was the world's only reusable large liquid propellant rocket engine. In clusters of three engines each, it provided the primary power to place NASA space shuttles and crews into earth orbit for the entire 30-year Space Shuttle flight program—1981 to 2011.

It was a daunting undertaking to design and develop the engine with performance requirements that had never before been achieved. The SSME is a high chamber pressure (over 3,000 pounds per square inch) rocket engine that burns LOX and LH_2 at a mixture ratio of 6 pounds of LOX for every pound of LH_2. It produces a rated thrust of 470,000 pounds (vacuum), with a specific impulse greater than 453 pounds of thrust per pound of propellant per second. This very high efficiency is achieved with the use of a *staged combustion cycle*, wherein a portion of the propellants, partially combusted at a fuel-rich mixture ratio, is used to drive the high-pressure turbopump turbines before being completely burned in the main combustor.

The turbine drive gases are produced in two "preburners" to provide the power for the two high-pressure turbines. The gases then exit into the main fuel injector and are burned with the remainder of the propellants in the main combustion chamber (Fig. 4.1). This results in maximum propellant efficiency because all the propellant is used in the main combustor, and none is wasted by being dumped overboard from a low-pressure turbine exhaust

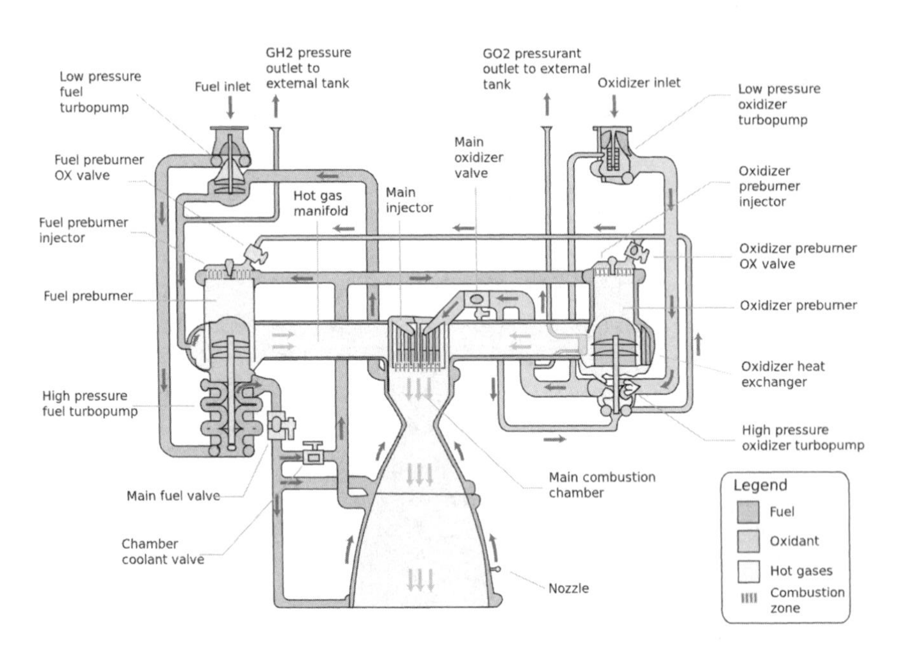

Fig. 4.1 An SSME propellant flow schematic showing how exhaust from the fuel and oxidizer turbopumps is burned in the main combustion chamber. (Rocketdyne/Pratt & Whitney, CP8_0931-4)

system, as was the case with all prior large liquid rocket engines. This improved efficiency is achieved at a significant cost in system pressures. With the turbines in series with the main combustor, the turbine exhaust pressure has to be higher than the main combustion chamber pressure. Although the turbines are designed for low-pressure ratio (approximately 1.5 to 1), the turbine inlet pressure has to be about 50 percent higher than the exhaust pressure to provide sufficient power. The preburners that provide the turbine drive gases have propellant injectors that require a minimum differential pressure to assure stable combustion. This further increases the required turbopump discharge pressures for the propellant pumps to as much as two and a half times the main combustion chamber pressure [2].

The oxidizer and fuel systems each contain a low-pressure turbopump, a high-pressure turbopump, and a preburner. To provide turbine power in the staged combustion cycle, 80 percent of the fuel (LH_2) is burned in the two preburners with 12 percent of the oxidizer. The turbine exhaust gases are then burned in the main combustion chamber (MCC) with the remainder of the propellants. The high combustion chamber pressure combined with the amplification effect of the staged combustion cycle made this engine a quantum jump in rocket engine technology and created a significant challenge to the contractor and government team charged with its design and development.

The LH_2 enters the engine at the low pressure fuel turbopump (LPFTP) inlet at a pressure of 30 psia and is increased in pressure by the 15,000 RPM turboinducer to over 250 psia. This pressure is required to prevent cavitation of the high pressure fuel turbopump (HPFTP). This three-stage centrifugal pump, operating at 35,000 RPM, further increases the pressure to over 6,000 psia. The LH_2 is then divided into three separate flow paths. Approximately 80 percent of the fuel flows to the two preburners; half of this, however, is used to cool the thrust chamber nozzle and is then mixed with the other half before entering the preburners. The remaining 20 percent of the fuel is used in the major component cooling circuit. The LH_2 is first routed to the MCC, where it provides coolant for the main combustion process by flowing through 390 milled slots in the copper alloy combustor. Having been converted to an ambient temperature gas by the MCC, the fuel is then routed to the LPFTP, where it is used as the power source for the partial admission single stage impulse turbine, which drives the LPFTP. The Space Shuttle then uses a small portion (0.7 pounds per second) of this gas to pressurize the main hydrogen tank, whereas the rest of it is used to cool the major hot gas system structure (hot gas manifold) and, finally, the main injector baffles and faceplates before being consumed in the MCC.

The LOX enters the engine at the low pressure oxidizer turbopump (LPOTP) inlet at a pressure of 100 psia and is increased in pressure by the 5,000 RPM turboinducer to over 400 psia. This pressure is required to prevent cavitation of the high pressure oxidizer turbopump (HPOTP). The dual inlet single stage centrifugal main impeller, operating at almost 30,000 RPM, further increases the pressure to about 4,500 psia. Most of the LOX is then routed through the main oxidizer valve to the coaxial element main injector of the MCC. A small amount of LOX (1.2 pounds per second) is routed through an engine-mounted heat exchanger and conditioned for use as the pressurant gas for the Space Shuttle's main oxidizer tank. The remainder of the LOX is ducted back into a smaller boost impeller on the same shaft to increase the pressure to as much as 8,000 psia. This provides enough pressure to allow the use of throttle valves to control the LOX flow rate into the two preburners. Thrust control is achieved by closed loop throttling of the oxidizer preburner (OPB) side, and mixture ratio control is accomplished by closed loop control of the fuel preburner (FPB) side. The throttle valves are controlled by an engine-mounted computer known as the main engine controller (MEC). A built-in recirculation flow path provides power for the six-stage axial flow hydraulic turbine that drives the LPOTP. A LOX flow rate of approximately 180 pounds per second is supplied from the discharge side of the main impeller. After passing through the turbine, this LOX is mixed with the discharge flow of the LPOTP and thereby returned to the HPOTP inlet.

The two preburners produce a hydrogen-rich steam that is used to power the two high-pressure turbines that drive the HPFTP and the HPOTP. Combustion of these gases is completed in the MCC.

THE REQUIREMENTS

The engine design requirements began being finalized in May 1972 with the continuation of 1971's fact-finding negotiation. Over 250 separate issues were identified and resolved in two months. With the NASA selection of the Orbiter contractor—Space Division of North American Rockwell Corporation—negotiations could begin to define the physical, functional, and electronic interfaces between the engine and the Orbiter. The first such meeting took place at Rocketdyne on August 10, 1972. In a series of technical meetings throughout the rest of 1972, fact-finding and interface issues were sufficiently resolved between the various contractor and NASA organizations to enable the baseline release of the two major design requirements documents.

The Interface Control Document (ICD), containing SSME design requirements relating to engine/vehicle interfaces, was released on February 9, 1973 [3]. These requirements included engine envelope; weight and center of gravity; dimensions, tolerances, and structural capabilities of all physical interfaces; electrical power, frequencies, and phase requirements; computer command and data formats and failure responses; and engine environment and performance requirements. The Contract End Item (CEI) specification [4] was released on May 10, 1973. The CEI specification contained detailed requirements for engine checkout, prestart, start, operation, and shutdown;

engine service life and overhaul requirements; design criteria for thermal, vibration, shock, acoustic, and aerodynamic loads; material properties, traceability, and fabrication process control; control system redundancy requirements; and required safety factors. Few changes were made in these requirements after the baseline release; however, three changes that came about later as a result of further Space Shuttle system definition are worthy of mention:

1. The original life requirement was for 100 missions and 27,000 seconds, including 6 exposures at the "Emergency Power Level" (EPL) of 109 percent. A change was made to maximize the allowed number of such exposures within the existing design. With the redefined shuttle, 27,000 seconds was equivalent to 55 missions. A fatigue analysis concluded that if the total number of missions was reduced to 55, then no limit need to be placed on the number of exposures at 109 percent. Because of this change, EPL was renamed Full Power Level (FPL).
2. The engine mixture ratio was to have been controlled by vehicle command to any value from 5.5 to 6.5. As the Space Shuttle mission was refined, this requirement was first reduced in range to 5.8 to 6.2 and then eliminated altogether in favor of a fixed mixture ratio of 6.0. To take advantage of this, the engine design was modified by reducing various system resistances; as a result, system pressures and turbine operating temperatures were reduced.
3. Early in 1978, a definitive shuttle trajectory analysis revealed that throttling all the way to the 50 percent power level during the period of maximum aerodynamic loading was not required. The Minimum Power Level (MPL) was raised from 50 percent to 65 percent, which allowed further system resistance reductions in subsequent engines.

A series of design verification specifications (DVS) was developed. The DVS contained all of the engine design requirements derived from the ICD, CEI, contract statement of work, and other sources such as company design standards and good industry practice. The engine-level requirements were contained in DVS-SSME-101. The engine component DVSs had similar identifications. Each detailed requirement was listed, its source was identified, and the methods of verification (proof that the design meets the requirement) and validation (proof that the requirement is valid) were specified. The methods to be employed for verification and validation were analysis, hardware inspection, laboratory or bench tests, subsystem hot-fire tests, and engine hot-fire tests. Emphasis was placed on obtaining the required proofs at the lowest possible level. These requirements formed the basis for the SSME development program until well into the flight program. Individual DVS task completions were used as benchmark control points or gates to allow continuation of the program for certain critical preplanned activities. The most significant of these was the first flight of the Space Shuttle, for which 991 DVS tasks had to be closed. After the DVS program was completed (after the first flight), a total of 4566 laboratory tests and 1418 subsystem hot-fire tests had been completed [5].

THE FIRST 10 YEARS

The first 10 years of SSME development resulted in an engine configuration designated as the First Manned Orbital Flight (FMOF) configuration. Engine testing for this phase of development began on May 19, 1975, with the first Integrated Subsystem Test Bed (ISTB) test and culminated with the flight of STS-1 on April 12, 1981. Figure 4.2 shows the FMOF development history along with the accumulated firing times to first flight.

GETTING STARTED

Engine development testing was planned to be conducted at the NASA rocket test site in Mississippi and to begin late in 1974. The Mississippi Test Facility (MTF) near Gulfport, Mississippi, had been used for static testing the Saturn (Apollo) launch vehicle stages, and the Saturn test facilities were modified, at NASA direction, to accommodate the SSME. MTF was later

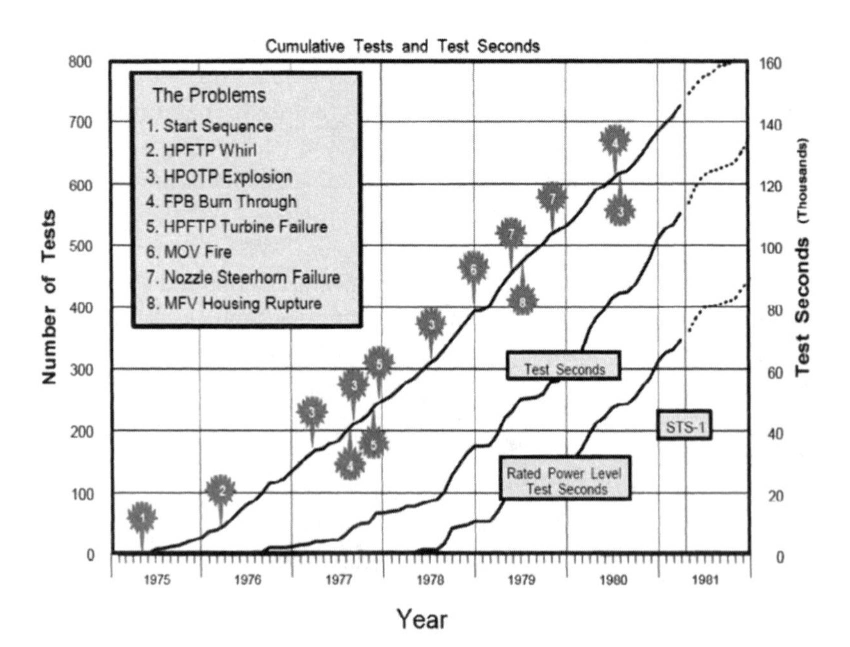

Fig. 4.2 The SSME development history leading to the FMOF configuration. (Rocketdyne/Pratt & Whitney, CP8_0931-8)

renamed the National Space Technology Laboratory (NSTL) and, more recently, the Stennis Space Center (SSC). In the meantime, component and subsystem testing was also planned for the Coca area of the Rocketdyne Santa Susana Field Laboratory (SSFL) at Chatsworth, California. Existing test facilities were to undergo major modifications to accommodate the turbopumps and combustion devices as well as various combinations of components arranged in subsystems. During 1973 and early 1974, the Coca construction project encountered unforeseen difficulties that eventually led to a schedule slip of about six months. At the same time, procurement delays, weight reduction design changes, and required structural improvements caused fabrication of major components to fall behind schedule [6].

In the summer of 1974, the SSME program was realigned under the leadership of Bob (J. R.) Thompson as the MSFC SSME project manager and Norm Reuel as the Rocketdyne vice president and program manager. Program schedules were adjusted by about six months, and increased management emphasis was provided to assure timely completion of the remaining development tasks. The most significant of these involved the ISTB.

The ISTB was originally planned as a "bobtail" engine. It was to be made up of the four turbopumps and two preburners with associated plumbing and controls, but without a thrust chamber assembly (TCA). (The TCA consists of the main injector, the main combustion chamber, and the nozzle.) The control system included all the required valves and actuators. However, during the fact-finding negotiations of July 1971, it was agreed that a shortened version of the TCA would be added to the ISTB. The shortened TCA was to have an area ratio (throat area divided by exit area) of 35 to 1 rather than the flight configuration of 77.5 to 1. The ISTB, then, became essentially an engine assembly with the program realignment, and activation of the ISTB test facility planned for the Coca test area (Coca 1C) was deferred in favor of testing the ISTB at NSTL, with the first test scheduled for May 1975.

A special management team was formed to determine and implement system and operational changes that would ensure this very key objective would be achieved. The team was headed by Dom Sanchini as associate program manager, and included Ted Benham as the manufacturing project manager and Ed Larson as the engineering project manager. The team investigated in detail the production release and fabrication system being used. With the concurrence and help of MSFC project management, quality and engineering changes were made to simplify paperwork and to provide a quick turnaround for hardware modifications without sacrificing quality or configuration control. This was achieved largely by assigning 25 top design engineers to on-the-floor manufacturing support with authority for on-the-spot approval of material review dispositions and design change rework modifications. The chief engine designer, Bob Crain, supervised the ISTB assembly. When the

ISTB was delivered to the test site, Sanchini was appointed vice president and program manager.

The Management and Budget Office of the White House had selected the ISTB's first full-up ignition test date as one of the major Space Shuttle program milestones by which this office would monitor the program's progress and health. This very important milestone was achieved on schedule. The ISTB was installed in NSTL test stand A1, and a countdown demonstration test (Test 901-001) was conducted on May 19, 1975. After five short exploratory ignition tests, the full thrust chamber ignition test was conducted on June 23, 1975. The engine development test program was underway.

COMPONENT TESTING

Within the program realignment of 1974, it was decided that the first article of each major component would be allocated to the ISTB. This action would accelerate engine testing and the discovery of any potential major system problems, but doing so would delay the beginning of the component test program until after the second article had been assembled. The component test program began in the same month as the engine test program (May 1975) with the low-pressure turbopumps (LPOTP and LPFTP). The two high-pressure turbopumps (HPOTP and HPFTP) began testing three months later, in August 1975. The test program planned for the Coca area, however, began with ignition system tests and progressed to preburners and then preburners with the MCC, and culminated, finally, with the TCA (MCC with a 35 to 1 nozzle) in August 1975.

Component tests were used to great advantage in the design verification program delineated in the DVSs. Most of the problems that were encountered, however, arose because of the complexity of the test facilities rather than the discovery of component failure modes. The test facilities were designed to accept various combinations of components arranged in subsystems and used facility devices (usually servo-controlled valves) to simulate the engine environment. The turbopump test stand had approximately 2,000 valves, including 24 that were servo-operated. Preburner propellants were supplied from a 14,000 psi system, with valves weighing as much as 5 tons. One of the more significant problems occurred on Coca-1A early in 1976. The oxidizer subsystem, which consisted of an LPOTP, HPOTP, and OPB (actually a half powerhead, which included the preburner), was being tested. At 19 seconds into Test 740-007, a facility rotary flowmeter failed, releasing flowmeter blades into the LOX flow stream [7]. The blades initiated a fire at a downstream throttle valve, which burned, causing a decrease in flow resistance. The decrease in flow resistance caused enough of a change in the operation of the HPOTP that it cavitated, lost axial thrust control, and began to rub internally. This resulted in a major fire that caused significant damage to the components and the facility. A similar failure occurred on a fuel subsystem test on Coca-1B the following year. Test 745-018 experienced a major fire that began with a fire in a facility throttle valve caused by cavitation-induced erosion [8].

With the advent of engine and component testing and its attendant loss of hardware, it soon became evident that the planned hardware was inadequate to support the scheduled test program and to keep up with the attrition realized from the development problems. This deficiency was to remain with the program for many years. The component test program, if pursued as originally planned, would have drained valuable resources from the engine test program to develop the complicated test facilities. The NASA administrator, Robert Frosch, stated in testimony to the Senate Subcommittee on Science, Technology, and Space, "...we have found that the best and truest test bed for all major components, and especially turbopumps, is the engine itself" [9]. The Coca area test facilities were gradually phased out from November 1976 to September 1977, largely because there were not enough resources to pursue an aggressive component test program in addition to the engine test program.

LESSON LEARNED

Concurrent development of a system with one or more of its subsystems appears to have a significant schedule advantage; however, in systems with significant failure rates, hardware attrition will threaten both system and subsystem schedules, becoming two programs vying for the same assets. Additional failures may exist because of complex subsystem test facilities.

ENGINE START SEQUENCE

The first hurdle that had to be overcome in the engine test program was to learn how to safely start and shut down the engine. Five years of analysis had produced sophisticated computer models that attempted to predict the transient behavior of the propellants and engine hardware during start and shutdown [10]. With these models, the basic control concepts were defined, and initial sequences were developed. The models had shown that the engine was sensitive to small changes in propellant conditions and that timing relative to opening the propellant valves was critical.

Expecting difficulties, a cautious step-by-step plan was developed to explore the start sequence in small time increments. Using this approach, it required 19 tests, 23 weeks, and 8 turbopump replacements to reach 2 seconds into an eventual 5-second start sequence. It took an additional 18 tests, 12 weeks, and 5 turbopump replacements before momentarily touching MPL. A safe and repeatable start sequence was eventually developed by making maximum use of the

MEC to control the propellant valve positions. Without the precise timing and positioning allowed by the MEC, it is doubtful that a satisfactory start could have been developed. The times specified in the following start sequence description refer to the initial flight engine configuration. Later configurations required minor modifications to the sequence timing.

The SSME starting sequence is diagrammed in Fig. 4.3. When a start command is received, the Main Fuel Valve (MFV) is immediately ramped to its full open position in two-thirds of a second. This enables the LH_2 to fill the downstream system and begin to power the high-pressure turbines. The latent heat of the hardware imparts enough energy to the hydrogen to operate as an "expander-cycle" engine for the early part of the start sequence. This eliminates the need for any auxiliary power to initiate the start sequence; however, it also creates a thermodynamic instability, referred to as the *fuel system oscillations*. When the cold LH_2 begins to flow into the thrust chamber nozzle, the hardware latent heat causes the hydrogen to expand rapidly, creating a flow blockage and momentary flow reversal. The result is a pulsating fuel flow rate with an unstable pressure oscillation at a frequency of approximately 2 Hz. The oscillations continue to increase in magnitude with dips (reductions in pressure) occurring at approximately 0.25, 0.75, and 1.25 seconds, until the establishment of MCC chamber pressure causes it to stabilize after 1.5 seconds. Events before stabilization had to be made to conform to the idiosyncrasies of the fuel system oscillations.

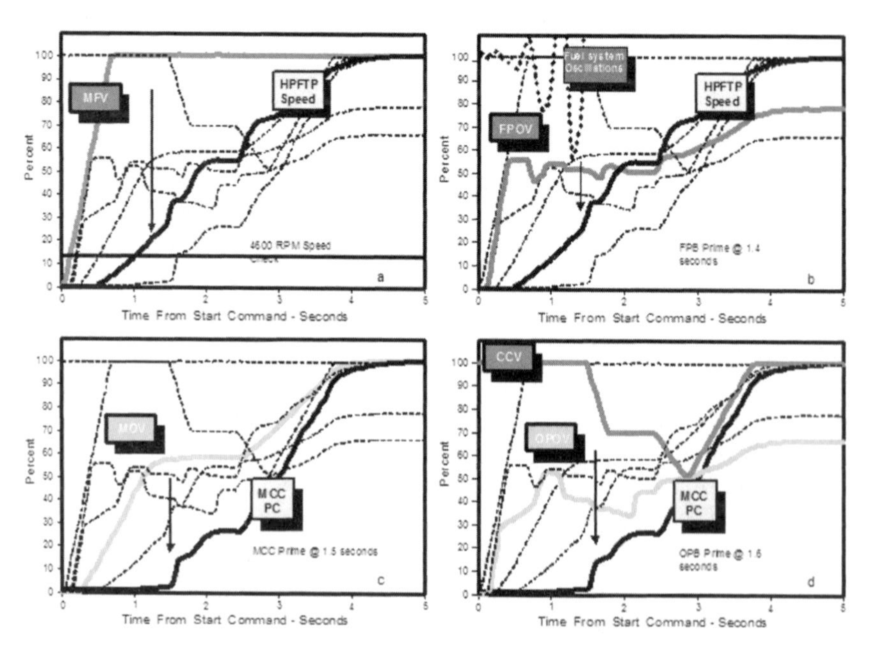

Fig. 4.3 An SSME start sequence showing the behavior of key engine parameters during the first five seconds. (Rocketdyne/Pratt & Whitney, CP8_0931-9)

Simultaneously with the opening of the MFV, electrical power is provided to the spark plugs in the augmented spark igniters (ASI) included in each of the three combustors. The ASI will then ignite the combustors when both fuel and oxidizer are present in the proper mixture ratio. The fuel is provided first by the MFV being opened, and then the oxidizer is provided later for each combustor separately through the three oxidizer valves. Each valve has an ASI LOX supply line that allows LOX to flow to the ASI upon initial valve motion (about 5 percent). The proper mixture ratio for ignition is achieved by the second dip in pressure caused by the fuel system oscillations.

After the MFV starts to open, the three oxidizer valves are separately subjected to a series of position commands intended to precisely control the oxidizer system's priming times for the three combustors. Priming is the process of filling the system with liquid, as with an old hand-cranked water pump. An oxidizer system is said to be "primed" when it is filled with liquid down to the combustor such that the flow rate entering the injector is equal to the flow rate leaving the injector to be burned in the combustor. This event generally results in a rapid rise in combustion chamber pressure. The target priming times for the three combustors are a tenth of a second apart; FPB prime is at 1.4 seconds, MCC prime is at 1.5 seconds, and OPB prime is at 1.6 seconds. Although part of the valve positioning is accomplished under a limited form of closed loop control, it is merely a convenient method of commanding the valves to a predetermined position and therefore can be treated as if it were all done as open loop commanded positions as a function of time.

The first oxidizer valve to be commanded is the Fuel Preburner Oxidizer Valve (FPOV). After a delay of 0.100 seconds, the FPOV is ramped to 56 percent open at its maximum slew rate. At 0.72 seconds, the FPOV is given a "notch" command to close about 10 percent and then reopen. This is done to compensate for the second pressure dip caused by the fuel system oscillations and to avoid damaging temperature spikes in the HPFTP turbine. During this dip, the FPB is ignited, and the additional power causes a slight acceleration in the HPFTP speed. Just before the third fuel system oscillation pressure dip, the FPOV is given another notch command, which is maintained throughout the priming sequence.

A safety check is made at 1.25 seconds to assure that the HPFTP speed is high enough to safely proceed through the priming sequence. The speed must be high enough at MCC prime to be able to pump hydrogen through the downstream system against the back pressure rise created by the MCC prime or an engine burnout will occur because of the resulting oxygen-rich combustion. It was determined from test experience that if the speed were to be less than 4,600 RPM at 1.25 seconds, it would likely be too low at MCC prime to maintain pumping capability. The engine must be shut down at 1.25 seconds because if the speed is discovered to be too low later in the start sequence, there is insufficient time to react and shut down safely.

When the FPB prime occurs at 1.4 seconds, there is a rapid rise of pressure at the inlet to the HPFTP turbine. Because the turbine back pressure is not provided until MCC prime, this pressure rise causes a high turbine pressure ratio and a significant acceleration in the HPFTP speed. The higher HPFTP speed is desirable for a cool fuel-rich start; however, the turbine back pressure must be applied (MCC prime) soon to prevent a runaway condition.

MCC prime is primarily controlled by of the way the Main Oxidizer Valve (MOV) is positioned. After an initial delay of 0.200 seconds, the MOV is slowly ramped to just under 60 percent open. This combination of time delay, ramp rate, and position provides a LOX flow rate that causes MCC prime to occur at 1.5 seconds and creates an engine system balance that will produce a safe low mixture ratio (between 3 and 4) for the stabilized operation just before activating the closed loop thrust control system at 2.4 seconds. When MCC prime occurs at 1.5 seconds, it causes a rapid rise in MCC chamber pressure, which, because it increases the turbine back pressure, acts as a break to decelerate the HPFTP.

The Oxidizer Preburner Oxidizer Valve (OPOV) is used to control OPB prime. Its initial opening is after a delay of 0.120 seconds; however, the opening retracts only the valve inlet seal, which is designed to provide sufficient oxygen to ignite the ASI and to have a small leakage flow into the OPB injector. The valve is designed so that the major flow path does not start to open until an indicated position of 46 percent. The slow ramp has no effect on the OPB LOX flow rate except to delay initiation of LOX flow until 0.84 seconds, when the main flow path through the valve starts to open. This flow path is partially open for about a third of a second before it recloses and the OPB is again run on valve leakage flow. The timing for this opening is scheduled to provide sufficient oxygen to allow the ASI to ignite the OPB before the second fuel oscillation pressure dip recovers and causes a significant decrease in mixture ratio. The next opportunity for ignition would be about half a second later. With valve leakage flow, OPB prime occurs at 1.6 seconds and causes an increase in drive power to both high-pressure turbines. The power increase stabilizes at about 2 seconds, with the MCC chamber pressure at approximately 25 percent of Rated Power Level (RPL). During this time, the chamber coolant valve, which was fully open at the start, is throttled down to 70 percent to force additional coolant flow through the MCC. The engine is allowed to run at this condition until 2.4 seconds to assure stable operation. The additional 0.4 seconds allow for and absorb normal variations in propellant pressures and temperatures.

By using the engine-mounted sensors, the MEC verifies proper ignition and operation of the three combustors at 1.7 seconds and again at 2.3 seconds. If no malfunctions are discovered, the closed loop thrust control system is activated at 2.4 seconds. The MEC compares the measured MCC chamber pressure to a preprogrammed chamber pressure ramp to RPL and modulates the OPOV in an attempt to zero out any differences. During this time, the

FPOV is simply moved by the MEC with position changes that are proportional to the amount of OPOV movement, and the Coolant Control Valve (CCV) is commanded open at a rate commensurate with the commanded chamber pressure ramp rate. Because of the engine dynamic response characteristics, the resulting chamber pressure lags behind the command by about 0.200 seconds. At 3.8 seconds, the closed loop mixture ratio control system is activated using the FPOV to adjust the fuel flow rate until the commanded mixture ratio is achieved. At 5 seconds, the engine has achieved stabilized operation at RPL with a mixture ratio of 6.

The engine design characteristics placed significant constraints on the start sequence. The priming sequence is the most critical. Very high (damaging) temperature spikes occur if any combustor prime coincides with the pressure dips caused by the fuel system oscillations. The timing of the sequences relative to each other is also critical. If the FPB prime were late or the MCC prime early, the insufficient fuel pump speed would cause very LOX-rich operation with major burning of the engine hardware. If the OPB prime were early or the MCC prime late, a rapid acceleration of the HPOTP could lead to its destruction. Because of the very compact design of the high-pressure pumps (the highest horsepower to weight ratio ever achieved), the very low inertia causes the pumps to accelerate and decelerate extremely quickly under abnormal conditions. If only the normal operating torque were applied to the HPOTP without the fluid load applied (gas in the pump or in cavitation), the turbopumps could accelerate from a dead stop to a destructive overspeed condition in less than a tenth of a second. The acceleration rate under this condition is almost 400,000 RPM per second [11].

The initial start sequence development tests on the ISTB were limited to starting to MPL (then 50 percent of RPL). The first test to achieve MPL was Test 901–037, a 3.36-second start transient test that was conducted at the end of January 1976. The first test to achieve stabilized operation with the closed loop mixture ratio control system activated was Test 901-042, on March 8, 1976. Operation at RPL was not achieved until January 1977 (Test 901-095). Although the ISTB start development tests resulted in a start sequence that would allow the ground test program to continue, the final start sequence was not arrived at until the end of 1978. The operation of the preburner valves evolved over time to better compensate for variations in external conditions and in response to specific problems as they occurred.

Initial engine tests were limited to 50 percent power level to comply with required proof test logic. The basic requirement is that for all pressure vessels, a proof test must be demonstrated at a pressure 20 percent higher than the maximum projected operational pressure. Proof pressures are determined early in the design process, and high turbine efficiencies and low-pressure losses create low-proof pressure. The staged combustion cycle magnifies optimistic values.

LESSON LEARNED

Establishing system proof test requirements based on optimistic component performance imposes a test limit on the system, and this limit requires at least one component design turnaround time to recover.

HPFTP SUB-SYNCHRONOUS WHIRL

On March 12, 1976, four days after the first stabilized test on the ISTB, Test 901-044 was scheduled for a 65-second exploratory test at 50 percent power level with 1 second at 65 percent. The test was terminated at 45.2 seconds because axial thrust was lost in the HPFTP. After the test, the HPFTP was bound up and could not be rotated with the turbopump torque tool (a normal post-test checkout).

It was later found that the HPFTP turbine end bearings had failed, causing this condition. A review of the test data revealed two major abnormalities. The HPFTP turbine gas temperature increased almost 200 R during the test. This and other measurements indicated a significant loss of turbine efficiency during the test. In addition, high-frequency vibration measurements on the HPFTP indicated a large amplitude vibration at a frequency of about half of the fuel pump speed. To expedite solving this problem, a combined Rocketdyne-MSFC team was formed under the leadership of Matt Ek, Rocketdyne vice president and chief engineer, and Otto Goetz, MSFC's leading turbomachinery expert. The team was ultimately expanded to include the foremost experts in the field of rotordynamics; the experts came from industry, government, and the academic community in the United States and Great Britain [12].

The HPFTP sub-synchronous whirl was a violent instability that caused the rotor to gyrate in the direction of normal rotation at a frequency of about half of the pump speed. This caused a forward precession of the rotor, which was actually an orbiting of the normal rotating axis. Being a true instability, the whirl was self-initiating and would usually start when the pump speed exceeded twice the first critical speed of the rotating assembly, with an inception frequency equal to the first critical speed (originally about 8,500 RPM). The amplitude would increase rapidly, and within six cycles, with the bending of the rather flexible rotor, the normal clearances would be breached, and internal rubbing would occur at many locations. With clearances closed and bearing supports bottomed out, the system's stiffness increased significantly, preventing a further increase in amplitude (limit cycle) and raising the first critical speed and, therefore, the whirl frequency. Bearing loads in the limit cycle condition were higher on the turbine end than on the pump end by a factor of three, and a significant number of turbine bearing failures occurred.

The team pursued a multi-disciplined approach, which included historical research, literature surveys, mathematical models, and consultations with universities and other companies with related knowledge or experience.

A vigorous test program included laboratory, component, subsystem, and engine tests. Twenty-two potential drivers were identified and analyzed; however, the team eventually concluded that two factors were far more significant than all the others.

The most significant destabilizing effects were hydrodynamic cross-coupling of the pump interstage seals combined with the low natural frequency of the rotating assembly. A series of design changes attacked these effects and worked to decrease cross-coupling drivers, to provide damping at the seals, and to increase the rotor critical speeds by stiffening the shaft and bearing supports. Over 10 months, the whirl inception speed was gradually increased from 18,000 RPM (below MPL) to above 36,000 RPM, which allowed whirl-free operation to above RPL. With the whirl problem eliminated, the HPFTP was capable of supporting the engine test program in mid-January 1977. For the first time, the SSME could be tested for extended durations at rated thrust [13].

HPOTP EXPLOSIONS

LOX pump explosions are nightmarish events in rocket engine development programs. The cost in program resources is quite severe because the turbopump assembly and surrounding hardware are usually lost to the program for any future use. Even more significant, though, is the fiendish nature of the failure. Once a fire has been ignited in the high-pressure LOX environment, it readily consumes the metals and other materials that make up the hardware. In most cases, the part that originated the failure is totally destroyed, leaving no physical evidence as to the failure cause. Program management is often left in a quandary as to what to do to prevent the failure from happening again. This leads to a process of speculating on possible failure causes and fixing everything that it could be.

In the time between when the HPFTP whirl problem was solved and the first shuttle flight, the SSME program experienced four HPOTP explosions. Two of them were caused by internal design flaws that had to be rectified [14]; the other two, although they did not represent design problems, did significantly impact the program resources in terms of available hardware and required recovery time [15].

Other failures during FMOF development testing included an MOV fire, an MFV housing rupture, two HPFTP turbine failures, and two instances of a fuel preburner body burn through.

THE GOALS

The DVS program was planned to verify all design requirements in a logical fashion, using certain key task completions as benchmark control points or gates that constrained the continuation of the program for some of the more critical activities. Superimposed on that program were other significant milestones that various NASA and other government agencies established as aids in tracking the general health of the SSME and shuttle flight engines. The most significant test milestone was established in terms of total accumulated test duration of the single engine ground test program (excluding cluster tests). John Yardley, associate administrator for the Space Transportation System, set a goal of 65,000 seconds as representing a sufficient level of development maturity to consider the engine flight worthy. NASA headquarters considered the achievement of this goal to be a flight constraint. The goal of 65,000 seconds was reached on March 24, 1980, during a test on Engine 2004.

The average test duration had increased significantly from year to year, which made it possible to accelerate the accumulated test seconds and to achieve the 65,000-second goal in the required time period. The dramatic increase in average test duration was possible because the development problems were being solved and increasing confidence allowed more tests of longer duration to be scheduled.

The original SSME Program Plan included a Preliminary Flight Certification (PFC) demonstration test program to be conducted before the first flight. Specific requirements for the PFC evolved gradually during the program, with the final requirements being established in early 1980. The PFC was defined in terms of a unit of tests called *cycles*. Each cycle consisted of 13 tests and 5000 seconds of test exposure, which included simulations of normal and abort mode flight profiles. It was required to conduct two PFC cycles on each of two engines of the flight configuration to certify that configuration for 10 shuttle missions. The PFC demonstration required 100 percent successful tests. If any test was shut down because of an engine problem, the PFC cycle did not count and had to be started over from zero.

Similar certification plans were developed and approved for each significant component or system change for the entire Space Shuttle program.

The First Flight

The first four flight configuration engines were assembled and acceptance tested in the first half of 1979. Engine acceptance testing included 1.5-second start verification, a 100-second calibration firing, and a 520-second flight mission demonstration test. Engine 2004 was allocated to the PFC demonstration program, and Engines 2005, 2006, and 2007 were installed in the Orbiter *Columbia* for the initial Space Shuttle flight. Several Shuttle program problems (such as Orbiter tile replacement) ensued, causing the first flight to be delayed.

During this time, significant changes were made to the three flight engines as a result of the test problems previously discussed. Because of the number and complexity of the changes, it was decided to repeat the final engine acceptance test. Engines 2005, 2006, and 2007 were removed from the Orbiter and shipped to the engine test site at NSTL. In June 1980, all three engines successfully completed a 520-second flight mission demonstration test and were subsequently reinstalled in the Orbiter *Columbia* [16].

A successful 20-second Flight Readiness Firing (FRF) was conducted on February 20, 1981. All three main engines were operated simultaneously at RPL with the entire Space Shuttle, including the SRB, on the launch pad in the launch attitude. The normal launch sequence was used, including starting the main engines at T minus 6.6 seconds (staggered by 0.120 seconds). Liftoff was precluded by not igniting the SRBs (normally at T minus zero). The FRF had been planned as the final "all-up" verification that the engines and all interfacing systems were capable of satisfactory operation. Engine performance was within expected limits, and post-test hardware inspections, leak tests, and other required checkouts were satisfactorily completed. The engines were ready for flight. The Space Shuttle era began at 0700 hrs (EST) on Sunday morning, April 12, 1981, with the flawless launch of the Space Shuttle *Columbia* on her maiden flight, STS-1. It was exactly 20 years after the world's first manned space flight, which carried cosmonaut Yuri Gagarin into earth orbit aboard the Russian spacecraft Vostok I. *Columbia* achieved her predicted orbit and remained there for two days. Americans (John Young and Bob Crippen) were back in space after an absence of almost six years. The launch was the first time in history that an all-new rocket launch system carried people on its first flight [17].

THE SECOND DECADE

The launch of STS-1 was a glorious ending to a decade of technical, logistical, and managerial difficulties involving thousands of specialists at Rocketdyne and MSFC. The FMOF configuration SSME had been developed and certified for Space Shuttle flights at 470,000 pounds of vacuum thrust and proven successful with the launch of STS-1 in April 1981. Then the bar was raised. The FPL configuration engine had to operate at 512,300 pounds of thrust—9 percent higher. This had been the original requirement, but development and certification at the higher power level had been postponed until after STS-1.

Development of the additional 9 percent thrust capability was more difficult than anyone had imagined. Seven major failures had occurred by the end of August 1982 (Fig. 4.4). Hardware funding was woefully inadequate for the task at hand, causing delays, encouraging reuse of poorly performing components, and requiring that heroic repairs to salvage damaged parts be performed. The SSME was certified and then de-certified for the higher thrust level. Programmatic problems led to the SSME program being split into two separate

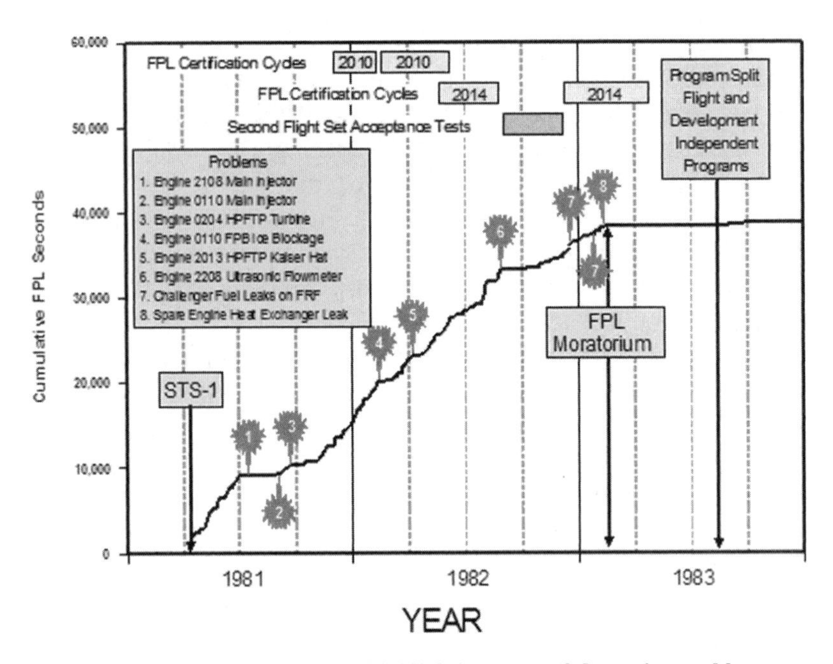

Fig. 4.4 The struggle to achieve FPL, highlighting some of the major problems encountered when test firing at FPL. (Rocketdyne/Pratt & Whitney, CP8_0931-40)

and equal programs—flight and development. The tragedy of *Challenger* in January 1986 caused the two programs to be reunited. The aftermath of *Challenger* saw significant improvements in the SSME and included a NASA contract with Pratt & Whitney to design and develop replacements for the two most troublesome components—the high-pressure turbopumps.

FPL Moratorium

SSME performance requirements were established to satisfy a variety of NASA, commercial, and military payloads. Initial flights were to require each engine to produce a vacuum thrust of 470,000 pounds. This was known as 100 percent power level or RPL. Later, payloads that would require a vacuum thrust of 512,300 pounds were envisioned. This was known as 109 percent power level or FPL. SSME structural design criteria and component performance requirements were based on FPL operating parameters. The program development plan was to achieve 100 percent/109 percent capability and to conduct a PFC test demonstration at 100 percent, followed by a full flight certification (FFC) test demonstration at 109 percent.

Engine testing began on May 19, 1975. Performance deficiencies limited the first engine to 50 percent power level. Redesigns for improved performance were planned and executed; however, progress was much slower than initially anticipated. It took 20 months before the engine was capable of a

5-second burn at 100 percent. It was a full three years, until May 1978, before the 10th engine ran a mission duration of 520 seconds at 100 percent. Low performance combined with engine failures in the summer of 1977 led to concerns by NASA and contractor management that the task of developing 109 percent capability in the time allotted would be too difficult to achieve. Some felt that even attempting that goal might jeopardize the engine's ability to support the planned flight program.

In February 1978, Mike Malkin, NASA's Space Shuttle program director, took decisive action to alleviate this concern. In a directive to Bob Lindstrom, MSFC shuttle projects manager, and J. R. Thompson, MSFC SSME project manager, Malkin stated, "The SSME development program has shown that as power level and run duration increase, new problems are uncovered that tend to prevent planned accumulation of test time" [19]. He directed that "the shuttle program be reoriented to concentrate engine development at the rated power level." FPL development was delayed until after the first flight.

FPL DEVELOPMENT

FPL development began while *Columbia* STS-1 was still in orbit. On April 13, 1981, at the Santa Susana Field Laboratory (SSFL) test stand A-3, a test to run over 200 seconds at FPL was attempted on FPL Engine 0110 (Test 750-131). On April 27, 1981, Engines 0110 and 0204 both demonstrated satisfactory operation at FPL On the same day, FPL Engine 2108 was delivered to NSTL test stand A-1 for installation.

The SSME program had three test stands, and each one contained an FPL engine for the first time. FPL development was underway, with each test stand starting to accumulate about 1000 seconds a month of FPL operation.

FPL development would exact a toll on test hardware. In under five months, all three of the original FPL engines experienced major failures, and then others followed. New and old failure modes were uncovered in rapid order. The first to surface was an old one.

This failure involved 600 small tubes, called LOX posts, that were used in the main injector to transfer LOX from the LOX inlet through the fuel side to the combustion zone. In 1978, two engines of the FMOF configuration suffered main injector LOX post failures. This failure mode was considered eliminated once some flow shields were added and the tip of some posts was changed to a stronger material. Two years later, the failure occurred on an FMOF configuration engine after 1080 seconds of operation at 109 percent power level.

Two more LOX post failures occurred on FPL configuration Engines 2108 and 0110, in July and September 1981, respectively. Data obtained from these tests were sufficient to adequately characterize the dynamic environment of the LOX posts. Converting to the stronger material posts with flow shields proved to eliminate main injector LOX post failures. That configuration was tested for the next 10 years, including 77 flights. Almost half a million seconds of test time were accumulated on 47 engines without a failure before the injectors were phased out with a new and improved powerhead design.

Three weeks after the last main injector LOX post failure, a catastrophic failure of the HPFTP occurred; this was attributed in part to a significant repair of the FPB injector.

Three additional catastrophic failures occurred in 1982: Engine 0110, caused by water left in the FPB; Engine 2013, attributed to the immature design of an experimental nut; and Engine 2208, caused by a braze joint failure between the main LOX duct and an experimental ultrasonic flowmeter.

Program Reassessment and Realignment

The catastrophic failure of Engine 2013 (Kaiser Hat failure) led to a series of reviews and investigations by various teams and committees over the next year. On the day after the failure, General Abrahamson asked Walt Williams, NASA chief engineer, to conduct a special review of the SSME program. He requested that a small team of experts not directly involved in the SSME program review the significant programmatic implications of the Engine 2013 failure. The review was to include an assessment of the maturity of the FPL configuration engine and the certification process.

Williams chose his team from among the country's elite scientists: Seymour Himmel, retired from the Lewis Research Center and a member of the prestigious Aerospace Safety Advisory Panel (ASAP); Eugene Covert, director of the Gas Turbine and Plasma Dynamics Laboratory and a professor at Massachusetts Institute of Technology; Maxime Faget, the chief designer of Apollo and retired from the JSC; Richard Weiss, chief scientist at USAF Rocket Propulsion Laboratory; and Thomas Morgan, Lieutenant General, USAF (Ret.). Walter Dankhoff, director of engine programs at the Office of Space Transportation Systems (OSTS), served as the executive secretary for the team.

The Williams team conducted hearings at Rocketdyne April 26–27, 1982, and at MSFC May 18–19. A third meeting was held at KSC on June 23, and the team findings were forwarded to General Abrahamson on July 2. These findings included that the HPOTP was unsafe to fly at FPL, the HPFTP turbine gas temperature was too high, and a long-term engine improvement program needed to be developed, starting with the high-pressure turbopumps. The team also concluded, "The review of procedures and facilities used in the production of the SSME and quality assurance practices used was very satisfactory" [33]. Williams returned in August and again in December.

Challenger Flight Readiness Firing and Leaks

Space Shuttle program requirements included an FRF for each new Orbiter before its first flight. The FRF was planned as the final "all-up" verification of the main engines, fully integrated with the Orbiter and external tank integrated propulsion systems. *Challenger*, the second new Space Shuttle Orbiter, was erected on the launch pad with the full Space Shuttle, including the SRB. Engines 2011, 2015, and 2012 were installed in engine positions ME-1, ME-2, and ME-3, respectively. Shortly after midnight on Saturday, December 18, 1982, propellant loading was initiated as part of a normal countdown for a Space Shuttle launch. The countdown proceeded through all three main engines operating at 100 percent power level, except that when T-0 went by (at 0800 hrs), the SRB was not ignited, and liftoff did not occur. The engines were allowed to run for an average time of 23 seconds before being commanded to shut down.

The FRF went well, with a few minor engine problems and one very serious indication. A hazardous gas detector mounted in the enclosed engine compartment indicated that, while the engines were firing, hydrogen was present in the compartment atmosphere at a concentration of 4,600 parts per million [18]. There must have been a fuel leak inside the compartment.

The engine compartment contained an inert atmosphere, provided by a gaseous nitrogen purge, which vented out as the atmospheric pressure decreased with altitude. During the flight, some outside air would flow into the compartment while nitrogen vented out. A hydrogen leak of the magnitude suggested could build to an explosive mixture because oxygen was now present. The leak had to be found and fixed [19].

Engine leak tests failed to determine the location or cause of the fuel leak [20]. Another FRF was planned. Joe Lombardo, MSFC propulsion division director, undertook the task of planning instrumentation for the FRF rerun, with the intent of isolating the leak. Seventeen hydrogen detectors were located in and around the engine compartment. Ten hazardous gas "grab bottles" were placed in the compartment, and two more were placed outside at the vent doors.

The second FRF was run on Tuesday, January 25. Again, the leak could not be found. Another exhaustive leak check was ordered. This time, the pressure was raised from 25 to 40 psig, and most joints were not bagged. It called for both mass spectrometer and soap bubble leak tests and included external surfaces of the components rather than just joints. Several small leaks were identified but not the one in question. The leak was found four days later and had been caused by a crack through the parent metal in the underside of a tee section, which transitioned from a toroidal fuel manifold on Engine 2011. The outlet elbow had been broken in a tooling accident. The normal fabrication sequence for the tee section was to machine weld the manifold, tee, and elbow together as a completed manifold assembly and then to weld the manifold to the

MCC. It was estimated that to remove the manifold and replace it with a totally new manifold would cause a schedule delay of four months. After much consternation, a repair procedure that would cause only a six-week delay was chosen. The tee section with the elbow was cut away from the rest of the manifold, and the MCC and a new tee/elbow assembly was manually welded in [21].

The MCC repair leak on Engine 2011 had far-reaching implications. Investigations would raise the question of whether safety was compromised for schedule, and this would lead to a major SSME reorganization at both MSFC and Rocketdyne. *Challenger*'s leaks led to another series of investigations. In the first week of March, a Rockwell corporate audit team was convened. The team included Joe McNamara and Charlie Feltz, both long-term executives of North American Aviation and Rockwell International, and they were closely familiar with Rocketdyne. At the same time, Jim Kingsbury along with a team of handpicked MSFC experts moved into Rocketdyne to monitor the day-to-day activity.

During the spring of 1983, armed with findings and recommendations from a variety of sources, MSFC and Rocketdyne replanned the SSME program. Lindstrom presented the plan to the NASA administrator on May 19, 1983, along with a request for a significant budget increase [22]. The final plan was presented on August 12.

During the August 12 meeting at NASA headquarters, Lindstrom announced that the SSME project had been realigned as two separate independent project offices [23]. Rocketdyne implemented a similar reorganization. Future design changes were to be incorporated as block changes. The *Challenger* configuration engines were designated as Block I, Phase I, and this was defined as the baseline configuration for continued flight at a maximum power level of 104 percent. The certification engines would continue to test in a certification extension program. Block I, Phase II, was to have improved high-pressure turbopumps for 109 percent flight, with certification to begin early in 1985 and ready for flight in 1986. Block II, Phase I, was planned as an increased margin configuration with a larger MCC throat area to reduce system pressures and a new powerhead with improved hot gas flow characteristics. It was planned to fly in 1988. Block II Phase II was planned as a contingency configuration with all new high-pressure turbopumps and would start development in 1988—if needed. The plan included design studies for all new high-pressure turbopumps in response to a May 27 letter from General Abrahamson to MSFC center director, Bill Lucas. The general directed that other American companies be solicited for support in designing improvements for turbopumps. Over the next several months, assets and responsibilities were divided between the two programs, and the funding level was increased to support Lindstrom's plan. The Block I nomenclature was soon dropped from both programs, and the development program was referred to simply as Phase II. Block I would later be used to define a more advanced configuration.

STS-26 (51-F) ENGINE 2023 IN-FLIGHT ERRONEOUS SHUTDOWN

On July 29, 1985, STS–26 (51-F) was launched carrying Spacelab 2, with the engine power level set at 104 percent. At 5 minutes, 43 seconds after liftoff, 3 minutes before the engines were scheduled to shut down, the engine controller on Engine 2023 confirmed an unsafe over-temperature operation of the HPFTP turbine and commanded the engine to shut down. It then dutifully notified the Orbiter computer of the action taken. The remaining two engines were kept running for an extra minute, staying at 104 percent rather than the normal throttle down to limit acceleration. At 9 minutes, 40 seconds, the LOX low-level indicator activated, warning of imminent depletion, and the Orbiter computer commanded the shutdown of the remaining two engines. The vehicle had reached sufficient velocity to achieve orbit, but not the planned orbit. It was 50 miles low and officially designated as an abort to orbit (ATO). The orbit was adjusted to the desired dimensions by several boosts by the orbital maneuvering system, and all mission objectives were completed.

Flight controllers on the ground realized that the engine shutdown was caused by instrumentation failures. When the channel B HPFTP temperature sensor on Engine 2021 voted for shutdown at 499 seconds, all redlines were inhibited for the remainder of the flight because controllers feared another erroneous shutdown.

The high temperature indication that caused the shutdown of Engine 2023 was not real. Two hot gas temperature sensors failed in such a way that the engine controller's computer logic concluded that the temperature was over the limit.

The RES7004-71 temperature sensor was the seventh iteration of the same basic design and was used to measure high-pressure turbine gas temperatures beginning with STS-3. The outer shell consisted of two cylinders and a mounting flange. The sensing element was contained in a 2-inch-long, 0.385-inch-in-diameter cylindrical shield welded to one side of a four-bolt mounting flange. The electrical wiring and connections were contained in a cylindrical external housing welded to the other side of the mounting flange. The heart of the sensing element was a 5-inch-long platinum wire, just a 1500th of an inch in diameter wound in a coil around a 0.059-inch-in-diameter sheath (coaxial tube) containing two 0.012-inch platinum lead wires. The ends of the coil were attached to the two lead wires and covered with a 0.090-inch-indiameter tip. An internal tube of crushable magnesium oxide and an external coating of aluminum oxide provided electrical insulation.

The coaxial tube penetrated the mounting flange and opened into the housing where the two lead wires were terminated in an electrical connector at the end of the housing. A three-wire harness connected to the engine controller where the wires were used to complete a Wheatstone bridge. A change in temperature caused a change in resistance of the tiny coil, which produced a voltage change in the Wheatstone bridge. Potting the interior with epoxy foam provided electrical insulation and physical shock protection.

Penetration of the mounting flange by the coaxial tube left a potential leak path, which was sealed by brazing around the tube (90 percent platinum and 10 percent rhodium) with a silver alloy. This sometimes resulted in liquid metal embrittlement of the tube. Mechanical and thermal loads during operation caused the brittle material to crack and allow hot gas (steam) leakage into the coaxial tube. Leakage would permeate through the magnesium oxide and reduce the resistance between the platinum lead wires and the tube. The end result was a temperature drift problem much more severe than that produced by charred foam. The –81 sensor addressed this problem both by changing the braze material to a gold alloy that would not cause liquid metal embrittlement of platinum/rhodium and by reducing the operating stress by more than half.

Microscopic examination of failed units revealed cracks initiated at grain boundaries. Some of the platinum had been metallurgically degraded during manufacturing, where grain enlargement caused some wire to have a grain size exceeding the wire's diameter. Samples from five new lots of wire were obtained from the sensor supplier and tested for the grain enlargement. Only one lot proved to be proper, and the entire lot was put on hold and reserved for making the –81 sensor. The failure rate decreased dramatically, but failures still occurred. Later studies concluded that the sensing element recrystallization temperature was within the operating range of the temperature being measured. This basic design concept was doomed in a temperature environment that could exceed 1,750 R [24]. Three years later, the –91 sensor was introduced and work began on a quadruple thermocouple design that would eliminate the tiny wire, replacing it with a much more robust system. The use of thermocouples for this measurement was discarded in the design definition phase because it would require the addition of a known reference temperature and would not meet the sensor accuracy requirement. In this case, the specified sensor accuracy requirement was much more stringent than needed to perform its function.

LESSON LEARNED

Accuracy requirements for instrumentation intended for use in redline circuits are traditionally overstated to be those that can be achieved rather than those that are required to perform the function. Highly precise measurements should be replaced with highly robust ones to prevent erroneous shutdown.

THE TRAGEDY OF CHALLENGER

STS-33 (51-L), the 10th and final flight of *Challenger*, was launched on January 28, 1986. At 73 seconds after liftoff, a failure of a solid rocket motor O-ring caused the destruction of the vehicle and the loss of the crew. The Space Shuttle fleet was grounded until the SRB joints were redesigned and proven safe, under the NRC's surveillance. The SSME program used the time for a massive bottom-to-top review of the engine. A design requirements review (DRR) was followed by a design certification review (DCR). The DRR board review was completed in July 1987 and resulted in 50 change requests to JSC and MSFC. The DCR board review was completed in February 1988, and 51 actions were assigned to resolve issues with design certification.

The Failure Modes and Effects Analysis (FMEA) was a formal contractual document that contained descriptions and analysis of all potential credible engine failures, their causes, and the resulting effect on the vehicle. An associated Critical Items List (CIL) contained the rationale for accepting (retaining) each potentially catastrophic failure in terms of the hardware design, processing, inspection, and test. A Hazards Analysis (HA), derived from the FMEA/CIL, categorized each hazard as controlled by design, by inspection or other action, or as an accepted risk.

These documents were prepared in accordance with NASA Handbook 5300.4, were approved for the program before the first Space Shuttle flight, and were in effect for STS-33 (51-L). Commission recommendation III called for a review of these analyses on all shuttle elements to "...identify those items that must be improved prior to flight to ensure mission safety." On March 24, Admiral Truly, associate administrator for space flight, directed that the FMEA/CIL/HA review be a complete rebaselining with a complete reapproval process [25]. The SSME program undertook a complete reconstruction of the documents in much more detail than ever done before. Two years later, the FMEA/CIL and HA were approved with much more information—fewer than 200 pages of analysis were replaced by almost 3000. The reassessment had identified 18 required changes to design, prelaunch inspection, and controller software.

A structural audit conducted from April 1987 to April 1988 required 148 structural analyses. A detailed weld assessment conducted at the same time addressed retention rationale for 3058 welds. A review of 11,000 Unsatisfactory Condition Reports (UCRs) found 1283 UCRs that were closed without recurrence control. These were further analyzed to assure adequate flight rationale. Six thousand UCRs were reassessed relative to the appropriate FMEA failure modes and causes, and subsequent trend analyses did not uncover any problems not previously recognized [26].

RETURN TO FLIGHT, STS-26R

Exhaustive reviews were conducted of the launch processing requirements at KSC. The Operational Maintenance Requirements and Specifications Document (OMRSD) and the resulting Operations and Maintenance

Instructions (OMI) were reviewed for accuracy and completeness. The OMRSD/OMI were reconciled with the rebaselined FMEA/CIL and resubmitted for approval. A 76-page cross index correlates individual FMEA failure codes to OMRSD requirements specified in the CIL. Prelaunch guidelines known as Launch Commit Criteria (LCC) were also reconciled with the FMEA/CIL and were modified for changes that the margin review board identified. The LCC were also resubmitted for approval.

While much of the post-*Challenger* recovery operation was being planned and structured, the SSME program was reorganized. The two programs, flight and development, were reunited as a single SSME program in August 1986—three years after being split into two programs.

Even though this was the seventh flight of *Discovery*, the launch preparations and review activity were as complete and thorough as if the launch system had been all new. All three *Discovery* engines completed acceptance tests just before Christmas 1987, and individual formal acceptance reviews were completed by the middle of January. Engine installation took place in the Orbiter Processing Facility (OPF), which was *Discovery*'s home during the *Challenger* stand-down. Installation was completed on January 24, 1988.

Two additional formal reviews were conducted in May and June before *Discovery*'s rollout from the OPF to the Vehicle Assembly Building (VAB) on June 21. In the VAB, *Discovery* was mated to the solid rocket booster/external tank (SRB/ET) stack, and then the complete Space Shuttle vehicle was rolled out to the launch pad on the fourth of July.

The engine configuration description in the FRF test readiness review identified 71 engine design changes that had been incorporated since the *Challenger* disaster. Fifty-five Engineering Change Proposals (ECPs), each of which specified verification requirements, approved and tracked the 71 changes. All requirements were completed, and Rocketdyne prepared formal verification complete reports and NASA approved them.

The engine power level for STS-26R was 104 percent. There were 20 Space Shuttle flights from *Challenger*'s first flight to the last one. Sixteen of those flights were at a power level of 104 percent, whereas the other four, with reduced payloads, were at 100 percent. A NASA risk assessment concluded that having to conduct flight trajectory analyses with two different power levels presented an opportunity to make a mistake. The Space Shuttle program baselined the power level at 104 percent for all future flights.

BUILDING MARGIN

Immediately after the *Challenger* disaster, a Rocketdyne SSME Margin Review Board was established to review and approve SSME changes to reduce risk and increase margin. Board membership included engineering, program office, and quality. The board approved changes for the next flight, changes to be phased in after the next flight, and a long-range plan to develop an SSME with significant margin improvements. Two of the engine operating goals established for the margin engine were that the component operating environment at 104 percent would be less severe than the existing 100 percent power level and that the environment at FPL would be less severe than the existing 104 percent power level [27]. The margin engine would be known as the Block II configuration and included a complete redesign of the MCC, powerhead, and both high-pressure turbopumps. Except for maintaining major component interfaces, all were "clean sheet" designs [28].

LARGE THROAT MCC

The large throat main combustion chamber (LTMCC) was first envisioned as a potential for up-rating the SSME to a higher thrust and second as a method of increasing operating margin. It was carried as a back burner program even before the first Space Shuttle flight [29]. Before the difficulties of FPL development became apparent, design was authorized in July 1981 as a backup for the FPL engine. During the difficult struggle to certify the SSME for FPL, it gained impetus as a potential for increased margin. It was recommended by Williams and the ASAP in the 1983 program reassessment, and it was included in Lindstrom's August 1983 program realignment.

The LTMCC had a favorable impact on the operating environment of the entire engine. Increasing the throat diameter by 6 percent (a little over half an inch) caused the chamber pressure to decrease by 270 psi for the same propellant flow rate. This effect was compounded upstream of the chamber, with system pressures being reduced as much as 380 psi. This in turn caused reductions in the LOX and fuel high-pressure pump speeds of 600 and 700 RPM. Power requirements were reduced 2,000 and 3,000 horsepower, with turbine temperature reductions of 100 and 160 degrees. The combined impact on component operating environments was equivalent to a thrust de-rating of about 7 percent.

The new combustion chamber contained more than just a larger throat. Other features of the LTMCC were aimed at increased margins and incorporation of lessons learned from decades of SSME testing. The inlet and outlet manifolds were changed from welded forgings to integral castings. In all, 46 welds were eliminated, of which 28 were classified in the FMEA/CIL as criticality one. Increasing the number of coolant channels from 390 to 430 enhanced combustion chamber cooling.

Warm hydrogen gas in the outlet manifold had always caused concern about hydrogen environmental embrittlement (HEE) of the metal. To prevent this, the inside of the outlet manifold and duct had been copper plated. The LTMCC outlet manifold and duct were fabricated with a material that is not susceptible to HEE, eliminating the concern and the copper plating. Additional design changes addressed improved producibility, cost savings, and schedule reduction.

Incorporation of the LTMCC into the SSME was delayed for several years because of concerns about engine performance and then combustion stability. Several new components under development were expected to undergo a significant increase in weight. Increased robustness meant heavier components. The weight of a two-duct powerhead designed by Rocketdyne was to increase by 180 pounds. A new HPOTP and a new HPFTP, referred to as alternate turbopumps (ATs), were expected to gain 190 and 290 pounds, respectively. Including the LTMCC would have resulted in a gain of 95 pounds and a loss of engine efficiency [30]*. The increased throat area caused a decrease in the expansion area ratio, resulting in a lower specific impulse. This combination of higher weight and lower performance would result in an unpalatable reduction in payload capability.

Analysis had indicated that recovering the parasitic fuel flow associated with combustion stability aids could restore the performance loss. The main injector baffle in the powerhead and acoustic cavities at the forward end of the main combustion chamber required extra fuel for cooling. Eliminating these devices would allow more efficient use of the propellants. Decades of experience with LOX/hydrogen combustion and coaxial element injection had strongly indicated that the design was inherently stable. This was not universally accepted, however, because it had not been proved beyond a shadow of a doubt. The accepted manner of proving combustion stability is to detonate bombs inside the combustion zone and to show that the disturbance is quickly damped out. A plan was set in to motion to fabricate the required test hardware to conduct multiple bomb tests to prove stability. Commitment to incorporate the LTMCC was delayed until after the stability tests were completed.

An interim configuration LTMCC was fabricated and assembled with Engine 0208 early in 1988. The engine was installed in test stand A-3 for three tests to demonstrate that the increased throat diameter had the desired effect. It was then removed and shipped to MSFC for additional testing on a new test stand called the Technology Test Bed (TTB).

PHASE II+ POWERHEAD

The powerhead contained all the elements for the staged combustion cycle. The FPB provided the hot gas to drive the HPFTP turbine, which exhausted into the hot gas manifold (HGM). On the other side, the OPB provided the drive gases for the HPOTP turbine, which flowed into the HGM through a heat exchanger, which converted liquid oxygen to gaseous oxygen, to be used to pressurize the external oxygen tank. The HGM collected the hydrogen-rich steam exhausted by the turbines and transferred it to the injector through five four-inch-in-diameter transfer tubes. A hurricane of hot gas then flowed

*Conversation with Roger Belt, Rocketdyne weight engineer.

through the forest of 600 LOX posts, went through the fuel elements at the bottom of each post, and was burned in the main combustion chamber.

Two transfer tubes were used on one side of the injector for the HPOTP, and three were used on the other side for the HPFTP. The required turbine flow rate for the HPFTP was over twice that required for the HPOTP. Because the flow area was only 50 percent greater, this created a much higher gas velocity on the HPFTP side.

A series of engine problems became apparent in the second half of 1981, shortly after FPL development began. These problems were related to high-pressure fuel turbine hot gas flow velocity and turbulence. Bob Crain, SSME chief designer, reacted by establishing a design team to pursue a long-range redesign of the hot gas flow path, which would "...not be limited by schedule constraints" [30]. His team was formed in early December 1981, and six months later, after another HPFTP catastrophic failure, he had the full support of the program office.

All of these elements were improved during a 10-year design and development program, which concluded with a much-improved powerhead known as the *two-duct powerhead* [31]. Engines with the new powerhead were identified as Phase II+ configuration, which was never approved for flight. The two-duct powerhead that was combined with a new HPOTP became the Block I flight configuration, which would fly in 1995.

ALTERNATE TURBOPUMPS (ATS)

During the restructuring activity after the *Challenger* disaster, MSFC awarded a contract to Pratt & Whitney (on August 18, 1986) for the design and development of replacement turbopumps for the HPOTP and the HPFTP [32]. Component testing of both Alternate Turbopump Development (ATD) turbopumps began in the spring of 1990 on Pratt & Whitney test stand E-8. Almost a year later, both turbopumps had been subjected to an engine test. The HPOTP/AT was first tested on Engine 0213 (Test 904-108) on February 16, 1991, and the HPFTP/AT was first tested on the same engine (Test 904-119) on May 31, 1991. Significant features of the HPOTP/AT compared to the heritage HPOTP are as follows:

- The number of rotating parts was reduced from 50 to 28.
- The number of welds was reduced from 300 to 4.
- The bearings were made from silicon nitride (ceramic).

Significant features of the HPFTP/AT compared to the heritage HPFTP are as follows:

- Welded sheet metal was replaced by precision investment castings.
- It has single crystal turbine blades.
- It has a strong pump inlet housing (factor of two on burst).

After incorporation, the AT designation was dropped. HPOTP/AT was referred to as the Block I HPOTP, and HPFTP/ATD was referred to as the Block II HPFTP.

CONFIGURATIONS AFTER CHALLENGER

During the 25 years after the *Challenger* disaster, significant safety improvements were incorporated in five major component changes made uniformly with time (Fig. 4.5). The phase II configuration was established with the 71 changes required by the *Challenger* return to flight margin review board. The first flight of phase II engines was STS-26R, on September 29, 1988. The Block I HPOTP was combined with the Phase II+ Powerhead to create the Block I engine configuration. The first flight of Block I engines was STS-77, on May 19, 1996.

A planned Block II configuration was delayed by problems with the Block II HPFTP. Because the Block II configuration included the LTMCC, which would also be delayed, it was decided to create an interim configuration using the heritage HPFTP and the LTMCC. It was designated as Block IIA. The first flight of Block IIA engines was STS-89, on January 22, 1998. The Block II HPFTP was combined with the LTMCC to create the Block II

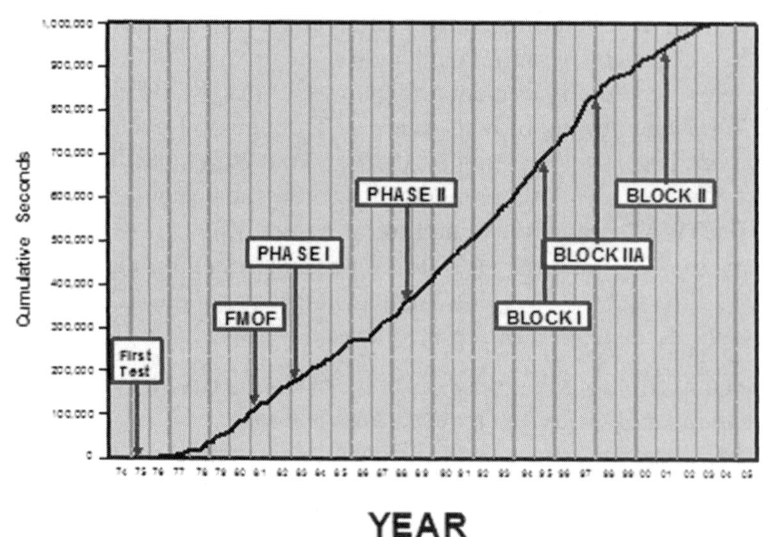

Fig. 4.5 Major milestones in the SSME test program. (Rocketdyne/Pratt & Whitney, CP8_0931-69)

engine configuration. The first flight of Block II engines was STS-110, on April 8, 2002.

With STS-110, the Pratt & Whitney high-pressure turbopumps were completely integrated into the SSME program. In August 2005, Pratt & Whitney acquired Rocketdyne.

ADVANCED HEALTH MONITORING SYSTEM (AHMS)

After 25 years of Space Shuttle flights, the engine controller was upgraded to an AHMS, which included a digital engine model with the capability to isolate failure sources and take some action. Although it was never fully activated, one feature was incorporated for a significant reliability improvement: A high-pressure turbopump vibration monitor was added with the capability to determine if an indicated turbopump failure is real or erroneously indicated by the instrumentation. Before this device was incorporated, the probability of a false indication of a major turbopump problem was approximately equal to the probability of a real failure. The first flight of the AHMS was STS-118, on August 8, 2007.

APPENDIX: SSME MANAGEMENT

NASA MSFC Project Mgr.		Rocketdyne Program Mgr.	
Bill Brown	1971–1974	Paul Castenholz	1971–1974
J. R. Thompson	1974–1982	Norm Reuel	1974–1975
Jud Lovingood	1982–1983	Dom Sanchini	1975–1983
Joe Lombardo (D)	1983–1986	Steve Domokos (D)	1983–1985
Bill Taylor (F)	1983–1986	Bob Paster (D)	1985–1986
Joe Lombardo	1986–1989	Jerry Johnson (F)	1983–1986
Jerry Smelser	1989–1992	Bob Paster	1986–1989
Boyce Mix (A)	1992	Byron Wood	1989–1992
Otto Goetz	1992–1994	Al Hallden	1992–1997
Gerry Ladner	1994–1997	John Plowden	1997–2000
George Hopson	1997–2003	Jim Paulsen	2000–2011
Gene Goldman	2004–2006		
Tim Kelley (A)	2006–2007		
Jerry Cook	2007–2011		

NASA MSFC Chief Engineer		Rocketdyne Chief Engineer	
Jerry Thomson	1971–1986	Willy Wilhelm	1971–1974
Jud Lovingood (A)	1986–1988	Matt Ek	1974–1977
Otto Goetz	1988–1992	Paul Fuller	1977–1983
Dennis Gosdin (A)	1992–1993	Byron Wood (D)	1983–1986
Rick Bachtel	1993–1995	Paul Dennies (F)	1983–1984
Gary Lyles	1995–1996	Al Hallden (F)	1984–1986
Len Worlund	1996–2002	Byron Wood	1986–1989
Ron Tepool	2002–2008	Al Hallden	1989–1992
Katherine Van Hooser	2008–2011	Fred Nitz	1992–1996
		Jim Paulsen	1996–1997
		Dan Adamski	1997–2009
		Doug Bradley	2009–2011

(D) Development, (F) Flight, (A) Acting

REFERENCES

[1] Release No. 71-119, Public Affairs Office, George C. Marshall Space Flight Center, National Aeronautics and Space Administration, Huntsville, Alabama, 13 Jul. 1971.

[2] Wilhelm, W. F., "Space Shuttle Orbiter Main Engine Design," *Society of Automotive Engineers Transactions*, Vol. 81, 1972, Paper 72-0807.

[3] 13M15000F, "Space Shuttle Orbiter Vehicle /Main Engine Interface Control Document," National Aeronautics and Space Administration, 9 Feb. 1973.

[4] CP320R0003B, "SSME Contract End Item Specification," Rocketdyne Division, Rockwell International, 10 May 1973.

[5] "RSS-8503-2, SSME Program Development Plan Approved 2 August 1982," Rocketdyne Division of Rockwell International, 3 Nov. 1982.

[6] RSS–8570-4, "SSME Award Fee Performance, September 1973–February 1974," Rocketdyne Division of Rockwell International, 9 Apr. 1974.

[7] Larson, E. W., "SSME Accident-Incident Report Coca-1A Test 740-007, Incident of 4 February 1976," RSS-8595-6, Rocketdyne Division of Rockwell International, 31 Mar. 1976.

[8] Spencer, E. G., "SSME Accident-Incident Report Coca-1B Test 745-018, SB-1 Valve Fire," RSS-8595-12, June 29, 1976, Rocketdyne Division of Rockwell International, Sep. 1977.

[9] Statement for the Record by Dr. Robert A. Frosch, Administrator, National Aeronautics and Space Administration, to the Subcommittee on Science, Technology and Space, of the Senate Committee on Commerce, Science and Transportation, United States Senate, 31 Mar. 1978.

[10] Seitz, P.E., and Searle, R. F., "Space Shuttle Main Engine Control System," Paper 73-0927, National Aerospace Engineering and Manufacturing Meeting, Society of Automotive Engineers, 16–18 Oct. 1973.

[11] Nelson, R. L., Unpublished SSME Transient Model Analysis Results, Rocketdyne Division of Rockwell International, n.d.

[12] Childs, D. W., "Transient Rotordynamic Analysis for the Space Shuttle Main Engine High Pressure Turbopumps," ASEE–NASA Summary Faculty Fellowship Program Final Report, University of Alabama, 1973.

[13] Ek, M. C., "Solution of the Sub-synchronous Whirl Problem in the High Pressure Hydrogen Turbomachinery of the Space Shuttle Main Engine," Paper 78-1002AIAA/SAE 14th Joint Propulsion Conference, Las Vegas, Nevada, 25–27 July 1978.

[14] Wood, B. K., "SSME Accident-Incident Report, Engine 0010 Test 901-284 High Pressure Oxidizer Fire," RSS-8595-22, Rocketdyne Division of Rockwell International, 15 Jan. 1981.

[15] Ek, M. C., "SSME Accident-Incident Report, Test 901-120 High Pressure Oxidizer Fire," RSS-8595-15, Rocketdyne Division of Rockwell International, 1978; Ek, M. C. "SSME Accident-Incident Report Test 901-136 High Pressure Oxidizer Fire," RSS-8595-13, Rocketdyne Division of Rockwell International, 20 Mar. 1978.

[16] Johnson, J. R., and Colbo, H. I., "Space Shuttle Main Engine Progress Through the First Flight," *AIAA 78-1373, AIAA/SAE 17th Joint Propulsion Conference*, Colorado Springs, Colorado, 27–29 July 1981.

[17] Thompson, J. R., "SSME Project Development and Production Guidelines," SA51-243, George C. Marshall Space Flight Center, National Aeronautics and Space Administration, Huntsville, Alabama, 21 May 1980.

[18] Picarella, J. L., and Fuller, P. N., "OV099 FRF Test Summary," SSME-83-0251, Rocketdyne Division of Rockwell International, 6 Jan. 1983.

[19] Malkin, M. S., "Teletype RUEANAT0234 SSME Development Planning (MH 78-18)," National Aeronautics and Space Administration, 15 Feb. 1978.

[20] RAR KLO-250R1, "Post FRF–2 Leak Check to Challenger Engines," Rocketdyne Division of Rockwell International, 28 Jan. 1983.

[21] Erv Eberle and Ross Mckown, "Engine 2011 Main Combustion Chamber Manifold Crack," BC 83-52, Rocketdyne Division of Rockwell International, 25 Feb. 1983.

[22] Lindstrom, R. E., NASA MSFC Manager, "Space Shuttle Projects Budget Briefing to the NASA Office of Space Flight, Aug. 1982.

[23] Lindstrom, R. E., "SSME Engine Program Realignment," Briefing, Marshall Space Flight Center, 12 Aug. 1983.

[24] Hill, A., "SSME Fast Response T/C Status," Briefing, Rocketdyne Division of Rockwell International, 11 Mar. 2002.

[25] Fletcher, J. C., NASA Administrator, Response to June 13, 1986, letter from the President on implementation of the Commission's recommendations, NASA Historical Reference Collection.

[26] "BC 87-219 Aerospace Safety Advisory Panel Review," Rocketdyne Division of Rockwell International, 16 Sep. 1987.

[27] "BC 86-153 Margin Engine Benefits Analysis," Rocketdyne Division of Rockwell International, Sep. 1981.

[28] "BC 87-69 SSME Review for Gene Covert and Walt Williams," Rocketdyne Division of Rockwell International, 17 Apr. 1987.

[29] "P10222 Enhanced Cooled MCC," Rocketdyne Division of Rockwell International, 13 July 1981.

[30] Crain, B., "Powerhead Flow Path Redesign," IL ED-0156-3040, Rocketdyne Division of Rockwell International," 4 Dec. 1981.

[31] Biggs, B., and Gosdin, D. to (consultant) Lombardo, J., "Phase II + Powerhead Investigation," Un-numbered briefing, Rocketdyne Division of Rockwell International, 28 Mar. 1991.

[32] Ryan, B., and Gross, L., "Developmental Problems and Their Solution for the Space Shuttle Main Engine Alternate Liquid Oxygen High-Pressure Turbopump: Anomaly or Failure Investigation the Key," NASA Technical Paper 3553, May 1995.
[33] Chief Engineer to Associate Administrator for Space Transportation Systems, Certification of SSME for operations above 100%, NASA, July 2 1982.

Chapter 5

Protecting the Body: The Orbiter's Thermal Protection System

Dennis R. Jenkins

Developing a thermal protection system (TPS), often called a *heat shield*, was one of the larger challenges facing scientists and engineers as they began seriously looking at lifting-reentry vehicles during the 1950s. In selecting a concept, researchers had to consider the physical, mechanical, chemical, and economic characteristics of the material and the vehicle [1]. Initially, almost all heat-shield research centered on various ablative and radiative metallic materials, and with one exception, it was not until well into the Space Shuttle conceptual development cycle that scientists and engineers seriously considered other materials.

All of the manned orbital space capsules used ablative heat shields, but during the 1950s and early 1960s, most scientists and engineers assumed any advanced TPS would be metallic. One of the larger problems that researchers uncovered was that most metals tended to rapidly oxidize when exposed to high temperatures. The oxidation quickly compromised the material's strength and thermal properties. As a result, researchers expended a great deal of effort trying to find a long-lasting coating that would provide adequate protection for the metal, not affect its mechanical properties, and not have a negative impact on the desired emittance properties. By 1964, researchers concluded there was no "ideal" coating for any given material because each application had different needs. Nevertheless, they had made significant progress toward understanding specific silicide and aluminide coatings for columbium, molybdenum, tantalum, and tungsten. The suborbital Mercury capsules used heat sinks, much like the ICBM warheads of the era [2].

Interestingly, early in the search for a reusable lifting reentry vehicle, a few engineers briefly investigated a ceramic tile concept similar to that ultimately used by the Space Shuttle. During 1951, Lawrence D. "Larry" Bell (1894–1956), Walter R. Dornberger (1895–1980), and Krafft A. Ehricke (1917–1984) were working at the Bell Aircraft Company in Buffalo, New York. One of their projects was a reusable lifting reentry vehicle known variously as BoMi, Project MX-2276, and RoBo; the concept eventually formed the basis for the ill-fated Dyna-Soar program.

In 1954, Bell researchers investigated covering the MX-2276 glider with lightweight bricks, each measuring 10 inches square. The bricks were made from Sil-O-Cel®, first developed by the Celite Products Company (now Celite Corporation) in 1924 and still available in various forms. The bricks had a density of 23.5 pounds per cubic foot, 86 percent of which was silicon dioxide derived from diatomaceous earth. These were generally similar to the firebricks used by Babcock & Wilcox for boilers installed on large ships. The bricks contacted the airframe only along the outer edges, with the beveled center filled with Fiberfrax insulation. A single center attachment consisting of a large countersunk ceramic pin and an Inconel bolt secured each brick. The bricks were spaced from their neighbors with a gap sufficient to accommodate thermal expansion and structural strains. A thin ceramic coating covered the outer surfaces of the bricks for aerodynamic smoothness and waterproofing. It was, essentially, the solution ultimately chosen for the Space Shuttle [3].

It took a while, but NASA Langley finally noticed the ceramic concept and awarded Bell a small contract in August 1965 for continued evaluation. By now, Lockheed had been working on a similar concept for several years under contract to NASA Ames. James N. Krusos led the effort at Bell, which ran through October 1966. Bell researchers selected porous ceramics because of their "advantageous insulation and weight characteristics." The perceived advantages of ceramics included their (1) ability to maintain a smooth and stable outer mold line, (2) need for little or no refurbishment, (3) lightweightness, and (4) relatively low cost, primarily because of their reusability [4].

Oddly, considering that the silicon dioxide Sil-O-Cel fire bricks investigated in 1954 weighed only 23.5 pounds per cubic foot, Bell now estimated that a fused silica ceramic would weigh between 34 and 68 pounds per cubic foot. Combined with support panels and insulation, the TPS weighed 3.28 pounds per square foot of surface area. An active cooling system in the underlying structure weighed another 0.49 pounds per square foot, for a total of 3.77 pounds per square foot. This compared favorably to a contemporary ablative heat shield at 4.0 pounds per square foot [5].

The researchers came away seemingly impressed, thinking ceramics held "promise." They cautioned, however, "that realization of the benefits of highly refractory and stable heat shields requires further refinement of structural analysis and material technologies." The researchers believed ceramics were attractive from a weight standpoint and in regard to component simplicity and refurbishment. The ceramics also maintained a constant outer mold line and vehicle smoothness, unlike ablative heat shields, which changed shape and became rough during entry [6].

The Space Shuttle

Ablators were the departure point for many of the TPS concepts during the Space Shuttle Phase A studies because they were well understood and proven. Researchers studied various materials, including silica/silica-fiber composites; mixed inorganic or organic composites with silica, nylon, or carbon fiber-reinforced resins (phenolics, epoxies, and silicones); and carbon- or graphite-based materials. These ablators had densities ranging from 10 pounds per cubic foot (using microballoons to keep the material lightweight) to 150 pounds per cubic foot (for high-heat protection). Researchers evaluated both charring and noncharring materials, but the need to maintain a precise outer mold line for aerodynamic efficiency usually mandated charring ablators.

One of the most promising ideas to come out of the Phase A studies was to mount the TPS so that it could be easily removed from the vehicle. This permitted technicians to service the systems and structure under the TPS, and allowed the refurbishment of the TPS itself, without taking the entire vehicle out of service for a prolonged period. In this concept, the TPS was mounted on panels that were installed over the basic vehicle structure with an appropriate insulator between. When the vehicle returned from orbit, technicians at the launch site would replace the panels; the old heat shield went to a shop for refurbishment and reuse on a future mission.

Dyna-Soar had eschewed ablators in favor of a metallic heat shield, and this seemed an elegant solution to most engineers. The use of radiative metallic panels covering a cool structural shell with a layer of insulation between would work for all but the hottest regions of the vehicle. Metals are intrinsically durable and therefore capable of extensive reuse, but they have oxidation and strength limitations. A material's ability to resist oxidation largely determines its temperature limitations, and, unfortunately, most of the exotic superalloys tended to oxidize quickly when exposed to high temperatures. At the time, coatings appeared to permit an upper-limit temperature of about 2500°F for 100 cycles, but actual real-world data was lacking [7].

During the follow-on Phase B studies, new materials, including hardened compacted fibers (HCFs) and oxidation-inhibited carbon-carbon, which were neither ablative nor metallic, presented engineers with an entirely different set of options. The most common HCF investigated during these studies was called mullite, although it was not truly mullite, which was a rare chemical found only on the Isle of Mull, off the west coast of Scotland. The glass and steel industries had long used mullite as a refractory material because it exhibited good high-temperature strength, adequate thermal-shock resistance, and excellent thermal stability. Because natural mullite was rare and expensive, scientists developed methods of fabricating synthetic mullite ceramics [8].

Researchers believed that HCF was a promising candidate for space shuttle applications because of its availability and temperature-overshoot capability. The HCF shingles were also lighter than equivalent ablators but somewhat heavier than most metallic concepts. Typical HCF materials were relatively soft, were extremely porous, and had inherently low emittance values. Adhesive bonding had to be used because the shingles would not support mechanical fasteners due to structural weakness. Researchers believed the HCF shingles were potentially useful for 100 flights in the range of 2000 to 3000°F. However, as with the superalloys, researchers needed to develop suitable coatings to protect the HCF before it could be successfully used on any space shuttle. In many ways, the mullite shingles were an indication of things to come [9].

The various studies also investigated a relatively new class of materials, all of which benefited significantly from the investment in Dyna-Soar. Researchers believed that oxidation-inhibited, fiber-reinforced, carbonaceous products offered the potential for a reusable, reasonably cost-efficient, high-temperature-resistant material. Carbon or graphite fibers, usually in the form of cloth, reinforced the carbonaceous material, and the resultant material offered an unsurpassed strength-to-weight ratio at high temperatures. It should be noted that these materials, like the reinforced carbon-carbon actually used on the Space Shuttle, were not insulators and that the back side of the material was essentially as hot as the front side. What the materials provided was a means of creating the appropriate outer mold line on the hottest parts of the vehicle (the nose and the wing leading edge), but some sort of insulating material still needed to protect the primary structure. This insulator was usually a fibrous micro-quartz material characterized by low strength, low density, and low thermal conductivity. Materials typical of this class include Dynaflex®, Micro-Fiber®, and Q-Felt® batt insulation.

Reusable Surface Insulation (RSI)

However, researchers were developing another material, although it was similar in many respects to the artificial mullite shingles investigated during Phase B and the conceptual descendant of the Sil-O-Cel firebricks. The

Lockheed Missiles & Space Company in Palo Alto, California, was making progress in the development of a ceramic RSI concept and by December 1960 had applied for a patent for a reusable insulation material made of ceramic fibers. The first potential use came in 1962, when Lockheed manufactured a 32-inch-diameter radome for the Apollo spacecraft. This radome was made from a filament-wound shell and a lightweight layer of internal insulation cast from short silica fibers. However, the Apollo design changed, and the radome never flew. Chemist Robert M. Beasley (1927–1990) led the development of the ceramic tiles ultimately used on the Space Shuttle. Initially, Beasley conceived of the idea while working at the Corning Glass Works in New York, but he did not initiate any real development until he moved to Lockheed during 1960 (Fig. 5.1).

By 1965, researchers at Lockheed had developed LI-1500, the first of what became the space shuttle tiles. This "Lockheed Insulation" (hence the "LI") material had a density of 15 pounds per cubic foot (the "15"), was 89 percent porous, could survive repeated cycles to 2500°F, and appeared to be truly reusable. A sample flew on an Air Force reentry test vehicle during 1968, reaching 2300°F with no apparent problems.

The material was brittle, with a low coefficient of linear thermal expansion, and, therefore, Lockheed could not cover an entire vehicle with a single piece of it. Rather, engineers expected to install the material in the form of tiles, generally 6 × 6-inch squares. The tiles had small gaps between them to permit relative motion and to accommodate the deformation of the metal structure under them because of thermal effects. A tile could crack as the metal skin under it deformed, so engineers decided to bond the tile to a felt pad and then bond the felt pad to the vehicle. Both bonds used a room temperature vulcanizing (RTV) adhesive. It was a breakthrough, although everybody approached the concept with caution and a certain reluctance. Oddly, nobody seemed to remember the firebricks Bell had proposed in 1954.

Fig. 5.1 An early LI-15 tile (before it was redesignated LI-1500) that was tested at the MSC in Houston. (NASA)

PHASE C/D DEVELOPMENTS

After long and heated battles within NASA, the DoD, the White House, the OMB, and Congress, President Nixon finally approved the development of the Space Shuttle on January 5, 1972. NASA issued the space shuttle request for proposals (RFPs) on March 17, 1972, to Grumman/Boeing, Lockheed, McDonnell Douglas/Martin Marietta, and North American Rockwell. The RFP required the use of an unspecified reusable TPS, although it allowed ablative material or other special forms of TPSs where beneficial to the program. The orbiter TPS was to be capable of surviving an abort scenario from a 500-mile circular orbit. Unsurprisingly, because the RFP required the contractors to use the NASA-developed MSC-040C concept, the four proposed vehicles looked remarkably similar [10].

Grumman was not completely confident with the Lockheed-developed RSI and proposed adding a layer of ablative material around the crew compartment and the orbital maneuvering system/reaction control system (OMS/RCS) modules during the early development flights to guard against a possible RSI burn-through [11].

Lockheed proposed LI-1500 tiles for the inboard part of the leading edge, using the more expensive carbon-carbon only on the outboard sections. Lockheed believed its TPS could provide 100 normal operational entries at 2500°F or a single-contingency entry at 3000°F [12].

McDonnell Douglas proposed mullite HCF shingles glued to the bottom and side of the orbiter over a thin layer of foam for strain relief. Ablator covered the upper surfaces of the fuselage, wing leading edges, and vertical stabilizer. For some reason, the mullite material was composed of "shingles," whereas the Lockheed RSI used "tiles" [13].

North American was also not completely convinced that the Lockheed-developed RSI would mature quickly enough for the early orbital flights and proposed using HCF shingles manufactured from aluminum silicate, although North American continued to monitor the progress of the silica tiles because they were lighter and more durable. The leading edges and nose cap were reinforced carbon-carbon (RCC) developed by Ling-Temco-Vought (LTV) based on research for the Dyna-Soar program. There was no substantial use of ablator anywhere on the vehicle [14].

ORBITER CONTRACT AWARD

After the stock market closed on July 26, 1972, NASA announced that the Space Transportation Systems Division of North American Rockwell had been awarded the $2.6 billion contract to design and build the Space Shuttle orbiter [15].

Given the progress Lockheed was making developing its RSI, NASA and Rockwell asked the Battelle Memorial Institute to evaluate mullite and

RSI—an evaluation that ultimately favored the Lockheed product. Interestingly, NASA and Rockwell believed that the leeward (top) side of the orbiter would not require any thermal protection. However, in March 1975, researchers from the Air Force Flight Dynamics Laboratory (AFFDL) conducted a briefing for space shuttle engineers on the classified results of the ASSET, PRIME, and BGRV programs. This data indicated leeward-side heating was not particularly severe but easily exceeded the 350°F capability of the aluminum skin. To alleviate this concern, Rockwell proposed to cover areas that never exceeded 650°F, including the upper surfaces of the wings and aft fuselage and payload bay doors, with an ablative elastomeric reusable surface insulation (ERSI) bonded directly to the aluminum skin. Engineers expected, given the relatively benign environment, the ERSI could survive many flights with only minimal refurbishment [16].

Further study showed drawbacks to using the ablator and engineers soon replaced it with a combination of LI-900 tiles and Nomex felt blankets. Black LI-3000 and LI-2200 tiles protected the bottom of the fuselage and wings, the entire vertical stabilizer, and most of the forward fuselage where temperatures exceeded 650°F but were less than 2300°F. The tile thickness varied as necessary to limit the backface temperature to 350°F on the orbiter and 250°F on the OMS pods. As researchers further refined the heating models, the LI-3000 tiles gave way entirely to LI-2200 and finally to a combination of LI-2200 and LI-900 [17].

In the meantime, a problem developed with the tiles themselves. As researchers refined the flight profiles and better understood the aerodynamic environment, engineers began to question whether the tiles could survive the harsh conditions. By mid-1979, it had become obvious that the tiles in certain areas did not have sufficient strength to survive the tensile loads of a single mission. NASA immediately began an extensive search for a solution, which eventually involved outside blue-ribbon panels, various government agencies, academia, and most of the aerospace industry. As Space Shuttle Program Deputy Director LeRoy E. Day recalled, "There is a case [the tile crisis] where not enough engineering work, probably, was done early enough in the program to understand the detail—the mechanical properties—of this strange material that we were using" (Fig. 5.2). It was, potentially, a showstopper [18]. North American manufactured several test articles to evaluate the TPS and its underlying structure; constant damage eventually led NASA to replace the LI-2200 tiles with a small RCC "chin panel."

As engineers spent more time examining the problem, they realized the issue was not with the actual tiles but with how the tiles attached to the underlying structure. Analysis indicated that although each individual component—the tile, the felt strain-isolation pads (SIP) under the tile, and the two layers of adhesives—had satisfactory tensile ability, when combined as a system they lost about 50 percent of their strength. Engineers largely attributed this to

Fig. 5.2 The nose cap and first few feet of the forward fuselage. Note the area on the bottom of the fuselage just behind the nose cap is in the original configuration that used LI-2200 tiles. (NASA)

stiff spots in the SIP (caused by needling) that allowed the system strength to decline to as low as 6 pounds per square inch instead of the baseline 13 pounds per square inch [19].

In October 1979, engineers developed a densification process that involved filling voids between fibers at the inner mold line (next to the SIP pad) with a slurry consisting of Ludox (a colloidal silica made by DuPont) and a mixture of silica and water. The densified layer acted as a plate on the bottom of the tile, eliminating the effect of the local stiff spots in the SIP and bringing the total system strength back up to 13 pounds per square inch (Fig. 5.3) [20]. Early in the program, each vehicle had more than 30,000 individual tiles, although an increased use of blankets reduced this to just 24,000 at the end of the program.

Installing the tiles on the vehicle presented its own issues, and Rockwell quickly ran out of time while *Columbia* was in the manufacturing facility in Palmdale, California. NASA needed to present the appearance of maintaining a schedule, and moving *Columbia* from Palmdale to KSC was a very visible milestone. Rockwell had installed slightly more than 24,000 tiles in Palmdale, and had 6000 to go. Unfortunately, by now it appeared that technicians would have to remove all of the tiles so they could be densified.

The challenge was how best to salvage as many of the installed tiles as possible while ensuring sufficient structural margin for a safe flight. To overcome this almost insurmountable challenge, engineers developed the tile proof test. Technicians induced a stress over the entire footprint of each tile equal to 125 percent of the maximum flight stress experienced at the most critical point on the tile footprint. The proof-test device used a vacuum chuck attached to the tile, a pneumatic cylinder to apply the load, and six pads attached to surrounding tiles to react the load. Because any appreciable tile load might cause internal fibers to break, acoustic sensors placed in contact with the tiles monitored for internal fiber breakage. This testing proved the majority of the tiles were adequately bonded for flight because only 13 percent failed the proof test; those were replaced with densified tiles (Fig. 5.4) [21].

For the next 20 months, technicians at KSC worked 3 shifts per day, 7 days per week, testing, removing, and installing 30,759 tiles on *Columbia*. By the time the tiles were installed, proof tested, often removed and reinstalled, and then re-proof tested, the technicians averaged 1.3 tiles per person per week. In June 1979, Rockwell estimated 10,500 tiles needed to be replaced; by January 1980, technicians had installed more than 9000 tiles, but the number remaining had ballooned to 13,100 because additional tiles failed the proof test or were otherwise damaged. By September 1980, only 4741 tiles remained to be installed, and by Thanksgiving, the number was below 1000. It finally appeared that the end was in sight [22].

This initial debacle caused a great deal of embarrassment for NASA and Rockwell, although it did not seriously delay the first flight because the SSMEs had problems as well. Nevertheless, for the next 30 years, many people believed the tiles were a major maintenance issue for the Space Shuttle program; in fact, they proved to be remarkable durable and trouble free—other than the continual need for pull-tests and waterproofing. The next vehicle will

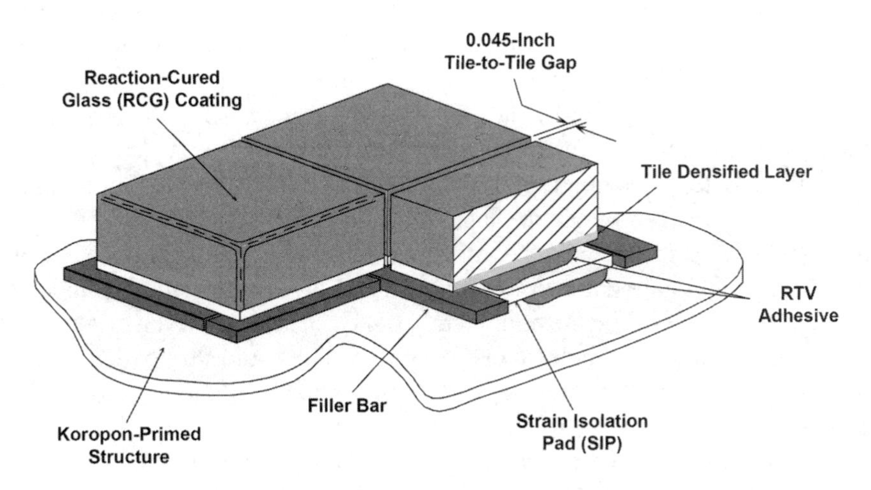

Fig. 5.3 A densified area on the bottom of the tile solved the early problems experienced by the tiles. This illustration shows the major components of the RSI system. (NASA)

Fig. 5.4 To verify that each tile was well bonded to the orbiter, technicians used this Rube Goldberg device, which used vacuum to gently pull on a tile while reacting the loads to the surrounding tiles. This is *Columbia* on October 3, 1979, in the Orbiter Processing Facility (OPF) at KSC. (NASA)

undoubtedly use something better, but for a first attempt at a reusable TPS, the tiles were a great success.

THE TPS

The TPS protected the aluminum structure of the orbiter during the atmospheric portion of flight on both ascent and entry. During a typical entry, the orbiter endured temperatures in excess of 2300°F and heat loads greater than 66,000 btu per square foot. The orbiter TPS also protected the structure from localized heating from the SSME, SRB, and OMS/RCS exhaust plumes during ascent. The TPS was primarily white on the upper surface and black on the lower surface to control on-orbit heating from solar radiation and to maximize heat rejection during entry [23].

In addition to protecting the structure from heat loads, the TPS outer mold line served as the aerodynamic shape of the vehicle. Rigid control of the step and gap between installed TPS components ensured a smooth surface; excessive steps and/or gaps could cause the boundary layer to transition from laminar to turbulent, resulting in significantly higher heat loads. Even minor steps and/or gaps could result in local overheating, which could slump (that is, melt and deform) tiles or permit subsurface plasma flow that, in turn, could degrade the TPS bond line or underlying structure [24].

The final TPS was composed of RCC, five types of tiles, two types of flexible insulation blankets, thermal barriers, gap fillers, thermal window panes, and thermal seals [25].

RCC

Where temperatures exceeded 2300°F, such as the wing leading edge and nose cap, the orbiter used a composite composed of pyrolyzed carbon fibers in a pyrolyzed carbon matrix with a silicon carbide coating called RCC. In addition, the external tank (ET) forward attach point also used an RCC arrowhead-shaped component because of the pyrotechnic shock environment of the ET separation mechanism. RCC weighed approximately 103 pounds per cubic foot and had an operating range of –250°F to 3000°F. The material had a flexural strength of 9000 psi and a tensile strength of approximately 4500 psi [26].

LTV (later part of Lockheed Martin) of Grand Prairie, Texas, manufactured the RCC, which cost almost $1 million per wing panel. The wing leading edge consisted of 22 RCC panels; because the wing profile changed from inboard to outboard, each RCC panel was unique. Contrary to popular perception, RCC was not an insulator, and the backface was essentially as hot as the frontface. Rather, RCC was simply a material that could withstand the appropriate aero and thermal loads; the underlying structure needed protection from extreme heat by other means [27]. The Leading Edge Structural Subsystem (LESS) was a complex piece of engineering that consisted of RCC panels and the hardware needed to attach them to the wing (Fig. 5.5).

RCC fabrication began with a graphite-saturated rayon cloth impregnated with a phenolic resin that was layed up as a laminate and cured in an autoclave. After curing, the laminate was pyrolyzed to convert the resin to carbon. The laminate was then impregnated with furfural alcohol in a vacuum

Fig. 5.5 RCC panel 8-L (the eighth panel on the left wing) shows the structure that supported the panel as well as the insulation behind it. (NASA)

chamber, cured, and pyrolyzed again to convert the furfural alcohol to carbon. This process was repeated three times until the desired properties were achieved [28].

To provide sufficient oxidation resistance to allow reuse, the outer layers of the RCC were converted to silicon carbide. To accomplish this, technicians dry-packed the RCC with material made of alumina, silicon, and silicon carbide and then placed it into an argon-purged furnace with a stepped-time-temperature cycle to 3200°F. A diffusion reaction occurred between the dry pack and carbon-carbon, and this reaction converted the outer layers to silicon carbide with a whitish-gray color. The piece was then sprayed with tetraethyl orthosilicate and sealed with a glossy overcoat. The RCC laminate was superior to a sandwich design because it was lightweight and rugged; it also promoted internal cross-radiation from the hot stagnation region to cooler areas, thus reducing stagnation temperatures and thermal gradients around the leading edge.

A series of floating joints attached the RCC panels to the wing leading edge to reduce loading caused by wing deflections. A T-seal, also made of RCC, between each panel allowed lateral motion and thermal-expansion differences between the RCC and the orbiter wing.

Because RCC was highly radiative and provided no thermal protection, the adjacent aluminum structure was protected by internal insulation. The forward wing spar and nose bulkhead were protected by a series of RSI tiles installed onto removable carrier panels and Inconel 601–covered Dynaflex batt insulation (called Incoflex) [29].

Five years before *Columbia*, beginning in 1998, engineers started modifying the leading-edge RCC to allow it to withstand larger punctures. The primary concern was impacts from micrometeorites and orbital debris. The original design allowed a 1-inch hole in the upper surface of any panel. But on the lower surface, no penetrations were allowed on panels 5 through 13 because any hole would allow heat from the hot gas flow during entry to quickly erode the 0.004-inch Inconel foil on the Incoflex insulators, exposing the leading-edge attach fittings and wing-front spar to the direct blast of superheated air. The upgrade included additional insulation able to withstand a penetration of up to 0.25 inch in diameter in the lower surface of panels 9 through 12 and up to 1 inch on panels 5 through 8 and 13. To achieve this, technicians wrapped Nextel 440 fabric insulation around the Incoflex insulators—one layer for panels 5 through 7 and 11 through 13, and two fabric layers for panels 8 through 10 (which had the highest-potential heating environment) [30].

Despite the modification, an ascent debris strike on an RCC panel on the left wing during STS-107 resulted in the destruction of *Columbia* during entry. A piece of foam from the bipod ramp on the external tank impacted the wing leading edge. The foam penetrated the panel, resulting in hot gas

Fig. 5.6 A test article used by the CAIB to prove foam could destroy RCC. (CAIB)

impinging on the internal structure during entry. Ultimately, the wing failed and *Columbia* broke up, killing her crew of seven. This led to a reevaluation of the relative strength of RCC, and subsequent testing revealed that it was not as robust as once thought, particularly to direct debris strikes. Nevertheless, it was one of the few materials that could withstand the entry environment and continued to be used until the end of the program, albeit with somewhat more inspection and analysis (Fig. 5.6).

RSI TILES

Tiles were manufactured from one of five substrate materials: LI-900, LI-2200, FRCI-12, AETB-8, or BRI-18. The lightest, LI-900, weighed 9 pounds per cubic foot and made up about 83 percent of the total tiles. Denser LI-2200 tiles weighed 22 pounds per cubic foot and were used on the undersurface of the vehicle, making up 2 percent of the tiles. FRCI-12 weighed 12.5 pounds per cubic foot and made up approximately 13 percent of the tiles, with 8-pound-per-cubic-foot AETB-8 and 18.5-pound-per-cubic-foot BRI-18 making up the remainder [31].

The tile substrate material and coating selection were dependent on the mechanical and thermal requirements of the particular location. For example, tiles located on the upper surface of the forward fuselage (nominally 0.75-inch-thick LI-900) experienced lower temperatures and required less strength than tiles on the nose landing gear door (2- to 3-inch-thick LI-2200, FRCI-12, or BRI-18). The thickness of the tiles varied with heat loads and outer mold line contour requirements from less than 1 inch to more than

5 inches. The substrate material was machined to the desired shape (usually a 6 × 6-inch square) before coating [32].

White LI-900 low-temperature reusable surface insulation (LRSI) tiles were used in selected areas of the fuselage, vertical stabilizer, upper wing, and OMS/RCS pods. These tiles protected areas where temperatures were between 750°F and 1200°F. As originally delivered, *Columbia* and *Challenger* made extensive use of LI-900 tiles on the sides of the fuselage and upper wing surfaces. Later vehicles used blankets in these areas, and *Columbia* was brought up to the same standard during Orbiter Maintenance Down Period (OMDP). *Challenger* was lost before any significant changes could be made.

LRSI tiles were of the same construction as LI-900 tiles and had the same basic characteristics as the high-temperature reusable surface insulation (HRSI) tiles but were cast thinner (0.2 to 1.4 inches). The tiles were produced in 8 × 8-inch squares and had a white reaction-cured glass (RCG) optical and moisture-resistant coating applied 10 millimeters thick to the top and sides. The coating was made of silica compounds with shiny aluminum oxide to obtain optical properties and provided on-orbit thermal control for the orbiter. The LRSI tiles were treated with waterproofing and installed on the orbiter in the same manner as the HRSI tiles. The LRSI tile had a surface emittance of 0.80 and a solar absorptance of 0.32 (Fig. 5.7).

Black HRSI tiles protected areas where temperatures were between 1200°F and 2300°F. The HRSI substrate consisted of low-density, high-purity, 99.8-percent amorphous silica fibers 1 to 2 millimeters thick. A slurry

Fig. 5.7 LI-900 tiles used extensively on the sides of the fuselage and upper wing surfaces of Columbia (and Challenger) as delivered while later vehicles used blankets in these areas. Columbia was later refitted.

containing fibers mixed with water was frame-cast to form soft, porous blocks onto which a colloidal silica binder solution was added. Sintering produced a rigid block that was cut into quarters and then machined to the precise dimensions required for individual tiles. HRSI tiles varied in thickness from 1 to 5 inches; generally, the tiles were thicker at the forward areas of the orbiter and thinner toward the aft end. The HRSI tiles withstood on-orbit cold soak conditions, repeated heating and cooling thermal shock, and extreme acoustic environments (163 decibels) at launch [33].

The HRSI tiles were coated on the top and sides with powdered tetrasilicide and sintered borosilicate RCG. The tiles were individually sprayed with 10 to 15 coats of the coating slurry to produce a final coating weight of 0.09 to 0.170 pounds per square foot. The tiles were heated to 2300°F for 90 minutes, resulting in a glossy black coating that had a surface emittance of 0.85 and a solar absorptance of about 0.85. Then, the tiles were waterproofed using heated methyltrimethoxysilane, rendering the fibrous silica material hydrophobic [34].

As an aside, an HRSI tile taken from a 2300°F oven could be immersed in cold water without damage. Also, surface heat dissipated so quickly that an uncoated tile could be held by its edges with an ungloved hand only seconds after removal from the oven while its interior still glowed red.

As LI-2200 tiles were replaced, new fibrous refractory composite insulation (FRCI) tiles were usually used instead. Researchers at NASA Ames developed these later in the program, and they were more durable and resistant to cracking than the earlier HRSI tiles. The tiles were essentially similar to the original HRSI tiles with the addition of a 3M Nextal (AB312—alumina-borosilicate fiber) additive. The resulting composite-fiber refractory material, composed of 20 percent Nextal and 80 percent pure silica, had entirely different physical properties than the original 99.8 percent silica tiles.

The black RCG coating on the FRCI-12 tiles was compressed as it was cured to reduce its sensitivity to cracking during handling and operations. In addition to the improved coating, the FRCI-12 tiles were about 10 percent lighter than the HRSI tiles for a given thermal protection. FRCI-12 tiles demonstrated a tensile strength at least three times greater than that of the HRSI tiles and an allowable temperature approximately 100°F higher.

Late in the program, Boeing Replacement Insulation (BRI-18) began to replace LI-2200 tiles around the landing gear and external tank doors. The Boeing Company in Huntington Beach, California, manufactured the raw substrate from silica mixed with other proprietary elements. The BRI-18 tiles made use of a toughened unipiece fibrous insulation (TUFI) coating developed at NASA Ames. TUFI was the first of a new type of composite known as *functional gradient materials*, in which the density of the material was relatively high on the outer surface and became increasingly lower within the insulation. Unlike the HRSI tiles, in which the tetrasilicide and borosilicate glass coating received little support from the underlying tile, the TUFI outer surface was fully integrated into the insulation, resulting in a more damage-resistant tile; TUFI tends to dent instead of shatter when hit [35].

Because of the constantly changing contours of the orbiter outer mold line, each tile needed to be manufactured to exacting dimensions to fit a specific location on the orbiter. There were two distinctly different types of tile machining, tracing a physical ("splash") model of the cavity on a stylus machine to produce a flight tile, and using a numerically controlled (NC) milling machine to create a tile based on a three-dimensional computer model [36].

Tracing the splash model almost always resulted in a tile that fit the cavity but often resulted in minor outer mold line changes that were not reflected back into the master database. The more precise NC machining was based on data from the master database, which ensured the outer mold line was precisely controlled. Technicians conducted a laser scan of the cavity on the orbiter and fed this data into the master database; oddly, however, this technique often did not produce a tile that fit as precisely as one the manual splash models produced. One of the disadvantages of NC machining was that the initial programming was time consuming, especially if sidewall jogs or nondesign features were required. This time was easily offset for a recurring replacement tile, such as a landing gear door corner tile or a tile adjacent to an RCS thruster that required frequent replacement. Another disadvantage of NC machining was the initial cost of the associated hardware, but this was ultimately offset by the time saved in modeling and the high quality of the finished product [37].

Technicians marked each tile with its part number and a unique serial number using a commercial high-temperature paint called Spearex. The part number was located on the forward or leading edge of the tile to assist in the correct orientation of the tile when it was installed. Because of discontinuities in the thin RCG coating, caused either by processing or flight damage, the tiles required re-waterproofing after each flight. Technicians pneumatically injected each tile with dimethylethoxysilane, which chemically reacted with the silica fibers of the base material and prevented water absorption [38].

Failure to re-waterproof the tiles could have resulted in increased weight from absorbed water or tile damage. The damage would be caused by the absorbed water freezing and subsequently contracting on orbit at cold soak temperatures below –70°F, inducing a fracture at the 1050°F isotherm. During entry, the ice would sublime (change from a solid to a vapor) and complete the failure of the tile at the region previously fractured [39].

Because the tiles could not withstand airframe-load deformation, stress isolation was necessary between the tiles and the orbiter structure. This was provided by SIPs made of Nomex felt. The SIP was bonded to the tiles and then bonded to the orbiter skin using silicon RTV adhesive. However, the SIP introduced stress concentrations at the needled fiber bundles, resulting in localized failure in the tile just above the RTV bond line. To solve this problem, the inner surface of the tile was densified to distribute the load more uniformly. The densification process used a Ludox ammonia-stabilized binder. When mixed with silica slip particles, it became cement, and when mixed with water, it dried to a finished hard surface. A silica-tetraboride coloring agent was mixed with the compound for penetration identification. The densification coating penetrated the tile to a depth of 0.125 inch, and the strength and stiffness of the tile and SIP system were increased by a factor of two [40].

Flexible Insulation Blankets

White blankets made of coated Nomex felt reusable surface insulation (FRSI) were initially used on the payload bay doors, portions of the upper wing surface, and parts of the OMS/RCS pods. The FRSI blankets protected areas where temperatures were below 700°F. The FRSI blankets nominally consisted of 3 × 4-foot, 0.16- to 0.40-inch-thick sheets, depending on the location. A white-pigmented 92-007 silicon elastomer coating provided aerodynamic-erosion protection and a moisture barrier. The density of the coated material was 5.4 pounds per cubic foot, and it was bonded directly to the orbiter by 0.20-inch-thick RTV silicon adhesive [41].

Nomex felt consisted of noncombustible, heat-resistant aromatic polyamide (aramid) fibers 2 deniers in fineness and 3 inches long. These were loaded into a carding machine that untangled the clumps of fibers and combed them to make a tenuous mass of lengthwise-oriented, relatively parallel fibers called a *web*. The cross-lapped web was fed into a loom, where it was lightly needled into a batt. Generally, two such batts were placed face to face and needled together to form felt. The felt then was subjected to a multineedle process that passed barbed needles through the fiber web in a sewing-like procedure that compacted the transversely oriented fibers into a pad. This needle-punch operation was repeated as many times as required to produce a felt product with specified physical properties. The needled felt was calendered to the proper thickness by passing through heated rollers at selected pressures. The calendered material was heat-set at 500°F to thermally stabilize the felt [42].

After the initial delivery of *Columbia*, NASA developed an advanced flexible reusable surface insulation (AFRSI) as a lightweight replacement for the more fragile LRSI tiles. AFRSI blankets protected areas where temperatures were below 1200°F and aerodynamic loads were minimal, such as the sides of the fuselage and portions of the upper wing surface. AFRSI was a quilted blanket consisting of quartz fibrous batt insulation and woven quartz fiber outer fabric quilted on 1-inch centers to a glass fiber inner fabric. The quilting was accomplished using a Teflon-coated quartz thread with three to four stitches per inch. The sewn quilted fabric blanket was manufactured in 3 × 3-foot squares with an overall thickness of 0.25 to 2 inches. The material

Fig. 5.8 AFRSI blankets. (Dennis R. Jenkins)

had a density of 8 to 9 pounds per cubic foot and was used where surface temperatures ranged from 750°F to 1200°F. The blankets were cut to the planform shape required and bonded directly to the orbiter by 0.20-inch-thick RTV silicon adhesive (Fig. 5.8) [43].

The first application of AFRSI was on the OMS pods of *Columbia* before STS-6. Originally, the OMS pods on *Columbia* and *Challenger* were covered with LI-900 tiles. AFRSI blankets replaced most of these tiles, although the aerodynamic environment on the front edge of the pod dictated that some tiles be retained. The LI-2200 tiles were added in response to a hot vortex that came off the leading edge of the double-delta portion of the wing. AFRSI blankets were subsequently used on *Discovery* and *Atlantis* to replace the majority of the LRSI tiles. The LRSI tiles on the mid-fuselage, payload bay doors, and vertical stabilizer of *Columbia* were replaced with AFRSI blankets during the *Challenger* stand-down. *Endeavour* was delivered with an even greater use of AFRSI, and the other orbiters migrated toward this configuration during normal maintenance as well as major overhauls. At this point, these blankets were used on the fuselage, upper elevons, upper wings, vertical stabilizer, and forward canopy surfaces of each orbiter. However, in an effort to save as much weight as possible for missions to the ISS, much of the AFRSI on the mid-fuselage and aft fuselage, payload bay doors, and upper wing surfaces of the three ISS orbiters was subsequently replaced by the lighter FRSI, saving up to 1400 pounds [44].

As of the last flight of *Atlantis*, in 2011, it (and its two OMS pods) was protected with 20,018 HRSI tiles, 3103 FRCI-12 tiles, 497 LRSI tiles, 325 AETB-18 tiles, 234 BRI-18 tiles, and 1472 blankets of various types. (These numbers varied slightly for each vehicle, with *Columbia* having additional HRSI-22 tiles on the vertical stabilizer [to protect the Shuttle Infrared Leeside Temperature Sensor (SILTS) pod] and additional HRSI-9 tiles on the upper wing leading-edge chines, and *Endeavour* having fewer tiles because of the extensive use of AFRSI.) Of these 24,177 tiles, 19,947 of them (83 percent) remained from the original build. Typically, approximately 20 tiles were damaged on each flight, and the average post-flight refurbishment consisted of about 200 tiles and 10 blankets. However, many of those were replaced for reasons unrelated to their failure, including needing to remove a tile (or blanket) to gain access to the structure or component it covered for maintenance, or replacing the tile (or blanket) with an improved version.

Other Materials

Thermal barriers prevented superheated air from flowing through the various penetrations in the orbiter, such as the landing-gear doors and crew hatch. The thermal barriers were tubular Inconel 750 wire mesh filled with Saffil alumina silica fibrous insulation wrapped into a ceramic sleeving with a Nextel AB312 alumina-borosilicate ceramic fabric outer cover. A black ceramic emittance coating was applied to the outer surface, and an RTV silicone adhesive attached the barrier to the structure. Thermal barriers were designed for service in areas seeing temperatures up to 2000°F (Fig. 5.9) [45]. Although most thermal barriers were used around penetrations such as the landing gear doors and crew hatch, similar material was used where the leading-edge RCC met the area protected with LI-2200 tiles.

Fig. 5.9 Where RCC panel 22 met the wingtip. (NASA)

Because the tiles were installed with a space between them, it was sometimes necessary to use gap fillers to ensure that heat did not seep between the tiles and impact the aluminum skin. Two types of gap fillers were used: the pillow/pad type and the Ames type. The pillow/pad gap filler was made of a ceramic fabric with alumina silica insulation with two layers of 0.001-inch-thick Inconel foil inside for support. The pad gap filler was stitched with quartz thread, and a black ceramic emittance coating was applied to the outer mold line. A coating of red RTV-560 was applied to the bottom surface as a stiffener to aid with installation into the tile-to-tile gaps. Approximately 5000 pad-type gap fillers, each 0.20-inch thick, were used on the lower fuselage, lower body flap, lower elevons, and vertical stabilizer. The Ames-type gap filler was made from Nextel ceramic fabric that was originally impregnated with a black silicone coating, although this was later replaced with a more durable ceramic coating [46].

Two types of internal insulation blankets were used: fibrous bulk and multilayer. The bulk blankets were fibrous materials with a density of 2 pounds per cubic foot and a sewn cover of reinforced-Kapton acrylic film. The cover material had 13,500 holes per square foot for venting. Acrylic-film tape was used for cutouts, patching, and reinforcements. Tufts throughout the blankets minimized billowing during venting. The multilayer blankets were constructed of alternate layers of perforated-Kapton acrylic-film reflectors and Dacron-net separators. There were 16 reflector layers in all, with the two-cover halves counting as two layers. The covers, tufting, and acrylic-film tape were similar to those used for the bulk blankets [47].

Flags and other markings were painted on the orbiter with a Dow Corning-3140 silicone elastomer, colored with pigments. It was basically the same paint used to paint automobile engines and would break down in temperature ranges between 800°F and 1000°F. Because of this, almost all markings were painted in relatively low-temperature areas of the orbiter.

CONCLUSIONS AND LESSONS LEARNED

Contrary to popular perceptions, the orbiter TPS proved to be quite robust and largely trouble free once the vehicle entered what passed for "operational" service. However, even before the final space shuttle design had been settled upon, some engineers worried about possible damage to the orbiter from debris separating from the booster (in the original two-stage concepts) or external tank (in the design actually built). This was less of a concern in most of the mainstream two-stage designs because the orbiter was mounted well forward on the fly-back booster, minimizing the chances that debris from the booster could impact anything critical (nose cap or wing leading edges) on the orbiter.

This became a greater issue in the final design with the decision to mount the orbiter well back from the nose of the external tank. Initially, the concern was that ice coming off the liquid oxygen tank could impact the orbiter, mostly because, as originally proposed in 1973, the ET did not use any insulation over much of its exterior. However, later that year, engineers decided the orbiter tiles were vulnerable to impact damage, and NASA banned ice and debris shedding from the ET. As early as 1979, NASA commissioned studies into the effects of foam debris on the orbiter, with one stating, "Failure of the HRSI coating was observed to be strongly dependent on the density and size of the projectile" [48].

The spray-on foam insulation used by the ET was mostly successful in eliminating ice debris, although the program accepted that small accumulations would be shed from certain brackets and other areas that could not realistically be insulated. But the solution turned into a problem when pieces of foam began falling off the ET. Eventually, of course, ET foam debris caused the loss of *Columbia* and her crew. The CAIB report offers a detailed discussion of foam, foam debris, and the normalization of deviance that surrounded the loss of *Columbia*.

Even after the loss of the orbiter, many engineers and managers refused to fully accept that something as light as foam could seriously damage the wing leading edge reinforced carbon-carbon, so much of the early investigation centered on possible tile damage, especially near the main landing gear doors.

Fig. 5.10 Even before the first flight, NASA investigated providing a method to repair the thermal protection system while on orbit. At the time, the concern was limited to repairing the HRSI tiles, and during 1980 researchers developed a derivative of MA-25S ablator (used on the X-15A-2) that could be applied to broken to missing tiles by astronauts during an EVA. (NASA)

Fig. 5.11 After the Columbia accident in 2004, NASA again looked at providing an on-orbit repair capability. The final system again used an ablator that looked much like the system originally developed in 1980. This time a capability was also developed to repair RCC, although it is unlikely it could have repaired a hole as large as the one that doomed Columbia. (NASA)

Eventually, of course, the laws of physics proved that a comparatively light piece of debris moving quickly could penetrate what seemed like a robust piece of RCC.

The return-to-flight brought many changes, including eliminating much of the questionable insulation on parts of the ET (bipod and PAL ramps) and the development of on-orbit TPS repair techniques (something that had been thought of before STS-1 but abandoned as impractical before the first flight). Even before the first flight, NASA investigated providing a method to repair the TPS while on orbit. At the time, the concern was limited to repairing the HRSI tiles, and during 1980, researchers developed a derivative of MA-25S ablator (used on the X-15A-2) that could be applied to broken or missing tiles by astronauts during an EVA. After the *Columbia* accident in 2004, NASA again looked at providing an on-orbit repair capability. The final system again used an ablator that looked much like the system originally developed in 1980. This time, a capability was also developed to repair RCC, although it is unlikely it could have repaired a hole as large as the one that doomed *Columbia*.

NASA rejected proposals for "robust RCC" and more damage-resistant tiles largely because President George W. Bush had announced in January

2004 that the Space Shuttle would be retired as soon as the assembly of the ISS was complete (Fig. 5.10 and Fig. 5.11) [49].

The orbiter TPS was the first reusable heat shield, and although it failed on STS-107, it proved to be a remarkable piece of engineering. On several occasions, it experienced significant damage yet provided sufficient protection for the orbiter to return home safely. One of these, on STS-27R, came when ablative insulating material from the right-hand SRB impacted *Atlantis* about 85 seconds into the ascent. During the post-landing inspection, NASA found 1 missing tile and more than 700 damaged tiles. Fortunately, the missing tile was located over a dense aluminum mounting plate for the L-band antenna, perhaps preventing a burn-through [50].

Although both tiles and RCC have found use on the Boeing X-37B and Lockheed Martin Orion, most researchers and engineers still appear to be fascinated with the concept of a metallic TPS for any future reusable vehicle.

REFERENCES

[1] Channon, S. L., and Barry, W. T., Aerospace Corporation, "Status of Reentry Vehicle Heatshields," AIAA-1967-1125.

[2] Gibeaut, W. A., and Bartlett, E. S., "Properties of Coated Refractory Metals," Battelle Memorial Institute report 195, Defense Metals Information Center, 10 Jan. 1964.

[3] "MX-2276: Advanced Strategic Weapon System," Interim Technical Report, Contract AF33(616)-2419, Bell report D143-945-011, 25 Oct. 1954, pp. 86–90.

[4] Krusos, J. N., "Study of Ceramic Heat Shields for Lifting Reentry Vehicles," NASA Langley contract NAS1-5370, Bell report CR-861, Aug. 1967; see also Krusos, J. N., "Ceramic Heat Shields for Lifting Re-Entry Vehicles," SAE-680643, 1 Feb. 1968.

[5] Krusos, "Study of Ceramic Heat Shields for Lifting Reentry Vehicles."

[6] Krusos, "Study of Ceramic Heat Shields for Lifting Reentry Vehicles."

[7] Hass, D. W., "Final Report: Refurbishment Cost Study of the Thermal Protection System of a Space Shuttle Vehicle," McDonnell Douglas Astronautics Co., Contract NAS1-10093, NASA report CR-111832, 1 Mar. 1971.

[8] Mullite fact sheet, located at http://www.azom.com/details.asp?ArticleID=925, accessed 1 October 2009.

[9] Hass, "Final Report: Refurbishment Cost Study."

[10] Nixon, R. M. "Space Shuttle Program Request for Proposal No. 9-BC421-67-2-40P," Statement by the President, 5 Jan. 1972, and NASA Press Release 72-4, 6 Jan. 1972; 17 Mar. 1972, pp. III-1 to III-2, IV-10, IV-12, IV-24.

[11] Proposal for Space Shuttle Program (multiple volumes), Grumman report 72-74-NAS, 12 May 1972.

[12] Proposal to NASA-MSC: Space Shuttle (multiple volumes), Lockheed Space & Missiles Company report LMSC-D157364, 12 May 1972.

[13] Space Shuttle Program (multiple volumes), McDonnell Douglas report MDC-E0600, 12 May 1972.

[14] Proposal for the Space Shuttle Program (multiple volumes), North American report SD72-SH-50-3, 12 May 1972.

[15] *Exploring the Unknown: Accessing Space*, Vol. IV, NASA SP-4407, Washington, DC, 1999, pp. 266–267. North American Rockwell became Rockwell International on 16 Feb. 1973.

[16] Cooper, P. A., and Holloway, P. F., "The Shuttle Tile Story," *Astronautics & Aeronautics*, Jan. 1981, pp. 24–34; "Technology Influence on Space Shuttle Development," Eagle Engineering Report 86-125C, 8 June 1986, pp. 6–4 and 6–5; Pless, W. M., "Space Shuttle Structural Integrity and Assessment Study, External Thermal Protection System," Lockheed report CR-134452/LG73-ER0082, June 1973.
[17] Pless, "Space Shuttle Structural Integrity and Assessment Study."
[18] First quote from Cooper and Holloway, "Shuttle Tile Story," p. 25. Day quote from an interview of LeRoy E. Day by John Mauer, 17 Oct. 1983, pp. 5–6, in the files of the JSC History Office, Houston, TX.
[19] Schneider, W. C., and Miller, G. J., "The Challenging 'Scales of the Bird'" (Shuttle Tile Integrity), Space Shuttle Technical Conference (CR-2342), Johnson Space Center, 28–30 June 1983, pp. 403–413.
[20] Schneider, W. C., and Miller, G. J., "Scales of the Bird," pp. 403–413. Probably the best known silane product is 3M Scotchgard™.
[21] Schneider and Miller, "Scales of the Bird."
[22] Hallion, R. P., and Young, J. O., "Space Shuttle: Fulfillment of a Dream," *Case VIII of The Hypersonic Revolution: Case Studies in the History of Hypersonic Technology, Vol. II, From Scramjet to the National Aero-Space Plane (1964–1986)*, USAF Histories and Museums Program, Washington, DC, 1998, p. 1,166.
[23] "Thermal Protection System Engineering Training Handbook," United Space Alliance, Kennedy Space Center, FL, 2008.
[24] "Thermal Protection System Engineering Training Handbook."
[25] *Shuttle Crew Operations Manual*, p. 1.2–15; STS-59 Press Kit, April 1994, p. 37; *Ames Astrogram* (newsletter), April 1, United Space Alliance report SFOC-FL0884, 1994; Redmond, C., "Improved Shuttle Tile to Fly on STS-59," NASA News Release 94-54, 31 Mar. 1994; "Shuttle Thermal Protection," LTV News Release V84-29, 1 Feb. 1990; Chin Panel, LTV News Release, 23 Mar. 1988; various material data sheets on TPS products, Lockheed Missiles and Space Company, Jan. 1984.
[26] Fleming et al. "A History of TPS Failures on Space Shuttle Orbiter."
[27] Fleming et al. "A History of TPS Failures on Space Shuttle Orbiter."
[28] Curry, D. M., Latchen, J. W., and Whisenhunt, G. B., "Space Shuttle Leading Edge Structural Development," 21st Aerospace Sciences Meeting, AIAA-1983-0483, 10–13 Jan. 1983; *Shuttle Crew Operations Manual*, p. 1.2–14.
[29] Fleming et al. "A History of TPS Failures on Space Shuttle Orbiter."
[30] Fleming et al. "A History of TPS Failures on Space Shuttle Orbiter."
[31] *Thermal Protection System Engineering Training Handbook*, United Space Alliance, Kennedy Space Center, FL, 2008.
[32] "*Thermal Protection System Engineering Training Handbook*," 2008.
[33] Fleming et al., "A History of TPS Failures on Space Shuttle Orbiter."
[34] Fleming et al., "A History of TPS Failures on Space Shuttle Orbiter."
[35] "New Tiles Mean Better Safety," http://www.nasa.gov/mission_pages/shuttle/behind scenes/new-tiles.html, accessed on 3 Oct. 2012.
[36] Laufenberg, K, *TPS Engineering Handbook*, United Space Alliance, Kennedy Space Center, FL, 2002.
[37] Gordon, M. P., *Space Shuttle Orbiter Thermal Protection System Processing Assessment*, Appendix A: Overview of the Space Shuttle Thermal Protection System, Rockwell International, 1995.
[38] Fleming et al. "A History of TPS Failures on Space Shuttle Orbiter."
[39] Laufenberg, *TPS Engineering Handbook*, 2002.
[40] *Thermal Protection System Engineering Training Handbook*, 2008.

[41] Fleming et al. "A History of TPS Failures on Space Shuttle Orbiter."
[42] Fleming et al. "A History of TPS Failures on Space Shuttle Orbiter."
[43] Fleming et al. "A History of TPS Failures on Space Shuttle Orbiter."
[44] STS-89 Orbiter Rollout Milestone Review presentation.
[45] Fleming et al. "A History of TPS Failures on Space Shuttle Orbiter."
[46] *Thermal Protection System Engineering Training Handbook*, 2008.
[47] *Shuttle Crew Operations Manual*, p. 1.2–14.
[48] Pessin, M. A., Lessons Learned from Space Shuttle External Tank Development: A Technical History of the External Tank," NASA-MSFC Second Generation RLV Program Office, 30 Oct. 2002; Rand, J. L. and Norton, D. J. "Effects of Soft Foam Insulation Impact," Texas Engineering Experiment Station report CR-160400, NASA contract NAS9-15962, 1979, p. 3.
[49] For tile repair, see, for example, Thermal Protection System Repair Kit Program, NASA contract NAS9-15970, General Electric Company Report 79SDR2310 (NASA CR-160418), 28 Nov. 1979; "Shuttle Orbiter TPS Flight Repair Kit Development," NASA contract NAS9-15971, McDonnell Douglas Astronautics Report MDC-G8261, Dec. 1979; "Thermal Protection System Flight Repair Kit," NASA contract NAS9-15969, Martin Marietta Corporation Report MCR-79-682, (NASA CR-160417), Dec. 1979.
[50] Harwood, W, "Legendary Commander Tells Story of Shuttle's Close Call," *Spaceflight Now*, Vol. 27, Mar. 2009; "STS-27R OV-104 Orbiter TPS Damage Review Team Summary Report," Volume I, TM-100355, Feb. 1989.

Chapter 6

Revolutionizing Electronics: Software and the Challenge of Flight Control

Nancy G. Leveson

Introduction

A mythology has arisen about the Space Shuttle's software, with claims being made about it being "perfect software" and "bug free" or having "zero defects [1]." All of these are untrue. But the overblown claims should not take away from the remarkable achievement by those at NASA, its major contractors (Rockwell, IBM, Rocketdyne, Lockheed Martin, and Draper Labs), and smaller companies such as Intermetrics (later Ares). All of them put their hearts into a feat that required overcoming what were tremendous technical challenges at that time. They did it using discipline, professionalism, and top-flight management and engineering skills and practices. We can learn many lessons that are still applicable to the task of engineering complex software systems, in both aerospace and other fields where software is used to implement critical functions.

Much has already been written about the detailed software design. This chapter will instead take a historical focus and highlight the challenges and how they were successfully overcome as the Space Shuttle development and operations progressed. The ultimate goal is to identify lessons that can be learned and used on projects today. The lessons and conclusions gleaned are necessarily influenced by the author's experience; others might draw additional or different conclusions from the same historical events.

To appreciate the achievements, we must first understand the state of software engineering at the time of the Space Shuttle's development and the previous attempts to use software and computers in spaceflight.

LEARNING FROM EARLIER MANNED SPACECRAFT

Before the Space Shuttle, NASA had managed four manned spacecraft programs that involved the use of computers: Mercury, Gemini, Apollo, and Skylab. Mercury did not need an onboard computer. Reentry was calculated by a computer on the ground, and the retrofire times and firing attitude were transmitted to the spacecraft while in flight [2].

Gemini was the first U.S. manned spacecraft to have a computer onboard. Successful rendezvous required accurate and fast computation, but the ground tracking network did not cover all parts of the Gemini orbital paths. Therefore, a ground computer could not provide the type of continuous updates needed for some critical maneuvers. The Gemini designers also wanted to add accuracy to reentry and to automate some of the preflight checkout functions. Gemini's computer did its own self-checks under software control during prelaunch, allowing a reduction in the number of discrete test lines connected to launch vehicles and spacecraft. During ascent, the Gemini computer received inputs about the velocity and course of the Titan booster so that it would be ready to take over from the Titan's computers if they failed, providing some protection against a booster computer failure. Switchover could be either automatic or manual, after which the Gemini computer could issue steering and booster cutoff commands for the Titan. Other functions were also automated, and, in the end, the computer operated during six mission phases: prelaunch, ascent backup, orbit insertion, catch-up, rendezvous, and reentry [3].

IBM, the leading computer maker at that time, provided the computer for Gemini. Both NASA and IBM learned a great deal from the Gemini experience because it required more sophistication than any computer that had flown on unmanned spacecraft to that date. Engineers focused on the problems that needed to be solved for the computer hardware, namely reliability and making the hardware impervious to radiation and to vibration and mechanical stress, especially during launch.

At the time, computer programming was considered an almost incidental activity. Experts wrote the software in low-level, machine-specific assembly languages. Fortran had been available for only a few years and was considered too inefficient for use on real-time applications, both in terms of speed and machine resources such as memory. The instructions for the Gemini computer were carefully crafted by hand using the limited instruction set available. Tomayko suggests that whereas conventional engineering principles were used to design and construct the computer hardware, the software development was largely haphazard, undocumented, and highly idiosyncratic: "Many managers considered software programmers to be a different breed and best left alone" [4].

The requirements for the Gemini software provided a learning environment for NASA. The programmers had originally envisioned that all the software needed for the flight would be preloaded into memory, with new programs to be developed for each mission [5]. However, the programs developed for Gemini quickly exceeded the available memory, and parts had to be stored on an auxiliary tape drive until needed. Computers had very little memory at the time, and squeezing the desired functions into the available memory became a difficult exercise and placed limits on what could be accomplished. In addition, the programmers discovered that parts of the software, such as the ascent guidance backup function, were unchanged from mission to mission. To deal with these challenges, the designers introduced *modularization* of the code, which involved carefully breaking the code into pieces and reusing some of the pieces for different functions and phases of flight, a common practice today but not at that time.

Another lesson was learned: the need for software specifications. McDonnell Douglas created a Specification Control Document, which was forwarded to the IBM Space Guidance Center in Oswego, to validate the guidance equations. Simulation programs, written in Fortran, were used for this validation.

Despite the challenges and the low level of software technology at the time, the Gemini software proved to be highly reliable and useful. NASA used the lessons learned about modularity, specification, verification, and simulation in producing the more complex Apollo software. In turn, many of the lessons learned from Apollo were the basis for the successful procedures used on the Space Shuttle. Slowly and carefully, NASA was learning how to develop more and more complex software for spacecraft.

After the success with Gemini, the software ambitions for Apollo stretched the existing technology even further. Although computer hardware was improving, it still created extreme limitations and challenges for the software designers. During Mercury, the ground-based computer complex that supported the flights performed one million instructions per minute. Apollo did almost 50 times that many—approaching one million instructions per second [6]. Today, the fastest computer chips can execute 50 billion instructions per second, or 50,000 times the speed of the entire Apollo computer complex. George Mueller, head of the NASA Office of Manned Space Flight, envisioned that computers would be "one of the basic elements upon which our whole space program is constructed" [7]. But first, the problems of how to generate reliable and safe computer software had to be solved. The handcrafted, low-level machine code of Gemini would not scale to the complexity of later spacecraft.

In 1961, NASA contracted with Draper Labs (then called the MIT Instrumentation Lab) to design, develop, and construct the Apollo guidance and navigation system, including software. The plans for using software in Apollo were ambitious and caused delays.

Much of the computing for Apollo was done on the ground. The final Apollo spacecraft was autonomous only in the sense it could return safely to Earth without help from ground control. Both the Apollo Command Module (CM) and the Lunar Exploration Module (LEM) included guidance and control functions, however. The CM's computer handled translunar and trans-Earth navigation, while the LEM's provided for autonomous landing, ascent, and rendezvous guidance. The LEM had an additional computer as part of the Abort Guidance System (AGS) to satisfy the NASA requirement that a first failure should not jeopardize the crew. Ground systems backed up the CM computer and its guidance system such that if the CM system failed, the spacecraft could be guided manually based on data transmitted from the ground. If contact with the ground was lost, the CM system had autonomous return capability.

Real-time software development on this scale was new to both NASA and MIT—indeed to the world at that time—and was considered to be more of an art than a science. Both NASA and MIT originally treated software as a secondary concern, but the difficulty soon became obvious. When testifying before the Senate in 1969, Mueller described problems with the Apollo guidance software, calling Apollo's computer systems "the area where the agency pressed hardest against the state of the art," and warning that it could become the critical path item delaying the lunar landing [8]. He personally led a software review task force that, over a nine-month period, developed procedures to resolve software issues. In the end, the software worked because of the extraordinary dedication of everyone on the Apollo program.

In response to the problems and delays, NASA created some important management functions for Apollo, including a set of control boards—the Apollo Spacecraft Configuration Control Board, the Procedures Change Control Board, and the Software Configuration Control Board—to monitor and evaluate changes. The Software Configuration Control board, chaired for a long period by Chris Kraft, controlled all onboard software changes [9].

NASA also developed a set of reviews for specific points in the development process. For example, the Critical Design Review (CDR) followed the preparation of the requirements definition, guidance equations, and engineering simulation of the equations and placed the specifications and requirements for a given mission under configuration control. The next review, a First Article Configuration Inspection (FACI), followed the coding and testing of the software and the production of a validation plan and placed the software under configuration control. The Customer Acceptance Readiness Review (CARR) certified the validation process after testing was completed. The Flight Readiness Review was the final step before clearing the software for flight. This review and acceptance process provided for consistent evaluation of the software and controlled changes, which helped to ensure high reliability and inserted much-needed discipline into the software development process. The control board and review structure became much more extensive for the Space Shuttle.

As with Gemini, Apollo's computer memory limitations caused problems, resulting in some features and functions being abandoned and tricky programming techniques being used to save others. The complexity of the resulting code made it difficult to debug and verify, which caused delays. MIT's distance from Houston also created communication problems and compounded the difficulty in developing correct software.

When it appeared that the software would be late, MIT added more people to the software development process, which simply slowed down the project even more. This basic principle that adding more people to a late project makes it even later is well known now, but it was part of the learning process at that time. Configuration control software was also late, leading to delays in supporting discrepancy reporting*. Another mistake made was to take shortcuts in testing when the project started to fall behind schedule.

The 1967 launch pad fire gave everyone time to catch up and fix the software, as the *Challenger* and *Columbia* accidents would for the Space Shuttle software later. The time delay allowed for significant improvements to be made. Howard "Bill" Tindall, who was NASA's watchdog for the Apollo software, observed at the time that NASA was entering a new era with respect to spacecraft software: no longer would software be declared complete to meet schedules, requiring the users to work around errors. Instead, quality would be the primary consideration [10]. The Mueller-led task force came to similar conclusions and recommended increased attention to software in future manned space programs. The dynamic nature of requirements for spacecraft should not be used as an excuse for poor quality, it suggested. Adequate resources and personnel were to be assigned early to this "vital and underestimated area" [11]. The Mueller task force also recommended ways to improve communication and to make coding easier.

Despite all the difficulties encountered in creating the software, the Apollo software was a success. Mueller attributed this success to the "extensive ground testing to which every subsystem was subjected, the simulation exercises which provided the crews with high fidelity training for every phase of the flights, and the critical design review procedures ... fundamental to the testing and simulation programs" [12].

In the process of constructing and delivering the Apollo software, both NASA and the MIT Instrumentation Lab learned a lot about the principles of software engineering for real-time systems and gained important experience

---

*Software *discrepancy* is the common NASA terminology for a software bug or error.

in managing a large real-time software project. These lessons were applied to the Space Shuttle. One of the most important lessons was that software is more difficult to develop than hardware; therefore,

- Software documentation is crucial.
- Verification must proceed through a sequence of steps without skipping any to try to save time.
- Requirements must be clearly defined and carefully managed.
- Good development plans should be created and followed.
- Adding more programmers does not lead to faster development.

NASA also learned to assign experienced personnel to a project early, rather than using the start of a project for training inexperienced personnel.

Skylab was the final large software effort before the Space Shuttle. Skylab had a dual computer system for attitude control of the laboratory and pointing of the solar telescope. The software contributed greatly to saving the mission during the two weeks after its troubled launch and later helped control Skylab during the last year before reentry [13]. The system operated without failure for over 600 days of operation. It was also the first onboard computer system to have redundancy management software. Learning from their previous experience, the software development followed strict engineering principles, which were starting to be created at that time. Software development was beginning to change from a craft to an engineering discipline.

Some unique factors in Skylab made the problem somewhat easier than it had been with Apollo. The software was quite small, just 16K words, and the group of programmers assigned to write it was correspondingly small. There were never more than 75 people, not all of whom were programmers, involved, and this minimized the problems of communication and configuration control, which were common in larger projects. Also, IBM assigned specialists in programming to the software in contrast to Draper Labs, which used spacecraft engineering experts. Draper and IBM had learned the fallacy of the assumption that it was easier to teach an engineer to program than to teach a programmer about spacecraft [14].

Limitations in memory were again a problem. The software resulting from the baseline requirements documents ranged from 9,000 words to 20,000 in a machine with only 16K words of memory. Engineers made difficult choices about where to cut. Memory limitations became the prime consideration in allowing requirements changes, which ironically may have actually contributed to the success of the software given the difficult problems that ensue from requirements changes.

Tomayko concludes that the Skylab program demonstrated that careful management of software development, including strict control of changes, extensive and preplanned verification activities, and the use of adequate development tools, results in high-quality software with high reliability [15].

However, the small size of the software and the development teams was also an important factor.

Challenges to Success for the Shuttle

The Space Shuttle design used computers in much more ambitious ways than previous spacecraft. The Space Shuttle onboard computers took the lead in all checkout, guidance, navigation, systems management, payload, and powered flight functions. These goals pushed the state of the art at the time, so NASA was required to create solutions to new problems. At the same time, NASA realized that conservatism was important to success. The result was often a compromise between what was desirable and what could be done with confidence.

The full scope of what needed to be accomplished was not recognized at first. NASA engineers estimated the size of the flight software to be smaller than that of Apollo. The cost was also vastly underestimated: Originally, NASA thought the cost for developing the Space Shuttle software would be $20 million, but the agency ended up spending $200 million just in the original development and $324 million by 1991 [16]. In 1992, NASA estimated that $100 million a year was being spent on maintaining the onboard software [58][†]. Note that the onboard software was only a small part of the overall software that needed to be developed. Four times as much ground support software was used in construction, testing, simulation, and configuration control.

NASA had learned from Apollo and the earlier spacecraft just how important software development was in spacecraft projects and that quality had to be a goal from the beginning. Software could not just be declared complete to meet schedules, requiring users to work around errors [17]. This goal was not completely realized, however, as witnessed by the number of user notes and software-related waivers during Space Shuttle operations[‡]. By STS-7, in June 1983, over 200 pages of such exceptions and their descriptions existed [18]. Some were never fixed, but the majority were addressed after the *Challenger* accident in January 1986, when flights were temporarily suspended.

[†]In 1992, an NRC committee studying the NASA Space Shuttle software process was told that the yearly cost for the flight software development contractors was approximately $60 million. Operation of the Shuttle Avionics Integration Lab (SAIL), which is used to test the flight software, required approximately $24 million per year. This total does not include the costs for the Space Shuttle Main Engine (SSME) software and other support contractors. See An Assessment of Space Shuttle Flight Software Development Processes, Committee for the Review of Oversight Mechanisms for Space Shuttle Software Development Processes, Aeronautics and Space Engineering Board, National Research Council, 1993.

[‡]User notes were provided to the astronauts to help them work around known software limitations and errors. Waivers were decisions on the part of the Space Shuttle program to recognize a condition, such as a known software error, as an acceptable risk. Thus, a condition that received a waiver was set aside and sometimes fixed at a later date when time and resources were available. Such conditions were not considered sufficient cause to hold up a flight.

When creating and maintaining the Space Shuttle software, NASA had to overcome many difficult challenges including continuing hardware and memory limitations, continually evolving software requirements, project communication and management, software quality, and computer hardware reliability. How NASA and its contractors met those challenges is described later. But first, to understand their solutions, some basic information about the Space Shuttle's software architecture is needed.

ONBOARD SOFTWARE ARCHITECTURE

There were three basic computer systems on board the Space Shuttle: a computer that controls the Space Shuttle Main Engines (SSME), the primary avionics computer, and a computer that backs up the primary computer.

The Main Engine Control (MEC) software, built by Rocketdyne and managed by Marshall Space Flight Center (MSFC), was a "first" in space technology. The Space Shuttle's three main liquid-propellant engines were the most complex and "hottest" rockets ever built [19]. The Space Shuttle's engines can adjust flow levels, sense how close to exploding they are, and respond in such a way as to maintain maximum performance at all times. This design would have been impossible without the use of a digital computer to control the engines. However, it also provided huge challenges to the software designers.

After studying the problem, Rocketdyne and Marshall decided to use a distributed approach. By placing controllers at the engines themselves, complex interfaces between the engine and vehicle could be avoided. In addition, a dedicated computer was the best way to handle the high data rates needed for active control. The designers also used a digital computer controller rather than an analog controller because software would allow for more flexibility. As the control concepts for the engines evolved over time, digital systems could be developed faster, and failure detection would be simpler [20].

The MEC software operated as a real-time system with a fixed cyclic execution cycle. The requirement to control a rapidly changing engine environment led to the need for a high-frequency cycle. Each major cycle started and ended with a self-test. It then executed engine control tasks, took sensor readings, performed engine limit monitoring tasks, and provided outputs. Next, it read another round of input sensor data and checked internal voltage, before finally performing a second self-test. Some free time was built into the cycle to avoid overruns into the next cycle. The primary avionics software was allowed direct memory access to engine component data to ensure that the main engine controller did not waste time handling data requests. NASA managers adopted a strict software engineering approach to developing and maintaining the MEC software, as they did for all the Space Shuttle software.

The second main onboard computer was the Primary Avionics Software System (PASS), sometimes referred to as the on-board data processing system (DPS). PASS provided functions used in every phase of a mission except for docking, which the crew performed manually [21]. The two main parts of the software were (1) an operating system and code that provided essential services for the computer (called the Flight Computer Operating System [FCOS]) and (2) the application software that ran the Space Shuttle. The application software provided guidance, navigation, and flight control; vehicle systems management (including payload interfaces); and vehicle checkout.

Because of memory limitations, the PASS was divided into major functions, called Operational Sequences (OPS), that reflected the mission phases (preflight, ascent, on-orbit, and descent). The OPS were divided into modes. Transition between major modes was automatic, but transition between OPS was normally initiated by the crew and required that the OPS be loaded from the magnetic tape-based Mass Memory Unit (MMU). Common data used by more than one OPS were kept in the computer memory continuously and were not overlaid [22].

Within each OPS, there were special functions (SPECs) and display functions (DISPs), which were supplemental functions available to the crew in additional to the functions being performed by the current OPS. Because SPECs and DISPs had lower priority than the regular OPS functions, they were kept on tape and rolled into memory when requested if a large OPS was in memory.

Originally, the FCOS was to be a 40-millisecond time-slice operating system[§], but a decision was made early to convert it into a priority-driven system. If the processes in a time-slice system get bogged down by excessive input/output operations, they tend to slow down the total process operation. In contrast, priority systems degrade gracefully when overloaded. The actual FCOS created had some features of both types: It had cycles similar to those in Skylab, but they could be interrupted for higher-priority tasks.

The PASS needed to be reconfigured from flight to flight. A large amount of flight-specific data was validated and installed into the flight software using automated processes to create a flight load. The flight load was tested and used in post-release simulations[¶]. About 10 days before flight, a small number of low-risk updates were allowed (after retesting). In addition, an automated process was executed on the day of launch to update a small set of environmental factors (such as winds) to adapt the software to the conditions for that particular day.

A third computer, the Backup Flight System (BFS), was used to back up the PASS for a few critical functions during ascent and reentry and for

§A time-slice operating system allocates predefined periods of time for the execution of each task and then suspends tasks unfinished in that time period and moves on to the next time slice.

¶A Titan carrying a Milstar satellite was lost in 1999 when a typo in a load tape that had inadvertently never been tested led to an incorrect attitude value being used by the flight control software.

maintaining vehicle control in orbit. This software was synchronized with the primary software so it could monitor the PASS. The BFS used a time-slice operating system, which led to challenges in synchronizing the PASS and BFS and ultimately to a delay of the first launch. If the primary computers failed, the mission commander could push a button to engage the backup software. A feature of this system was that the mission commander had to make a decision quickly about switching to the BFS—the BFS could only remain in a "ready" stage for a short time after failure of the PASS. It was not possible to switch back from the BFS to the PASS later. In addition, the BFS had to "stop listening" whenever it thought the PASS might be compromising the data being fetched so that the BFS would not also be polluted.

Originally, the BFS was intended to be used only during preoperational testing but was extended to STS-4 and later for the life of the Space Shuttle. It was never actually engaged in flight.

Support for developing the PASS software (including testing and crew training) was provided in a set of ground facilities. The Software Production Facility (SPF) provided a simulation test bed that simulated the flight computer bus interface devices, provided dynamic access and control of the flight computer's memory, and supported digital simulation of the hardware and software environments [23]. NASA defined SPF requirements early on, and all new capabilities required NASA approval. IBM maintained this facility, whereas Rocketdyne was responsible for the MEC testing.

After the software for a mission was finished being developed, testing and simulation continued at the Shuttle Avionics Integration Laboratory (SAIL), which was designed to test the actual Space Shuttle hardware and flight software in a simulated flight environment and with a full cockpit for human-in-the-loop testing and integration testing with other flight and ground systems. SAIL was used to verify that the flight software loads were compatible with hardware interfaces, the flight software performed as designed, and the flight software was compatible with mission requirements. Major contractors involved in the SAIL included Rockwell-Downey, Lockheed Martin, and Draper Labs.

This architecture was designed to overcome some of the limitations and challenges for real-time software at the time, as discussed next.

HARDWARE AND MEMORY LIMITATIONS

In Gemini and Apollo, important functions had to be left out of the software due to lack of adequate computer memory. NASA and its contractors struggled to overcome these limitations. The Apollo software, for example, could not minimize fuel expenditures or provide the close guidance tolerance that would have been possible if more memory had been available [24]. Much development time was spent simply deciding which functions could be eliminated and how to fit the remainder in memory.

Although computer memory had increased by the early 1970s from the tiny computers used in previous spacecraft, there were still severe limits on how many instructions could be in memory at the same time. The techniques developed to share memory created additional system complexity. As with the earlier spacecraft, much effort in the Space Shuttle software development was expended in reducing the size of the software that had to be resident in the computer at any given time.

The earliest onboard computers had only a 4K to 16K word (8-bit) memory. In comparison, the main memory in the PASS computers used for the first Space Shuttle flights contained 106K 32-bit words. The onboard code required 400K words of memory, however, including both instructions and data**. The operating system and displays occupied 35K words of memory at all times. With other functions that had to be resident, about 60K of the 106K was left for application programs. The solutions were to delete functions, to reduce execution rates, and to break the code into overlays, with only the code necessary to support a particular phase of the flight loaded into computer memory (from tape drives) at any time. When the next phase started, the code for that phase was swapped in.

The majority of effort went into the ascent and descent software. By 1978, IBM reduced the size to 116K words, but NASA headquarters demanded it be reduced to 80K. It never got down to that size, but it was reduced to below 100K by moving functions that could wait until later operational sequences. Later, the size increased again to nearly the size of the memory as changes were made.

Besides the effort expended in trying to reduce the size of the software, the memory restrictions resulted in other important limitations. For example, requests for extra functions usually had to be turned down. The limited functionality in turn affected the crew.

Space Shuttle crew interfaces are complex due to the small amount of memory available for the graphics displays and other utilities that would have made the system more useful and simpler for the users. As a result, some astronauts have been very critical of the Space Shuttle software. John Young, the chief astronaut in the early 1980s, complained, "What we have in the Shuttle is a disaster. We are not making computers do what we want." Flight trainer Frank Hughes also complained, saying, "The PASS doesn't do anything for us" when noting that important functions were missing. Both said, "We end up working for the computer rather than the computer working for us" [25].

**Compare this with the approximately 5 million lines of code on commercial aircraft today and 15 to 20 million lines on military aircraft. The ISS has 2.4 million lines of code onboard.

Some of the astronaut interface problems stemmed from the fact that steps usually preprogrammed and performed by the computers must be done manually in the Space Shuttle. For example, the crew must reconfigure PASS from ascent mode to on-orbit mode, and this process takes several minutes and needs to be reversed before descent. John Aaron, one of the NASA designers of the PASS interface, responded that management "would not buy" simple automatic reconfiguration schemes and, even if they did, there was no computer memory to store them [26].

The limited memory also required many functions to be displayed concurrently on a screen due to the large amount of memory required for such displays. As a result, many screens were so crowded that reading them quickly was difficult. Compounding this difficulty was the primitive nature of graphics available at the time. These interface limitations along with others added to the potential for human error.

Some improvements suggested by the astronauts were included when the Space Shuttle computers were upgraded in the late 1980s to computers with 256K memory. A further upgrade, the Cockpit Automation Upgrade (CAU), was started in 1999 but was never finished because of the decision to retire the Space Shuttle.

The astronauts took matters into their own hands by using small portable computers to augment the onboard software. The first ones were basically programmable calculators, but beginning with STS-9 in December 1983, portable microcomputers with graphics capabilities were used in flight to display ground stations and to provide functions that were impractical to add to the primary computers due to lack of memory.

The backup computer (BFS) software required 90K words of memory, so memory limitations were never a serious problem. However, memory limitations did create problems for the MEC computers and their software. The memory of the MEC computers was only 16K words, which was not enough to hold all the software originally designed for it. A set of preflight checkout functions were stored in an auxiliary storage unit and loaded during the countdown. Then, at T–30 hours, the engines were activated and the flight software was read from auxiliary memory. Even with this storage-saving scheme, fewer than 500 of the 16K words were unused.

The Challenge of Changing Requirements

A second challenge involved requirements. The Space Shuttle's software requirements were continually evolving and changing, even after the system became operational and throughout its 30-year operational lifetime. Although new hardware components, such as GPS, were added to the Space Shuttle during its lifetime, the Space Shuttle hardware was basically fixed, and most of the changes over time went into the computers and their software. NASA and its contractors made over 2,000 requirements changes between 1975 and the first flight in 1981. Even after the first flight, requirements changes continued. The number of changes proposed and implemented required a strict process to be used, or chaos would have resulted.

Tomayko suggests that NASA lessened the difficulties by making several early decisions that were crucial for the program's success: NASA chose a high-level programming language, separated the software contract from the hardware contract and closely managed the contractors and their development methods, and maintained the software's conceptual integrity [27].

USING A HIGH-LEVEL LANGUAGE. Given all the problems in writing and maintaining machine language code in previous spacecraft projects, NASA was ready to use a high-level language. At the time, however, there was no appropriate real-time language so a new one was created for the Space Shuttle program. HAL/S (High-order Assembly Language/Shuttle) [28, 29] was commissioned by NASA in the late 1960s and designed by Intermetrics[††].

HAL/S had statements similar to Fortran and PL/1 (the most prominent programming languages used for science and engineering problems at the time) such as conditional (IF) and looping (FOR or WHILE) statements. In addition, specific real-time language features, such as the ability to schedule and coordinate processes (WAIT, SCHEDULE, PRIORITY, and TERMINATE), were included. To make the language more readable by engineers, HAL/S retained some traditional scientific notation, such as the ability to put subscripts and superscripts in their normal lowered or raised position rather than forcing them onto a single line.

In addition to new types of real-time statements, HAL/S provided two new types of program blocks: COMPOOL and TASK. Compools are declared blocks of data that are kept in a common data area and are dynamically sharable among processes. While processes had to be swapped in and out because of the memory limitations, compools allowed the processes to share common data that stayed in memory.

Task blocks are programs that are nested within larger programs and that execute as real-time processes using the HAL/S SCHEDULE statement. The SCHEDULE statement simplified the scheduling of the execution of specific tasks by allowing the specification of the task name, start time, priority, and frequency of execution.

HAL/S was originally expected to have widespread use in NASA, including a proposed ground-based version named HAL/G ("G" for ground), but external events overtook the language when the DoD commissioned the Ada

[††]In the preface to the language specification document, the name "HAL" is described as being in honor of Draper Lab's J. Halcombe Laning.

programming language. Ada includes the real-time constructs pioneered by HAL/S, such as task blocks, scheduling, and common data. NASA adopted Ada rather than continuing to use HAL/S because commercial compilers were available and because the DoD's insistence on its use seemed to imply it would be around for a long time.

The use of a high-level programming language allowed top-down structured programming. Its use, along with improved development techniques and tools, has been credited with doubling Space Shuttle software productivity over the comparable Apollo development processes.

SEPARATING THE HARDWARE AND SOFTWARE CONTRACTS. IBM and Rockwell had worked together during the competition period for the orbiter contract. Rockwell bid on the entire spacecraft, intending to subcontract the computer hardware and software to IBM. To Rockwell's displeasure, NASA decided to separate the software contract from the orbiter contract. As a result, Rockwell still subcontracted with IBM for the computer hardware, but IBM had a separate software contract that Johnson Space Center (JSC) managed closely.

Tomayko suggests several reasons for why NASA made this unusual division [30]. First, software was, in some ways, the most critical component of the Space Shuttle. It tied the other components together, and, because it did not weigh anything in and of itself, it was often used to overcome hardware problems that would require extra mechanical systems and components. NASA had learned from the problems in the Apollo program about the importance of managing software development. Chris Kraft (at JSC) and George Low (at NASA headquarters), who were both very influential in the manned space program at the time, felt that Johnson had the software management expertise (acquired during the previous manned spacecraft projects) to handle the software directly. By making a separate contract for the software, NASA could ensure that the lessons learned from previous projects would continue to be followed and accumulated and that the software contractors would be directly accountable to NASA management.

In addition, after operations began, the hardware remained basically fixed, whereas the software was continually changing. As time passed, Rockwell's responsibilities as prime hardware contractor were phased out, and the Space Shuttles were turned over to an operations group. In late 1983, Lockheed Corporation and not Rockwell won the competition for the operations contract. By keeping the software contract separate, NASA was able to continue to develop the software without the extreme difficulty that would have ensued by attempting to change the software developers while it was still being developed.

The concept of developing a facility (the SPF) at NASA to produce the onboard software originated in a Rand Corporation memo in early 1970. This memo summarized a study of software requirements for Air Force space missions during the 1970s [31]. One reason for a government-owned and operated software "factory' was that it would be easier to establish and maintain security for DoD payloads, which could require special software interfaces and control. More important, it would be easier to change contractors, if necessary, if the software library and development computers were government owned and on government property. Finally, having close control by NASA over existing software and new development would eliminate some of the problems in communication, verification, and maintenance encountered in earlier manned spacecraft programs. The NASA-owned but IBM-run SPF had terminals connected directly to Draper Laboratory, Goddard Space Flight Center, Marshall Space Flight Center, Kennedy Space Flight Center, and Rockwell International. The SAIL played a similar role for prelaunch, ascent, and abort simulations as did the Flight Simulation Lab and the SMS for other simulations and crew training.

MAINTAINING CONCEPTUAL INTEGRITY. The ongoing vehicle development work did not allow NASA to employ an ideal software development process, where requirements are completely defined before design, implementation, and verification.

The baseline requirements were established in parallel with developing the Space Shuttle test facility. Originally, it was assumed that these tests would require only minor changes in the software. This assumption turned out to be untrue; the avionic integration and certification activities going on in Houston, at the Kennedy Space Center in Florida, and in California at Downey and Palmdale, and in other activities like the Approach and Landing Tests (ALTs)[‡‡], resulted in significant changes in the software requirements. In many cases, NASA and its contractors found that the real hardware interfaces differed from those in the requirements, operational procedures were not fully supported, and additional or modified functions were required to support the crew.

The strategy used to meet the challenge of changing requirements had several components:

- Rigorously maintained requirements documents.
- Using a small group to create the software architecture and interfaces and ensuring that their ideas and theirs alone were implemented (called "maintaining conceptual integrity").
- Establishing a requirements analysis group to provide a systems engineering interface between the requirements definition and software implementation worlds; this group identified requirements and design tradeoffs and communicated the implications of the trades to both worlds [33].

[‡‡]ALT (Approach and Landing Tests) were a series of taxi and flight trials of the prototype Space Shuttle *Enterprise* conducted in 1977 to test the vehicle's flight characteristics both on its own and when mated to the Shuttle Carrier Aircraft, before the operational debut of the Space Shuttle system. In January 1977, *Enterprise* was taken by road from the Rockwell plant at Palmdale to the Dryden Flight Research Center at Edwards Air Force Base to begin the flight test phase of the program.

This strategy was effective in accommodating changing requirements without significant cost or schedule impacts.

Three levels of requirements documents were created. JSC engineers wrote Levels A and B, whereas Level C, which was more of a design document, was Rockwell International's responsibility. John Garman created the Level A document, which described the operating system, application programs, keyboards, displays, other components of the software, and the interfaces to the other parts of the vehicle. William Sullivan wrote the Level B guidance, navigation, and control requirements, while John Aaron wrote the system management and payload specifications for the Level B document. James Broadfoot and Robert Ernull assisted them. Level B specifications differed from Level A in that they were more detailed in terms of what functions were executed when and what parameters were needed. Level B also defined what information was to be kept in the HAL/S COMPOOLS for use by different tasks. Level C, developed for the contractors to use in development, was completely traceable to the Level B requirements.

The very small number of people involved in the requirements development contributed greatly to the requirements' conceptual integrity and therefore to the success of the Space Shuttle's software development effort [34].

Early in the program, the Draper Lab was retained as a consultant to NASA on requirements development because Draper had learned the hard way on Apollo and had become a leader in software engineering. Draper provided a document on how to write requirements and develop test plans, including how to develop highly modular software. Draper also wrote some of the early Level C requirements as a model for Rockwell. Rockwell, however, added a lot of implementation detail to the Draper requirements and delivered detailed design documents rather than requirements. These documents were an irritation to IBM, which claimed that they told IBM too much about how to do things rather than just what to do. Tomayko interviewed some IBM and NASA managers who suspected that Rockwell, miffed when the software contract was taken away from them, delivered incredibly detailed requirements because they thought that if they did not design the software, it would not be done right [35]. In response, IBM coded the requirements to the letter, which resulted in exceeding the available memory by over two times and demonstrating that the requirements were excessive.

Rockwell also argued for two years about the design of the operating system, calling for a strict time-slice system with synchronization points at the end of each cycle. IBM, at NASA's urging, argued for a priority-interrupt driven design similar to the one used on Apollo. Rockwell, more experienced with time-slice operating systems, fought this proposal from 1973 to 1975, convinced it would never work [36]. Eventually, Rockwell used a time-slice system for the BFS, whereas IBM used a priority-driven system for the PASS. The difference between the two designs caused complications in the synchronization process. In the end, because the backup must listen in on PASS operation in order to be ready to take over if necessary, PASS had to be modified to make it more synchronous.

The number of changes in the software requirements was a continuing problem, but at least one advantage of having detailed requirements (actually, design specifications) very early was that it allowed the use of some software during the early Space Shuttle hardware development process. Because of the size, the complexity, the still evolving nature of the requirements, and the need for software to help develop and test the Space Shuttle hardware, NASA and IBM created the software using incremental releases. Each release contained a basic set of capabilities and provided the structure for adding additional functions in later releases. Seventeen interim releases were developed for the first Space Shuttle flight, starting in 1977. The full software capability was provided after the ninth release in 1978, but eight more releases were necessary to respond to requirements changes and to identified errors.

NASA had planned that the PASS would involve a continuing effort even after first flight. The original PASS was developed to provide the basic capability for space flight. After the first flight, the requirements evolved to incorporate increased operational capability and changing payload and mission requirements. For example, over 50 percent of the PASS modules changed during the first 12 flights in response to requested enhancements [37]. Among the Space Shuttle enhancements that changed the flight control requirements were changes in payload manifest capability, MEC design, crew enhancements, addition of an experimental autopilot for orbiting, system improvements, abort enhancements (especially after the *Challenger* accident), provisions for extended landing sites, and hardware platform changes (including the integration of GPS). The *Challenger* accident was not related to software, but it required changes in the software to support new safety features.

After STS-5, the maintenance and evolution process was organized by targeting requested changes to specific software Operational Increments (OIs). Software changes were generally made to correct deficiencies, to enhance the software's capabilities, or to tailor it to specific mission requirements. OIs were scheduled updates of the primary and backup software. Each OI was designed to support a specific number of planned missions. The OIs included required additions, deletions, and changes to thousands of lines of code. OIs were scheduled approximately yearly, but they could take up to 20 months to complete; therefore, multiple OIs were usually being worked on at the same time.

All requested changes were submitted to the NASA Shuttle Avionics Software Control Board (SASCB). The SASCB ranked the changes based on program benefits, including safety upgrades, performance enhancements, and cost savings. A subset of potential changes was approved for requirements development and placed on the candidate change list. Candidates on the list were evaluated to identify any major issues, risks, and impacts[§§] and then detailed size and effort estimates were created. Approved changes that fit within the available resources were assigned to specific OIs.

Once a change was approved and baselined, implementation was controlled through the configuration management system, which identified (a) the approval status of the change, (b) the affected requirements functions, (c) the code modules to be changed, and (d) the builds (such as operational increment and flight) for which the changed code was scheduled. Changes were made to the design documentation and the code as well as to other maintenance documentation used to aid traceability [38].

COMMUNICATION CHALLENGES

A third challenge involved project communication. As spacecraft grew in complexity and as more engineers and companies became involved, communication problems increased. The large number of computer functions in the Space Shuttle meant that no one company could do it all, which increased the difficulty in managing the various contractors and fostering the required communication. One of the lessons learned from Apollo was that having the software developed by the Draper Lab at a remote site reduced informal exchanges of ideas and created delays. To avoid the same problems, the developers of the Space Shuttle software were located in Houston.

In response to the communication and coordination problems during Apollo development, NASA had created a control board structure. A more extensive control board structure was created for the Space Shuttle. The results, actions, and recommendations of the independent boards were coordinated through a project baselines control board, which in turn interfaced with the spacecraft software configuration control board and the orbiter avionics software control board. Membership on the review boards included representatives from all affected project areas, which enhanced communication among functional organizations and provided a mechanism to achieve strict configuration control. Changes to approved configuration baselines, which resulted from design changes, requirements change requests, and discrepancy reports, were coordinated through the appropriate boards and ultimately approved by NASA. The project office performed weekly audits to verify consistency between approved baselines and reported baselines.

Finally, the review checkpoints, occurring at critical times in development, that had been created for Apollo were again used and expanded.

[§§]For example, impacts to the Mission Control Center, the Launch Processing System, procedures and training, or flight design requirements.

QUALITY AND RELIABILITY CHALLENGES

The Space Shuttle was inherently unstable, which means it could not be flown manually even for short periods of time during either ascent or reentry without full-time flight control augmentation [39]. There were also vehicle sequencing requirements for the SSME and solid rocket booster (SRB) ignition, launch pad release and liftoff operations, and SRB and external tank (ET) separation, which had to occur within milliseconds of the correct time. To meet these requirements, the Space Shuttle was one of the first spacecraft (and vehicles in general) to use a fly-by-wire flight control system [40]: In such systems, there are no mechanical or hydraulic linkages between the pilot's control devices and the control surfaces or reaction control system thrusters. Because sensors and actuators had to be positioned all over the vehicle, the weight of all the wire became a significant concern, and therefore multiplexed digital data buses were used.

The critical functions provided by digital software and hardware led to a need for high confidence in both. NASA used a fail-operational/fail-safe concept, which meant that after a single failure of any subsystem, the Space Shuttle had to be able to continue the mission. After two failures of the same subsystem, it still had to be able to land safely.

Essentially, there are two means for the "failure" of digital systems. The first is the potential for the hardware on which the software executes to fail in the same way that analog hardware does. The protection designed to avoid or handle these types of digital hardware failures is similar and often involves incorporating redundancy.

In addition to the computer hardware failing, however, the software (which embodies the system functional design) can be incorrect or can include behaviors that are unsafe in the encompassing system. Software is pure design without any physical realization and therefore "fails" only by containing systematic design defects. In fact, software can be thought of as *design abstracted away from its physical representation*. Software (when separated from the hardware on which it is executed) is pure design without any physical realization of that design. This abstraction reduces many physical limits in design and thus allows exciting new features and functions that could not be achieved using hardware alone to be incorporated into spacecraft. However, it also greatly increases potential complexity and changes the types of failure modes. With respect to fault tolerance, potentially unsafe software behavior always stems from pure design defects, so redundancy—which simply duplicates the design errors—is not effective. Although computer hardware reliability can depend on redundancy, dealing with software errors must be accomplished in other ways.

COMPUTER HARDWARE REDUNDANCY ON THE SHUTTLE

To ensure fail-operational/fail-safe behavior in the MEC, redundant computers were used for each engine. If one engine failed, the other would take over. Failure of the second computer led to a graceful shutdown of the affected engine. Loss of an engine did not create any immediate danger to the crew, as demonstrated in a 1985 mission where an engine was shut down but the Space Shuttle still achieved orbit [41]. Early in a flight, the orbiter could return to a runway near the launch pad. Later in the flight, it could land elsewhere. If the engine failed near orbit, it could still be possible to achieve an orbit and modify it using the orbital maneuvering system engines.

The redundant MEC computers were not synchronized. MSFC considered synchronizing them but decided the additional hardware and software overhead was too expensive [42]. Instead, they employed a design similar to that used in Skylab, which was still operating at the time the decision was made. Two watchdog timers were used to detect computer hardware failures, one watchdog incremented by a real-time clock and the other by a clock in the output electronics. The software reset both. If the timers ran out, a failure was assumed to have occurred, and the redundant computer took over. The timeout was set for less than the time of a major cycle (18 milliseconds).

The MEC had independent power, central processors, and interfaces, but the input/output devices were cross strapped such that if Channel A's output electronics failed, then Channel B's could be used by Channel A's computer.

Packaging is important for engine controllers because they are physically attached to an operating rocket engine. Rocketdyne bolted early versions of the controller directly to the engine, which resulted in vibration levels of up to 22 g and computer failures. Later, a rubber gasket was used to reduce the levels to about 3–4 g. The circuit cards within the computer were held in place by foam wedges to reduce vibration problems further [43].

When the original MEC hardware was replaced with a new computer with more memory, the new semiconductor memory had additional advantages in terms of speed and power consumption, but it was unable to retain data when power was shut off. To protect against this type of loss, the 64K memory was duplicated and each loaded with identical software. Failure of one memory chip caused a switchover to the other. Three layers of power also provided protection from losing memory. The first layer was the standard power supply. If that failed, a pair of 28-volt backup supplies, one for each channel, was available from other system components. A battery backup that could preserve memory but not run the processor provided a third layer of protection.

The upgraded PASS computers also used semiconductor memory, with its size, power, and weight advantages. Solutions had to be created to protect against the stored PASS programs disappearing if power was lost, although a different solution was devised for the PASS computer memory.

Like the MEC, PASS used redundancy to protect against computer hardware failures, but the designers used an elaborate synchronization mechanism to implement the redundancy. Again, the objective was fail-operational/fail-safe.

To reach this goal, critical PASS software was executed in four computers that operated in lockstep while checking each other. If one computer failed, the three functioning computers would vote it out of the system. If a second computer failed, the two functioning computers took over, and so on. A minimum of three computers is needed to identify a failed computer and to continue processing. A fourth computer was added to accommodate a second failure.

The failure protection did not occur in all flight phases. PASS was typically run in all four redundant computers during ascent, reentry, and a few other critical operations. During most orbital operations, the guidance, navigation, and control software was run on one computer while the system management software was run on a second computer. The remaining three computers (including the one running the BFS) were powered down for efficiency [44].

Even when all four computers were executing, depending on the configuration, each of the computers was given the ability to issue a subset of the commands. This partitioning might be as simple as each computer controlling a separate piece of hardware (such as the reaction control jets), or it could be more complex. Reallocation of some functions had to occur if one or more of the computers failed, which complicated the design. Input data also had to be controlled so that all the computers received identical information from redundant sensors even in the face of hardware failures [45].

Synchronization of the redundant computers occurred approximately 400 times a second. The operating system would execute a synchronization routine during which the computers would compare states using three cross-strapped synchronization lines. All of the computers had to stop and wait for the others to arrive at the synchronization point. If one or more did not arrive in a reasonable amount of time, they were voted out of the set. Once the voting was complete, they all left the synchronization point together and continued until the next synchronization point. Although the failed computer was automatically voted out of the set if its results did not match, the astronauts had to manually halt it to prevent it from issuing erroneous instructions. The capability to communicate with the hardware the failed computer was commanding was lost unless the DPS was reconfigured to pick up the buses lost by the failed computer [45].

The BFS ran on only one computer and therefore was not itself fault tolerant with respect to hardware failures. The only exception was that a copy of its software was stored in the mass memory unit so that another computer could take over the functions of the backup computer in case of a BFS computer failure.

The same software was run on the four independent computers so the hardware redundancy scheme used could not detect or correct software errors. In the 1970s (when the Space Shuttle software was created), many people believed that "diversity," or providing multiple independently developed versions of the software and voting on the results, would lead to very high reliability. Theoretically, the BFS was supposed to provide fault tolerance for the PASS software because it was developed separately (by Rockwell) from the PASS. In addition, a separate NASA engineering directorate, not the onboard software division, managed the Rockwell BFS contract.

In reality, using different software developed by a different group probably did not provide much protection. In the mid-1980s, Knight and Leveson showed that multiple versions of software are likely to contain common failure modes even if different algorithms and development environments are used [47]. Others tried to demonstrate that the Knight and Leveson experiments were wrong, but instead they confirmed them [48]. People make mistakes on the hard cases in the input space; they do not make mistakes in a random fashion.

In addition, almost all the software-related spacecraft losses in the past few decades (and, indeed, most serious accidents related to erroneous software behavior) involved specification or requirements flaws. not coding errors [49, 50]. In these accidents, the software requirements had missing cases or incorrect assumptions about the behavior of the system in which the software was operating. Often, the engineers misunderstood the requirements for safe behavior, such as an omission of what to do in particular circumstances or they did not anticipate or consider some special cases. The software may be "correct" in the sense that it successfully implements its requirements, but the requirements may be unsafe in terms of the specified behavior in the surrounding system, the requirements may be incomplete, or the software may exhibit unintended (and unsafe) behavior beyond what is specified in the requirements. Redundancy or even multiple versions that implement the same requirements do not help in these cases. If independently developed requirements were used for the different versions, there would be no way that they could vote on the results because they would be doing different things.

The BFS was never engaged to take over the functions of a failed PASS during a Space Shuttle flight. However, the difficulty in synchronizing the four primary computers with the BFS did lead to what has been called "The Bug Heard Round the World" [51], when the first launch was delayed because of a failed attempt to synchronize the PASS computers and the BFS computer. The BFS "listened" to all the inputs and some outputs to and from the PASS computers so it would be ready to take over if switched in by the astronauts. Before the launch of STS-1, the BFS refused to sync up with (start listening to) some of the PASS data traffic. The problem was that a few processes in the PASS were occurring one cycle early with respect to the others. The BFS was programmed to ignore all data on any buses for which it heard unanticipated PASS data fetches in order to avoid being polluted by PASS failures. As a result, the BFS stopped listening.

The Approach to Software Quality on the Shuttle

Because redundancy is not effective for requirements and design errors (the only type of errors that software has), the emphasis was on avoiding software errors by using a rigorous process to prevent introducing errors and extensive testing to find them if they were introduced.

Using the management and technical experience gained in previous spacecraft software projects and in ALT, NASA and its contractors developed a disciplined and structured development process. They placed increased emphasis on the front end of development, including requirements definition, system design, standards definition, top-down development, and creation of development tools. Similarly, during verification, they emphasized design and code reviews and testing. Some aspects of this process would be considered sophisticated even today, 30 years later. NASA and its contractors should be justly proud of the process they created over time. Several aspects of this process appear to be very important in achieving the high quality of the Space Shuttle software.

Extensive Planning before Starting to Code. NASA controlled the requirements, and NASA and its contractors agreed in great detail on exactly what the code must do, how it should do it, and under what conditions. That commitment was recorded. Using those requirements, extremely detailed design documents were produced before a line of code was written. Nothing was changed in the specifications (requirements or design) without the agreement and understanding of everyone involved. One change to allow the use of GPS on the Space Shuttle for example, which involved changing about 6,300 lines of code, had a specification of 2,500 pages. Specifications for the entire onboard software filled 30 volumes and 40,000 pages [52].

When coding finally did begin, top-down development was the norm, using stubs[¶¶] and frequent builds to ensure that interfaces were correctly defined and implemented first, rather than finding interface problems late in development during system testing. No programmer changed the code without changing the specification, so the specifications and code always matched.

Those planning and specification practices made it possible to maintain software for over 30 years without introducing errors when changes were necessary. The common experience in industry, where such extensive planning and specification practices are rare, is that fixing an error in operational software is very likely to introduce one or more additional errors.

[¶¶]Stubs are used in place of modules that have not yet been developed. They act as procedures and return default values so the software can execute before all the procedures are written and interfaces can be checked.

CONTINUOUS IMPROVEMENT. One of the guiding principles of the Space Shuttle software development was that if a mistake was found, not only should the original mistake be fixed but also whatever permitted the mistake in the first place. The process that followed the identification of a software error was to (1) fix the error, (2) identify the root cause of the fault, (3) eliminate the process deficiency that let the fault be introduced and not detected earlier, and (4) analyze the rest of the software for other, similar faults [53]. The goal was not to blame people for mistakes but to blame the process. The development process was a team effort; no single person was ever solely responsible for writing or inspecting the code. Thus, there was accountability, but accountability was assigned to the group as a whole.

CONFIGURATION MANAGEMENT AND ERROR DATABASES. Configuration management is critical in a long-lasting project. The PASS software had a sophisticated configuration management system and databases that provided important information to the developers and maintainers. One database contained the history of the code itself, showing every time it was changed, why it was changed, when it was changed, what the purpose of the change was, and what specification documents detailed the change. A second database contained information about the errors that were found in the code. Every error made while writing or changing the software was recorded, along with information about when the error was discovered, how the error was revealed, who discovered it, what activity was going on when it was discovered (testing, training, or flight), how the error was introduced into the software, how the error managed to slip past the filters set up at every stage to catch errors, how the error was corrected, and whether similar errors might have slipped through the same holes.

TESTING AND CODE REVIEWS. The complexity and real-time nature of the software meant that exhaustive testing was impossible, despite the enormous effort that went into the test and certification program. There were too many interfaces and too many opportunities for asynchronous input and output. However, the enormous amount of testing that went into the Space Shuttle software certainly contributed greatly to its quality.

Early error detection, starting with requirements, was emphasized and extensive developer and verifier code reviews in a moderated environment were used. It is now widely recognized that human code reviews are a highly effective way to detect errors in software, and these reviews appear to have been very effective in this environment too. Both the developers and the testers used various types of human code reviews. Testing in the SPF and SAIL was conducted under the most flight-like conditions possible (but see the STS-126 communications software error described in a later section in this chapter).

One interesting experience early in the program convinced NASA that extensive verification and code inspections paid off handsomely. For a year before STS-1, the software was frozen, and all mandatory changes were made using machine language patches. In parallel, the same changes were made in the STS-2 software. Later, it was determined that the machine language patches for STS-1 were of higher quality than the corresponding high-level language (HAL/S) changes in STS-2 [54]. This result seemed to defy common beliefs about the danger of patching software. Later, the difference was explained by the fact that nervousness about the patching led to the use of much more extensive verification for the patches than for the high-level language changes.

IBM's staffing cutbacks after 1985 provided another lesson about the importance of verification. At that time, IBM transitioned from long development time to shorter but more frequent operational increments. The result was less time spent on verification and the introduction of a significant number of software errors that were discovered in flight, including three that affected mission objectives and some Severity 1 errors***.

Learning from these experiences (and others), NASA and its contractors implemented more extensive verification and code inspections on all changes starting with STS-5.

There was some controversy, however, about the use of independent verification and validation (IV&V). Before the *Challenger* accident, all software testing and verification was done by IBM, albeit by a group separate from the developers. Early in 1988, as part of the response to the accident, the House Committee on Science, Space, and Technology expressed concern about the lack of independent oversight of the Space Shuttle software development [55]. A National Research Council committee later echoed the concern and called for IV&V [56]. NASA grudgingly started to create an IV&V process. In 1990, the House committee asked the GAO to determine how NASA was progressing in improving independent oversight of the Space Shuttle software development. The GAO concluded that NASA was dragging its feet in implementing the IV&V program they had reluctantly established [57]. NASA then asked another NRC study committee to weigh in on the controversy, hoping that this committee would agree with them that IV&V was not needed. Instead, the second NRC committee, after looking at the results that IV&V had attained during its short existence, recommended that it be continued [58]. After that, NASA gave up fighting it.

***Space Shuttle flight software errors are categorized by the severity of their potential consequences without regard to the likelihood of their occurrence. Severity 1 errors are defined as those that could produce a loss of the Space Shuttle or its crew. Severity 2 errors can affect the Space Shuttle's ability to complete its mission objectives. Severity 3 errors affect procedures for which alternatives, or workarounds, exist. Severity 4 and 5 errors consist of very minor coding or documentation errors. There is also a class of Severity 1 errors, called Severity 1N, which, although potentially life threatening, involve operations that are precluded by established procedures, are deemed to be beyond the physical limitations of the Space Shuttle, or are outside system failure protection levels.

SOFTWARE DEVELOPMENT CULTURE. A final important contributor to the software quality was the culture of the software development organizations. Software developers had a strong sense of camaraderie and a feeling that what they were doing was important. Many of them worked on the project for a long time, sometimes their whole career. They knew the astronauts, many of whom were their personal friends and neighbors. These factors led to a culture that was quality focused and that believed in zero defects.

Smith and Cusumano note, "These were not the hotshot, up-all-night coders often thought of as the Silicon Valley types" [59]. The Space Shuttle software development job entailed regular 8 a.m. to 5 p.m. hours, where late nights were an exception. The atmosphere and the people were very professional and of the highest caliber. They have been described as businesslike, orderly, detail oriented, and methodical. Smith and Cusumano note that they produced "grownup software and the way they do it is by being grown-ups" [60].

The culture was intolerant of "ego-driven hotshots": "In the Space Shuttle's culture, there are no superstar programmers. The whole approach to developing software is intentionally designed not to rely on any particular person" [61]. The cowboy culture that flourishes in some software development companies today was discouraged. The culture was also intolerant of creativity with respect to individual coding styles. People were encouraged to channel their creativity into improving the process, not violating strict coding standards [62]. In the few occasions when the standards were violated, such as the error manifested in STS-126 (see below), they learned the fallacy of waiving standards for small short-term savings in implementation time, code space, or computer performance.

Unlike the current software development world, many women were involved in Space Shuttle software development, many of them senior managers or senior technical staff. It has been suggested that the stability and professionalism may have been particularly appealing to women [63].

Observers have highlighted the importance of culture and morale on the Space Shuttle software development process. They have noted that during periods of low morale, such as during the early 1990s, when the PASS development organization went through several changes in ownership and management, personnel were distracted, and several serious errors were introduced. During times of higher morale and steady culture, the number of errors was reduced [64].

GAPS IN THE PROCESS

The software development process evolved and was improved over time, but gaps still existed that need to be considered in future projects. The second NRC study committee, created to provide guidance on whether IV&V was necessary, at the same time examined the Space Shuttle software process in depth, as well as many of the software errors that had been found, and it made some suggestions for improvement [65]. Three primary limitations were identified. One was that the verification and validation inspections by developers did not pay enough attention to off-nominal cases. A study sponsored by NASA had determined that problems associated with rare conditions were the leading cause of software errors found during the late testing stage [66]. The NRC committee recommended that verification activities by the development contractors include more off-nominal scenarios, beyond loop termination and abort control sequence actions, and that they also include more detailed coverage analysis.

A second deficiency the NRC committee identified was a lack of system safety focus by the software developers and limited interactions with system safety engineering. System-level hazards were not traced to the software requirements, components, or functions. The committee found several instances where a flight software manager signed off on potentially hazardous software issues and did not report them to the responsible people or boards.

A final identified weakness related to system engineering. The NRC committee studying Space Shuttle safety after the *Challenger* accident had recommended that NASA implement better system engineering analysis: "A top-down integrated systems engineering analysis, including a system-safety analysis, that views the sum of the STS elements as a single system, should be performed to identify any gaps that may exist among the various bottom-up analyses centered at the subsystem and element levels" [67].

The IV&V contractor (Intermetrics and later Ares) that was added after IV&V was introduced was, in the absence of any other group, doing this system engineering task for the software. The second NRC committee concluded that the most important benefits of the IV&V process that was forced on NASA and the contractors were in system engineering. By the time of the second NRC committee report, the IV&V contractor had found four Severity 1 problems in the interaction between the PASS and the BFS. One of these could have caused the shutdown of all the Space Shuttle's main engines, and the other three involved errors that could have caused the loss of the orbiter and the crew if the backup software had been needed during an ascent abort maneuver. The second NRC committee echoed the need for better systems engineering and system safety. Hopefully, this is a lesson NASA and others will learn for the future.

LEARNING FROM ERRORS

Some have misleadingly claimed that the process used to produce the Space Shuttle software led to perfect or bug-free software, but this was not the case. Errors occurred in flight or were found in other ways in software that had flown. Some of these errors were Severity 1 errors, which could have caused the loss of the Space Shuttle. During the standdown after the *Challenger* accident, eight PASS Severity 1 errors were discovered, and two others were found in 1985. In total, during the first 10 years of Space Shuttle flights, 16 Severity 1 errors were found in released PASS software, 8 of which remained in code used in flight. An additional 12 errors of Severity 2, 3, or 4 occurred during flight in this same period. None of these threatened the crew, but three threatened the mission, and the other nine were worked around [68]. In addition, the Space Shuttle was flown with known software errors: For example, 50 waivers were written against the PASS on STS-52, all of which had been in place since STS-47. Three of the waivers covered Severity 1N errors. These errors should not detract from the excellent processes used for the Space Shuttle software development, but they simply attest to the fact that developing real-time software is extremely difficult.

IBM and NASA were aware that effort expended on quality at the early part of a project would be much cheaper and simpler than trying to put quality in toward the end. They tried to do much more at the beginning of the Space Shuttle software development than in previous efforts, as had been recommended by Mueller's Apollo software task force, but it still was not enough to ensure perfection. Tomayko quotes one IBM software manager: "We didn't do it up front enough," the "it" being thinking through the program logic and verification schemes [69].

Obviously, none of the software errors led to the loss of the Space Shuttle, although some almost led to the loss of expensive hardware, and some did lead to not fully achieving mission objectives, at least using the software. Because the orbital functions of the Space Shuttle software were not fully autonomous, astronauts or Mission Control could usually step in and manually recover from the few software problems that did occur. For example, a loss was narrowly averted during the maiden flight of *Endeavour* (STS-49), on May 12, 1992, as the crew attempted to rendezvous with and repair an Intelsat satellite [70]. The software routine used to calculate rendezvous firings, called the Lambert Targeting Routine, did not converge on a solution because of a mismatch between the precision of the state vector variables, which describe the position and velocity of the Space Shuttle, and the limits used to bound the calculation. The state vector variables were double precision, whereas the limit variables were single precision. The satellite rescue mission was nearly aborted, but a workaround was found that involved relaying an appropriate state vector value from the ground.

Shortly before STS-2, during crew training, an error was discovered when all three SSMEs were simulated to have failed in a training scenario. The error caused the PASS to stop communicating with all displays, and the crew engaged the BFS. An investigation concluded the error was related to the specific timing of the three SSME failures in relation to the sequencing to connect the Reaction Control System (RCS) jets to an alternate fuel path. Consistent with the continuous learning and improvement process used, a new analysis technique (called Multi-Pass Analysis) was introduced to prevent the same type of problem in the future [71].

As another example, during the third attempt to launch *Discovery* on August 29, 1984 (STS-41D), a hardware problem was detected in the Space Shuttle's main engine number 3 at T–6 seconds before launch, and the launch was aborted. However, during verification testing of the next operational software increment (before the next attempt to launch *Discovery*), an error was discovered in the master event controller software related to SRB fire commands. This error could have resulted in the loss of the Space Shuttle due to the inability to separate the SRBs and ET. That discrepancy was also in the software for the original *Discovery* launch that was scrubbed because of the engine hardware problem. Additional analysis determined that the BFS would not have been of any help because it would not have been able to separate the SRBs either if the condition had occurred. The occurrence of the conditions that would have triggered the PASS error were calculated to be one in six launches. A software patch was created to fix the software error and to assure all three booster fire commands were issued in the proper time interval. The problem was later traced to the requirements stage of the software development process and additional testing and analysis were introduced into the process to avoid a repetition.

Another timing problem that could have resulted in failure to separate the ET was discovered right before the launch of STS-41D in August 1984. In the 48 hours before the launch, IBM created, tested, and delivered a 12-word code patch to ensure sufficient delay between the PASS computed commands and the output of those commands to the hardware.

With the extensive testing that continued throughout the Space Shuttle program, the number of errors found in the software did decrease over the life of the Space Shuttle, largely because of the sophisticated continuous improvement and learning process used in the software development process. A point was reached where almost no errors were found in the software for the last few Space Shuttle flights, although there were also fewer changes made to the software during that period [72].

However, a potentially serious software error did occur in April 2009, just two years before the Space Shuttle's retirement. The error manifested itself in flight STS-126 a few minutes after *Endeavour* reached orbit [73]. Mission Control noticed that the Space Shuttle did not automatically transfer two communication processes from launch to orbit configuration mode. Mission Control could not fix the problem during the flight, so they manually operated necessary transfers for the remainder of the flight. The pathway for this bug had been introduced originally in a change made in 1989 with a warning inserted in the code about the potential for that change to lead to misalignment of code in the COMPOOL. As more changes were made, the warning got moved to a place where it was unlikely that programmers changing the code would see it. The original change violated the programming standards, but that standard was unclear, and nobody checked that it was enforced in that case. Avoiding the specific error that was made was considered "good practice," but it was not formally documented, and there were no items in the review checklist to detect it. The SPF did not identify the problem either—testers would have needed to take extra steps to detect it. The SAIL could have tested the communication switch, but it was not identified as an essential test for that launch. Testing at the SAIL did uncover what hindsight indicated were clear problems in the communication handover problem, but the test team misinterpreted what happened during the test—they thought it was an artifact of lab setup issues—and no error reports were filed. Although *test as you fly and fly as you test* is a standard rule in spacecraft engineering, the difficult of achieving this goal is demonstrated by this specific escape, even given the enormous amounts of money that went into testing in the SAIL lab.

A final example is a software error that was detected during analysis of post-flight data from STS-79. This error resulted from a "process escape." Hickey notes that most of these errors can be traced to periods of decreasing morale among the IBM programming staff or pressures leading to decreased testing and not following the rigorous procedures that had been developed over the years [74].

In hindsight, it is easy to see that the challenges NASA and its contractors faced in terms of memory limitations, changing requirements, communication, and software and computer hardware quality and reliability were enormous, particularly given the state of technology at the time. Luckily, the Space Shuttle software developers did not have this hindsight when they started, and they went forward with confidence they could succeed, which they did spectacularly, in the manner of the U.S. manned space program in general.

LESSONS LEARNED, CONCLUSIONS, AND ANALOGIES

There can always be differing explanations for success (or failure) and varying emphasis placed on the relative importance of the factors involved. Personal biases and experiences are difficult to remove from such an evaluation. However, most observers agree that the process and the culture were important factors in the success of the Space Shuttle software as well as NASA's strong oversight, involvement, and control.

Lessons Learned

Oversight and Learning from the Past. NASA learned important lessons from previous spacecraft projects about the difficulty and care that need to go into the development of the software. These lessons included that software documentation is critical, verification must be thorough and cannot be rushed to save time, requirements must be clearly defined and carefully managed before coding begins and as changes are needed, software needs the same type of disciplined and rigorous processes used in other engineering disciplines, and quality must be built in from the beginning. NASA maintained direct control of the Space Shuttle software rather than ceding control to the hardware contractor and, in fact, constructed its own software development "factory" (the SPF). By doing so, NASA ensured that the highest standards and processes available at the time were used and that every change to human-rated flight software during the long life of the Space Shuttle was implemented with the same professional attention to detail.

Development Process. The development process was a major factor in the software's success. Especially important was careful planning before any code was written, including detailed requirements specification, continuous learning and process improvement, a disciplined top-down structured development approach, extensive record keeping and documentation, extensive and realistic testing and code reviews, detailed standards, and so on.

The Software Development Process. Culture matters. The challenging work, cooperative environment, and enjoyable working conditions encouraged people to stay with the PASS project. As those experts passed on their knowledge, they established a culture of quality and cooperation that persisted throughout the program and the decades of Space Shuttle operations and software maintenance activities.

System Safety and System Engineering. With the increasing complexity of the missions anticipated for the future and the increasing role of software in achieving them, another lesson that can be learned is that we will need better system engineering, including system safety engineering. NASA maintained control over the system engineering and safety engineering processes in the Space Shuttle and employed the best technology in these areas at the time. The two Space Shuttle losses are reminders that safety involves more than simply technical prowess, however, and that management can play an important role in accidents and must be part of the system safety considerations. In addition, our system and safety engineering techniques need to be upgraded to include the central role that software plays in our complex spacecraft systems. Unfortunately, the traditional hazard analysis techniques used in the Space Shuttle do not work very well for software-intensive systems [75].

CONCLUSIONS AND ANALOGIES

Beyond these lessons learned, some general conclusions and analogies can be drawn from the Space Shuttle experience to provide guidance for the future. One is that creating high-quality software is possible but requires a desire to do so and an investment of time and resources. Software quality is often given lip service in other industries, where often speed and cost are the major factors considered, quality simply needs to be "good enough," and frequent corrective updates are the norm.

Some have suggested that unique factors separated the Space Shuttle from other software development projects: one dedicated customer, a limited problem domain, and a situation where cost was important but less so than quality. However, even large government projects with a single government customer and large budgets have seen spectacular failures in the recent past. Examples are the new IRS software [76], several attempted upgrades to the Air Traffic Control system [77], a new FBI system [78], and even an airport luggage system [79]. The luggage system cost $186,000,000 for construction alone and never worked correctly. The other cited projects involved, for the most part, at least an order of magnitude higher costs than the baggage system and met with not much more success. In all of these cases, enormous amounts of money were spent with little to show for them. They had the advantage of newer software engineering techniques, so what was the significant difference?

One difference is that NASA maintained firm control over and deep involvement in the development of the Space Shuttle software. NASA used its experience and lessons learned from the past to improve its practices. With the current push to privatize the development of space vehicles, will the lesser oversight and control lead to more problems in the future? How much control will and should NASA exercise? Who will be responsible for system engineering and system safety?

In addition, software engineering is moving in the opposite direction from the process used for the Space Shuttle software development, with requirements and careful preplanning relegated to a less important position than starting to code. Strangely, in many cases, a requirements specification is seen as something that is generated after the software design is complete or at least after coding has started. Many of these new software engineering practices are being used by the firms designing new spacecraft today. Why has it been so difficult for software engineering to adopt the disciplined practices of the other engineering fields? Many software development projects still depend on cowboy programmers, "heroism," and less-than-professional engineering environments. How will NASA ensure that the private companies building manned spacecraft instill a successful culture and professional environment in their software development groups? Ironically, many of the factors that led to the success of the Space Shuttle software were related to limitations of computer hardware in that era, including limitations in memory that prevented today's common "requirements creep" and uncontrolled growth in functionality as well as requiring careful functional decomposition of the system requirements. Without the physical limitations that impose discipline on the development process, how can we impose discipline on our projects and ourselves?

The overarching question is how will engineers ensure that the hard learned lessons of past human space projects are conveyed to those who design future systems and that we are not, in the words of Santayana, condemned to repeat the same mistakes?

REFERENCES

[1] Fishman, C., "They Write the Right Stuff," Fast Company, 6:5–9 and 104–6, December 1996. Available online at http://www.fastcompany.com/online/06/writestuff.html; Singh, T. "Why NASA Space Shuttle Software Never Crashes: Bug-free NASA Shuttles," *The Geeknizer*, 17 July 2011.
[2] Tomayko, J., "Computers in Spaceflight: The NASA Experience," NASA Contractor Report CR-182505, 1988.
[3] Gemini sources, on-line at http://www.ibiblio.org/apollo/Gemini.html.
[4] Tomayko, "Computers in Spaceflight."
[5] Tomayko, "Computers in Spaceflight."
[6] Tomayko, "Computers in Spaceflight."
[7] Slotkin, A. L., *Doing the Impossible: George E. Mueller and the Management of NASA's Human Spaceflight Program*. Springer-Praxis, Chichester, UK, 2012.
[8] Slotkin, *Doing the Impossible.*
[9] Tomayko, "Computers in Spaceflight."
[10] Tomayko, "Computers in Spaceflight."
[11] Slotkin, *Doing the Impossible.*
[12] Slotkin, *Doing the Impossible.*
[13] Tomayko, "Computers in Spaceflight."
[14] Tomayko, "Computers in Spaceflight."
[15] Tomayko, "Computers in Spaceflight."
[16] GAO, "NASA Should Implement Independent Oversight of Software Development," GAO/IMTEC-91-20, 1991.
[17] Tomayko, "Computers in Spaceflight."
[18] Post-Challenger Evaluation of Space Shuttle Risk Assessment and Management, Aeronautics and Space Engineering Board, National Research Council, January 1988.
[19] Tomayko, "Computers in Spaceflight."
[20] Mattox, R. M., and White, J. B., "Space Shuttle Main Engine Controller," Report NASA-TP-1932 M-360, NASA, 1 Nov. 1981.
[21] GAO, "NASA Should Implement Independent Oversight of Software Development," GAO/IMTEC-91-20, 1991.
[22] Carlow, G. D., Architecture of the Space Shuttle Primary Avionics Software System, *Communications of the ACM*, Vol. 27, No. 9, September 1984, pp. 926–36.

[23] Garman, J. R., "Software Production Facility: Management Summary Concepts and Schedule Status," NASA Data Systems and Analysis Directorate, Spacecraft Software Division, 10 Feb. 1981, p. 12.
[24] Tomayko, "Computers in Spaceflight."
[25] Tomayko, "Computers in Spaceflight."
[26] Tomayko, "Computers in Spaceflight."
[27] Tomayko, "Computers in Spaceflight."
[28] *HAL/S Language Specification*, United Space Alliance, 2005.
[29] Martin, F. H., *HAL/S: The Avionics Programming System for the Shuttle*, AIAA, 315, 1977.
[30] Tomayko, "Computers in Spaceflight."
[31] Tomayko, "Computers in Spaceflight."
[32] Brooks, F., *The Mythical Man Month*, Addison-Wesley, Englewood Cliffs, NJ, 1973.
[33] Madden, W. A., and Rone, K. Y., "Design, Development, Integration: Space Shuttle Primary Flight Software System," *Communications of the ACM*, Vol. 27, No. 9, September 1984, 914–925.
[34] Tomayko, "Computers in Spaceflight."
[35] Tomayko, "Computers in Spaceflight."
[36] Tomayko, "Computers in Spaceflight."
[37] Tomayko, "Computers in Spaceflight."
[38] Hickey, J. C., Loveall, J. B., Orr, J. K., and Klausman, A. L., "The Legacy of Space Shuttle Flight Software," *AIAA Space 2011 Conference*, 2–29 Sep. 2011, Long Beach, California, 2011.
[39] Hickey, et al., "The Legacy of Space Shuttle Flight Software."
[40] Hickey, et al., "The Legacy of Space Shuttle Flight Software."
[41] Tomayko, "Computers in Spaceflight."
[42] Tomayko, "Computers in Spaceflight."
[43] Tomayko, "Computers in Spaceflight."
[44] Hickey, et al., "The Legacy of Space Shuttle Flight Software."
[45] Hickey, et al., "The Legacy of Space Shuttle Flight Software."
[46] Hickey, et al., "The Legacy of Space Shuttle Flight Software."
[47] Knight, J. C., and Leveson, N. G., "Experimental Evaluation of the Assumption of Independence in Multiversion Software," *IEEE Transactions on Software Engineering*, vol. SE-12, No.1, 1986, 96–109.
[48] Knight, J. C., and Leveson, N. G. "A Reply to the Criticisms of the Knight and Leveson Experiment," *ACM Software Engineering Notes*, January 1990.
[49] Leveson, N. *Safeware: System Safety and Computers*, Addison-Wesley Publishing Company, Englewood Cliffs, NJ, 1996.
[50] Leveson, N., *Engineering a Safer World*. MIT Press, Cambridge, MA, 2012.
[51] Garman, J. R., "The "Bug" Heard 'Round the World," *ACM SIGSOFT Software Engineering Notes*, October 1981, pp. 3–10.
[52] Hickey, et al., "The Legacy of Space Shuttle Flight Software."
[53] Keller, T. W., "Maintenance Process Metrics for Space Shuttle Flight Software," *Forum on Statistical Methods in Software Engineering*, National Research Council, Washington, DC, 11–12 Oct. 1993.
[54] Hickey, et al., "The Legacy of Space Shuttle Flight Software."
[55] Chairman, House Committee on Science, Space, and Technology, to NASA Administrator, 31 March 1988.
[56] Post-Challenger Evaluation of Space Shuttle Risk Assessment and Management, Aeronautics and Space Engineering Board, National Research Council, January 1988.

[57] GAO, "NASA Should Implement Independent Oversight of Software Development," GAO/IMTEC-91-20, 1991.
[58] An Assessment of Space Shuttle Flight Software Development Processes, Committee for Review of Oversight Mechanisms for Space Shuttle Flight Software Processes, Aeronautics and Space Engineering Board, National Research Council, 1993.
[59] Smith, S. A., and Cusumano, M. A., *Beyond the Software Factory: A Comparison of Classic and PC Software Developers*, Massachusetts Institute of Technology, Sloan School WP#3607=93\BPS, 1 Sept. 1993.
[60] Smith and Cusumano, *Beyond the Software Factory.*
[61] Smith and Cusumano, *Beyond the Software Factory.*
[62] Hickey, et al., "The Legacy of Space Shuttle Flight Software."
[63] Smith and Cusumano, *Beyond the Software Factory.*
[64] Hickey, et al., "The Legacy of Space Shuttle Flight Software."
[65] An Assessment of Space Shuttle Flight Software Development Processes, Committee for Review of Oversight Mechanisms for Space Shuttle Flight Software Processes, Aeronautics and Space Engineering Board, National Research Council, 1993.
[66] Hecht, H., "*Investigation of Shuttle Software Errors*," SoHar Incorporated, study prepared for Polytechnic University, Brooklyn, New York, and the Langley Research Center, Hampton, Virginia, under NASA Grant NAG1-1272, April 1992.
[67] Post-Challenger Evaluation of Space Shuttle Risk Assessment and Management, Aeronautics and Space Engineering Board, National Research Council, January 1988.
[68] An Assessment of Space Shuttle Flight Software Development Processes.
[69] Tomayko, "Computers in Spaceflight."
[70] An Assessment of Space Shuttle Flight Software Development Processes.
[71] Hickey, et al., "The Legacy of Space Shuttle Flight Software."
[72] Hickey, et al., "The Legacy of Space Shuttle Flight Software."
[73] NASA, "Shuttle System Failure Case Studies: STS-126, NASA Safety Center Special Study, NASA, April 2009.
[74] Hickey, et al., "The Legacy of Space Shuttle Flight Software."
[75] Leveson, *Engineering a Safer World.*
[76] Broache, A., "IRS Trudges on with Aging Computers," *CNET News*, April 12, 2007, http://news.cnet.com/2100-1028_3-6175657.html.
[77] Lewyn, M., "Flying in Place: The FAA's Air Control Fiasco," *Business Week*, April 26, 1993, pp. 87–90.
[78] Eggan, D. and Witte, G., "The FBI's Upgrade That Wasn't," *The Washington Post*, August 18, 2006, http://www.washingtonpost.com/wp-dyn/content/article/2006/08/17/AR2006081701485.html.
[79] Johnson, K., "Denver Airport Saw the Future. It didn't Work," *New York Times*, August 27, 2005.

Chapter 7

Flying the Shuttle: Operations from Preparation through Flight to Recovery

N. Wayne Hale

Introduction

The Space Shuttle was arguably one of the most complex machines ever built, and it launched some of the most complex payloads ever lofted into space. The enterprise was correspondingly complex to operate and maintain. Missions were expected to be completely successful in all planned objectives. This high expectation resulted in the development and exercise of excruciatingly meticulous procedures to deconstruct every aspect of each mission objective and payload to build comprehensive plans to meet those expectations.

In the mature phase of the Space Shuttle program, the flight preparation process was well defined, and all individuals in the process knew their roles exquisitely well. Workforce numbers fluctuated widely over time, with variation in the mature phase of the Shuttle program between 12,000 and 30,000 workers. This included both "prime" contractor personnel and government civil servants in a roughly constant ratio of 10 contractors for every civil servant. This number does not include myriads of workers at suppliers, parts vendors, and the like. Nor does that number include personnel who were not involved directly in the flight production process but who were working on such diverse operations as vehicle upgrades, infrastructure renewal, and other ancillary tasks. If the Space Shuttle was expensive to fly, this was largely because of the salaries paid to American workers to prepare and fly it.

The flight manifesting process started with NASA form 100. This was an application for a payload or an experiment to fly on the Space Shuttle. The form could come from a NASA center or organization (such as the Hubble Space Telescope program from Goddard Space Flight Center in Maryland), a commercial interest, an academic institution, or another government agency. The shortest time from filing a form 100 to flight was approximately 18 months but only for some very small experiments manifested late in the process. At the other extreme, some payloads languished over two decades waiting their turn to fly because priorities changed and accidents interrupted the process. Some applicants never flew.

PAYLOAD DEVELOPMENT

After being accepted as a potential shuttle payload, each potential flight customer would develop a Payload Integration Plan (PIP) detailing such things as the cargo weight, length, center of gravity, orbital parameter requirements, power requirements, cooling requirements, crew time requirements, and commanding requirements. The shuttle was an extraordinarily capable vehicle that could provide many services to complex payloads, including robotics and extra-vehicular-activity (EVA, or spacewalk) servicing. It was very onerous for many simple payloads that required little support or few interfaces to develop a PIP along with the other documentation required to fly on the shuttle. There were no shortcuts. A good lesson that the Space Shuttle program learned and that we can apply to future space haulers is that it should become easier for small and simple customers to get a ride. A mini-industry of interface organizations, such as the Goddard Space Flight Center hitchhiker program, developed to run interference for small, simple payloads. SpaceHab became a commercial agent to provide similar services (Fig. 7.1).

Human space flight requires extraordinary attention to safety, and every potential payload had to pass stringent safety reviews. These could be time consuming and difficult, and often they required the experiment or payload to be redesigned. For many small experiments, extensive safety analysis and documentation became an impediment to flying cargo into space on the shuttle.

The computers and associated electronics on the shuttle were designed and built in the early 1970s. As more sophisticated and less expensive electronics became available, new payloads were designed with modern electronics. However, to fly on the shuttle, the new, more capable electronics had to be dumbed down to the shuttle interface requirements so that they could provide for crew displays and controls, ground commanding, telemetry, and so forth. This obsolescence of shuttle avionics, driven by lack of funding to upgrade, became an obstacle to flying many payloads in space. Not only was this true for the flight systems, but the ground processing facilities and

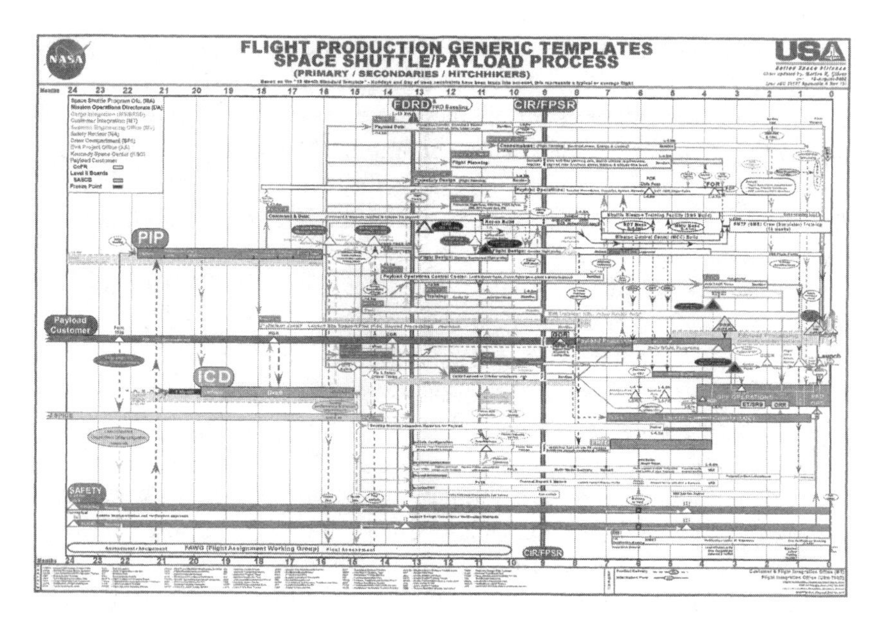

Fig. 7.1 Simplified Space Shuttle planning process. (Cargo Integration Office, Space Shuttle Program, NASA/JSC)

associated computer infrastructure also remained stubbornly obsolescent, complicating even the preflight test and checkout work for each payload.

Finally, to feed both the system's technical interface parts and the crew training organization, detailed information about the payload or cargo element had to be made available many months in advance. When complex experiments were being developed in academic settings, or new satellites were being built commercially, the designs had to be frozen early to provide the basic information required to feed the shuttle system. Late changes and upgrades could not be accommodated easily.

The NASA headquarters office for the shuttle (which went through various titles over time) prioritized the payloads and worked with the assessment team at JSC to put together a rough manifest for each flight, initially of the big payload elements only. This included assessments of the power requirements, the lift requirements, the center of gravity and physical space requirements in the payload bay, the mission trajectory (including pointing requirements), the mission duration, and consumables such as rocket fuel and cryogenic oxygen and hydrogen. All of these parameters were assessed to make certain that the proposed flight might actually be capable of being flown. After a rough mission manifest was established, the work transferred to the flight planning working group in Houston at the Shuttle program office. This office took the roughed-out manifest, applied the turnaround and scheduling constraints, and came up with a planning manifest for multiple flights [1]. In the meantime, the Flight Operations Integration office at JSC assigned a Flight Integration Manager and a Payload Integration Manager, jocularly known as the FIM and PIM. Monthly and, later on, weekly, meetings were held with representatives from all the operational areas to ensure a flight manifest that could be successfully accomplished was built. It was key to maximize return from each mission, so additional secondary payloads and mid-deck experiments were added to the flight's maximum capability. Mission planners at JSC started running assessment programs to ensure that the mass to orbit, Center of Gravity (cg), payload attach points, and other technical constraints all meshed. At this point, the flight duration was calculated, and an orbiter, the main engines, the ET, and a pair of solid rocket motors were initially assigned. The program manager, always in consultation with NASA headquarters, approved these assignments and the proposed launch date. The process was highly orchestrated, with high-level management reviews periodically examining the mission content's development; these reviews included the Flight Planning and Stowage Review (FPSR), the Cargo Integration Review (CIR), and the Flight Operations Review (FOR). At the start was a process leading to the baselining of the Flight Definition and Requirements Document (FDRD), the bible that all the flight preparation processes referred back to the program manager [1].

Early in the program, it was thought that Space Shuttle flights would become routine and that variations between flights would be minimal. In practice, this was not the case. Payloads were unique and were generally flown only once. Experiments were unique and even if flown again had modifications to the equipment, processes, and procedures. Every flight combined several major payload elements and mid-deck experiments. Very little in the flight manifesting and planning process was common across flights; every flight was unique, with its own list of objectives and their relative priority. In essence, each flight manifest was handcrafted, requiring a significant workforce to hand build the interface control documents (ICDs) for the payload accommodations (power, structural attachment, and the like), as well as the training for the crew and flight controllers. No flight was ever a carbon copy of any previous flight. Assumptions that space flight will become routine and nonvarying are not likely to become true in the near future. For future space systems, cost/schedule projections that depend on flying multiple similar flights should be considered highly suspect.

Meanwhile, at contractor facilities around the nation, the hardware necessary for the shuttle flight was in process. Lockheed-Martin took nearly three years from raw materials in the door to finished product rollout to build an ET in the Michoud facility just outside New Orleans. Alliant Tech Systems (ATK) shipped spent SRB case segments by railcar up to Clearfield, Utah, for a lengthy cleaning, inspection, and refurbishment process before they were transported to the remote Promontory, Utah, facility, where the rubbery solid propellant was poured into each case segment with great care. After curing and a myriad of inspections, the solid rocket motor segments were shipped by rail to KSC in Florida, where they were inspected and stored in the Rotation Processing and Surge Facility (RPSF) until final assembly in the Vehicle Assembly Building (VAB). Pratt Whitney Rocketdyne (PRW) removed the SSMEs from the shuttle orbiter after every flight for inspection at the Space Shuttle Main Engine Shop, adjacent to the Orbiter Processing Facility (OPF) and the VAB. Detailed teardown of the critical high-pressure pumps was accomplished at the PRW facility at West Palm Beach in southern Florida. New-build SSME pumps (or complete engines) were trucked from Canoga Park, California, to the Stennis Space Center in Bay St. Louis, Mississippi, for "green run" test firings before being certified for space flight. And United Space Alliance (USA) poured half a million man hours into turning around each Space Shuttle orbiter after flight in the OPF at Kennedy (Fig. 7.2). Every few years, each orbiter required an Orbiter Maintenance Down Period (OMDP) for maintenance, typically about three times longer than a standard flight turnaround. The OMDP accomplished major and invasive inspections, overhauls, in-depth maintenance, and upgrades. After the SRBs were recovered from the Atlantic Ocean, the parachutes, recovery systems, hydraulics systems, and more were cleaned after their salt water immersion and then reassembled and tested. All of these hardware production and refurbishment activities had to be choreographed to come together at the right time.

Space suits, cameras, tools, and robotic equipment all had to be refurbished for reuse, tested, and certified for space flight. Critical fluids such as cryogenic hydrogen, cryogen oxygen, and hypergolic rocket fuels had to be produced in chemical plants around the country and shipped to KSC. Pyrotechnics had to be carefully assembled and inserted into systems where

Fig. 7.2 SSME undergoing maintenance at KSC's Main Engine Shop. (Wayne Hale)

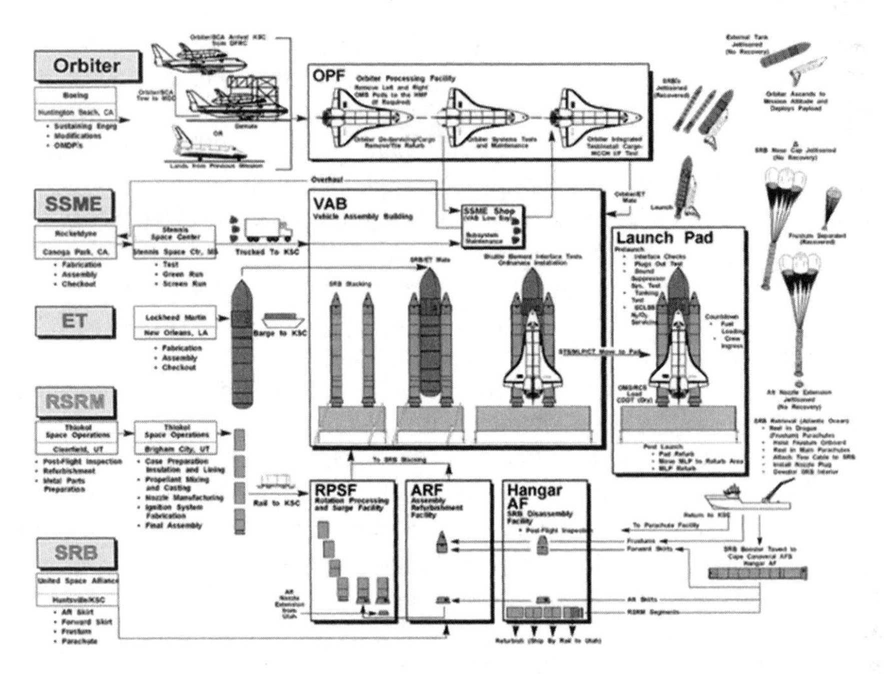

Fig. 7.3 Space Shuttle processing schematic. (Space Shuttle Program Office, NASA Lyndon B. Johnson Space Center)

their properly timed explosive action was vital. None of these processes were short, and all of them were expensive in terms of manpower and dollar outlays (Fig. 7.3).

Over the years, management paid a significant amount of attention to shortening many of these hardware preparation elements. In particular, the number of days an orbiter spent in the OPF was closely tracked, and from time to time, strong goals were set to minimize this time. An early “win” that reduced turnaround time was that the safety organizations agreed to allow hazardous hypergolic rocket fuels to be retained onboard the vehicles during refurbishment. This agreement came with significant operational controls, but overall turnaround time was reduced significantly (Fig. 7.4) [2].

MISSION PLANNING

Early in the program, flight rates for a 4-orbiter fleet were expected to exceed 40 flights a year, or 10 flights for each orbiter. This would have required very short OPF maintenance stays, about 20 days or fewer [3]. In practice, OPF turnaround times typically averaged 70 to 90 days no matter which management incentives were levied to reduce the time. The number of required days had several root causes. First, high flight rates never materialized; about six to eight flights a year for a three- or four-orbiter fleet was typical. More than 90 days were available for turnaround, and work expanded to fill that available time. Second, the optimistic estimates of "airliner-like maintenance" ignored the realities of space flight. Not only is the space flight environment much more demanding, but the design goals to minimize weight also required that all equipment designs sacrifice a level of robustness for weight savings. Third, rocket engines are significantly more complex and unforgiving than aviation jet turbine engines. The SSMEs required extensive and invasive inspections for every flight and frequently other significant maintenance work. This engine work was best accomplished with the engines off the orbiter, so the engines were removed after every flight and sent to the SSME shop for inspection and maintenance. Many times, components had to be replaced with factory-rebuilt components, which in their turn had required a hot fire or "green run" at the engine test stands at Stennis Space Center in Mississippi. With the excuse that "we have to pull the engines anyway," many more engineering requirements found their way into the orbiter turnaround process. Finally, after each of the Space Shuttle accidents, even though the orbiters had not been the root cause of those accidents, myriad additional safety checks were instituted during the "return to flight' process.

All of these causes were unavoidable with the state of technology for the Space Shuttle systems; all contributed to the lengthy and invasive turnaround timelines, and all increased costs and reduced the ultimate flight rate. Any new space launch systems that endeavor to be more economical or have a higher flight rate than the Space Shuttle will be compelled to make different design choices early in the vehicle's development so that it has more robustness and ease of maintenance at the cost of increased weight and decreased lift. Projections of turnaround costs and schedule for reusable space systems must be considered highly suspect until the system is actually flying and maintenance costs can be measured in reality. Exuberant and unwarranted

Fig. 7.4 Facility turnaround work in the OPF. (NASA photo no. KSC-2012-5803)

optimism about operations always dominates projections during early program development phases.

A lesson from the Space Shuttle program is that a reusable vehicle might be cheaper to operate than an expendable one only if the refurbishment and inspection costs can be reduced below the cost to build a new spacecraft. Refurbishment costs also depend on the flight rate, and a very low flight rate (fewer than 10 per year) does not lend itself to making a reusable vehicle cheaper than one that is expended each time.

However, one thing about the Space Shuttle was new every time it flew: the flight software, as we saw in Chapter 6. Software is an issue that famed aerospace executive Norman Augustine addresses: "...something that can be added to airplanes and other systems and that weighs nothing, yet is very costly, and violates none of the physical laws of the universe... is called software" [4]. Early treatises by Space Shuttle managers all stated that software would "soon be frozen" and that development costs would be eliminated. This never happened. Mandatory software changes were always required to accommodate various specialized payloads and hardware upgrades (for cost, schedule, performance, and safety reasons). In addition, software changes were needed merely to improve safety itself. Each flight required reconfiguration of the software for certain trajectory constants called "I-Loads." After the currently planned set of mandatory changes was implemented, management edict followed broken management edict to stop changing the Space Shuttle software. It was simply unrealistic to believe that software could be frozen on a complex, high-value asset such as a spacecraft, which was in operation for 30 plus years and which hosted complex payloads with different orbital parameters.

JSC's Mission Operations Directorate became responsible for all software changes because many constants and mission parameters that changed every flight were operationally required to meet mission plans (orbital inclination, day of launch, and orbital altitude all varied with every flight). It took time and resources to develop the software and the eight levels of testing to assure it worked properly. Patch sets often had to be developed for flights when late changes were directed from program management. These patches had to be verified and tested by hand. Every change associated with flight software was extremely costly despite many management programs over the years to "drain the swamp," or simplify the software reconfiguration processes. Among the significant interfaces that had to be developed for almost every payload was the command interface table. Sending a payload command from the obsolescent MCC to the geriatric shuttle flight avionics sometimes became a test of patience, determination, and ingenuity. Testing each command before flight became mind numbing for those involved.

Ground displays for the launch control center at KSC, the mission control center at JSC, and the Payload Operations Control Center at MSFC had to be developed uniquely to accommodate the different payload complements. Computer models of the payloads and their operations had to be coded and verified to be used in training well in advance of the flight.

Detailed instructions known as Orbiter Maintenance Requirements and Specifications Document (OMRSD) and detailed work documents had to be developed to guide technicians at KSC to properly install and check out even simplified payloads. This meant that a massive number of pages had to be generated for each flight to prepare work documents.

Mission Operations Directorate through the USA contractor workforce also performed the "flight cycle design." This was the detailed engineering and trajectory analysis for each flight. Results from this analysis defined such vital parameters as how much of each type of consumables to load and what trajectory to fly. Included was a variety of off-nominal analysis was performed for each flight. The goal of management was "one cycle to flight," which was rarely realized, typically because of significant manifest changes required by program management. Some flights went through as many as six flight design cycles because of launch date changes, manifest changes, and orbiter assignment changes. When the flight rate is low and the optimization required for each flight is high, as it always was for the shuttle, much analysis and engineering must be performed on each mission. Management edicts to minimize the man-hours required for computational analysis for flight cycles were ineffective in the face of the reality of changing manifests and headquarters-driven changing priorities.

Buried in the preparation process was the delivery of the training software loads to the various shuttle simulators—the Shuttle Mission Simulators, the part task trainers, and so on. All the interaction by and for each payload had to be validated and verified to mimic real life.

Once the training load had been delivered, it was time to think about integrated crew and flight controller training. Roughly a year before flight, a flight crew and flight control team were identified at JSC. For some complex payloads, an astronaut payload commander and a lead flight director were assigned much earlier, but for typical flights, assignments came one year before the launch.

When astronaut candidates (Ascan) are selected, they undergo a rigorous training process that familiarizes them with the generic aspects of space flight and the peculiarities of the shuttle (or other spacecraft). This basic training process typically takes 18 months and involves a significant number of personnel from various NASA centers. Specialized training such as that required for EVA or for Pilot-Astronaut can add significantly to that time. After basic training (termed –1000 and –2000 level training) was completed, an astronaut candidate was certified ready for flight assignment. This was not automatic, and depending on program needs, a lengthy assignment to various special projects on the ground awaited each newly minted astronaut. From selection to first flight took five years on average.

Upon being assigned to a flight, an astronaut crew started flight-specific training (–9000 level training). This typically was centered at JSC but with significant time scheduled at the payload developer's locations and other NASA centers. Assignment to an international flight (such as ISS) required training at foreign locations as well. In-depth training classes and training tools would be developed for each payload and experiment. Using these training tools, the crew trained repeatedly on how to operate the special equipment for the flight in normal mode and on what to do in emergencies or other off-nominal situations. Flights with extensive science payloads and SpaceLab or SpaceHab research flights could require significantly more payload and experiment training than a typical shuttle flight. Deployable satellites carried a significant overhead of training. Flights requiring rendezvous, robot arm operations, and the like required special flight consideration and design as well as training products.

A Crew Training Lead started coordinating the multicenter (sometimes multinational) training requirements and schedule even before a crew was assigned. The schedule could be quite complex depending on the availability of training resources such as simulator schedules, or overseas requirements (ISS training on European, Japanese, or Russian modules). Payload training was frequently accomplished by MSFC payload training organization. Other scheduling constraints such as mandatory public affairs events frequently made scheduling the crew and the flight control team training even more challenging. Changes to flights before the one in consideration drove even further schedule change because resources were shifted between various crews and their flights. Training schedules were updated weekly and frequently contained major shifts in activities.

Meanwhile, the lead Flight Director would convene a series of Flight Techniques meetings (or, for ISS flights, Joint Operations Panel meetings) to examine each operation required for the flight and to ensure the best procedures were developed for the crew and flight controllers so that the flight would be successful. Development of a Flight Specific Flight Rules Annex documented decisions as recommended by the Lead Flight Director and was approved by the Space Shuttle Program Manager and the ISS Program Manager. Complex missions that had multiple operations and payloads and that competed for time and crew involvement required complex rules to determine priorities under the various circumstances that might occur.

The lead Flight Activities officer started building the Crew Activity Plan (CAP). This detailed plan covered each minute of the flight with detailed directions for the crew and flight control team. Building a CAP involved interactive work with the attitude timeline requirements to ensure that the vehicle was properly pointed for thermal control reasons and power production, and to support deploys and retrievals, rendezvous, scientific operations, or a variety of other constraints.

The lead Payload Officer would ensure that the principal Investigator's requirements for operations of all payloads were properly documented in the crew checklists and Flight Rules annex.

At the Flight Techniques panel priorities were hammered out; new operations were examined closely, and detailed decisions of how to handle every conceivable contingency were written. This could be a hugely time consuming activity. Again, early in the Space Shuttle program, it was envisioned that if a malfunction occurred, the flight would simply be terminated early and, after repairs, the payload would be flown again. With the enormous costs associated with each Space Shuttle launch, this was extraordinarily unrealistic. The high cost of every flight, not to mention the risk to the crew and the vehicle, mandated that every flight be made as productive as possible. In the face of anomalies, trajectory changes, programmatic priority changes, or equipment problems, the crew and ground team would work quickly and efficiently to maximize the return on each flight. Because the flight planners and Flight Directors worked over the priorities and options for many months, they became experts in exactly what to do to replan a flight with reshuffled priorities. Failure was not an option. Like the scouts, a flight control team had to "be prepared" for any old thing.

Flight plans were developed in accordance with the Crew Scheduling Constraints— sleep shift limits were observed, and rest periods and off-duty days were sacrosanct—at least for the nominal plan before flight. During flight, if replanning was required, the flight planners frequently did what was necessary.

Meanwhile, the senior member of the training team, the Sim Sup (Simulation Supervisor) would attend all the meetings, the Flight Operations Panels, the Flight Techniques, the Crew Procedures Control Boards, and the Flight Rules Control Boards and would listen to the discussion of the objectives, the payload operations, and the contingency plans. Then, the Sim Sup and his evil team of training personnel would pore over all the procedures and rules looking for weak spots, cases that were overlooked, and problems that were not resolved properly. With this knowledge, the team would put together elegant training scenarios designed to stress the entire ground and flight team. And they did. Integrated simulations were sometimes like a chess match with masterminds vying for control, and sometimes like a roller derby with body slams, taunts, and insults. It was always entertaining. And important lessons were learned. Because the Shuttle Mission Simulator (SMS) used real flight computers and real flight software, and because the payload computer models were extremely high fidelity, problems that would not have been uncovered in any other way showed up. That is good in training, bad in flight. The teams would coalesce, appreciating each other's strengths and weaknesses, covering each other in tight situations.

In addition, all parts of the plan were tested during the training. The simulation teams were incredibly creative in posing excruciatingly difficult problems at the worst possible time in each training session. Their motto was that the flight should never be as hard as the simulations.

And it paid off; over half of all shuttle flights had to have major replanning in flight because of changing conditions or equipment failures. In spite of this, virtually every flight completed over 90% of all preplanned objectives.

During the development of the Space Shuttle, a partnership with the national defense and security agencies was critical. The shuttle was to loft all national security payloads once it was operational. In actual practice, the *Challenger* accident ended those agencies' dependence on the Space Shuttle as they looked back to expendable launch vehicles for their programs. Nevertheless, several national security missions continued on the Space Shuttle, and NASA had to accommodate their special needs. NASA was developed as a civilian space agency, and chiseled into its DNA was a strong relationship with the public through the media. Huge problems were caused by working with the restrictions of classification guides, need to know, and limited secure meeting rooms and telephones. Identifying who was cleared for a meeting became a huge obstacle when the results from a lawsuit removed security clearance information from the front of identification badges. NASA tried hard to comply with the need for secrecy, and the effectiveness of those processes is still classified. However, one serious incident could have resulted in the loss of a shuttle when the STS-27 mission was badly damaged by ascent debris—in a foreshadowing of the 2003 *Columbia* accident. The payload customer prohibited NASA from using its regular television downlink to inspect the tiles but allowed the use of a very low-resolution encrypted video system only. The result was that the mission controllers and engineers on the ground could not evaluate the damage accurately. Fortunately, the mission landed safely, with post-flight horror at the condition of the orbiter. All of the NASA human space flight organizations were glad when the last classified mission was complete.

Future human spacecraft will undoubtedly be involved in national security work at some point. The operational safety aspects for human spacecraft must never be sacrificed lightly. National security information risks must not outweigh the safety of crews and their highly valuable spacecraft.

A critical step in launch preparations occurred when NASA would have to negotiate with the Eastern Range (run by the U.S. Air Force) for a shuttle launch date. The Eastern Range did not play favorites with vehicles or programs; it operated on a strict "first come, first served" basis. If a launching organization reserved a date, in principle that organization could not be forced off it. If the shuttle had a tight launch window constraint, this could lead to some very high-level and intense negotiation! Generally, "customers" were granted two days—the planned launch date and one backup day—before they would have to stand down and allow another user to have a turn at launching. With the frequent launch date changes for the Space Shuttle, the program Launch Integration Manager was kept busy going on bended knee to the headquarters of the 45th Space Wing at Patrick Air Force Base, just down the road from KSC.

Finally, at about two weeks before the scheduled launch date, the Flight Readiness Review (FRR) loomed for the entire shuttle team. All elements, whether they made hardware or developed flight trajectories, stood up in front of the senior NASA management in a face-to-face meeting at KSC and proudly displayed their work. Questions were asked, formal votes were taken, and sometimes actions were assigned. However, when the chairman of the FRR signed the Certification of Flight Readiness (CoFR), the shuttle was ready to launch.

LAUNCH

With the shuttle assembled and on the launch pad, and the payload installed, what could be easier than counting down to launch? It never was easy. Technical problems in the old and complex ground system frequently thwarted launch plans. Technical problems with payloads could cause significant delays. The complexity of the flight vehicle left it vulnerable to small issues becoming showstoppers. And there was always the weather. Over half the shuttle flights experienced launch delays between the flight readiness review and the planned launch date. The typical culprit was the weather.

Future vehicles should be designed to be more resistant to weather problems. The shuttle's fragile thermal protection tiles required no flying through precipitation; even a small cloud in the wrong place could hold up a launch if the weather reconnaissance aircraft detected droplets in the cloud. Nevertheless, many weather requirements will remain a constraint for future rocket launches. Consider the problem of induced lightning strikes: As demonstrated on Apollo 12, the exhaust plume of a rocket is highly electrically conductive. Even when aircraft or other conventional vehicles could operate near slightly charged clouds with impunity, rockets will act as giant lightning rods, and a discharge can occur. Hardening a launch vehicle against a lightning strike is very difficult. It is likely that any sort of charging condition (such as thunderstorms in the area) will continue to constrain the launch of any rocket in the near future. Moreover, central Florida is one of the most active lightning areas in the world. Low clouds and low visibility restrict visual tracking with the range that safety organizations require; even if these conditions do not impede the launch vehicle directly, public safety will require good visibility and no low cloud decks in the launch area.

Then there is the problem of winds at the launch pad; strong winds could push a rocket into the launch pad structures, lightning suppression towers, or other obstacles. Winds will likely be restrictive at some level for all launch vehicles. In the last few weeks at the launch pad, a threat of a hurricane that might approach Florida's east coast could cause a retreat from the pad, bringing the shuttle back into the relative protection of the VAB. Only in Hollywood movies do rocketeers attempt to launch through the eye of a hurricane! Mortals who are more prudent will take reasonable precautions and fly another day. Of course, the shuttle had restrictive weather requirements for an abort landing back at KSC's Shuttle Landing Facility (SLF) runway. However, review of launch scrubs showed that most of the time, a weather delay was caused by both the vertical launch environment over the pad and the 20-minutes-later forecast for the abort landing site runway. Improvements in landing characteristics for future vehicles may only slightly increase the odds of beating Mother Nature when it comes to getting off the launch pad.

At KSC, a team comprised largely of the same skilled individuals who assembled and prepared the vehicle controlled the launch. The separation of the launch control team from the flight control team was more than geographic; it was cultural as well. When it came to interpreting telemetry data on launch day, it was vital to have intimate knowledge of how every part functioned, how it was installed, and what problems might have been incurred during prelaunch. The launch control team did not have to worry about the peculiarities of the mission content or the myriad options that might come into play during flight; they just had to deliver a perfectly functioning vehicle when the countdown clock ticked to T = 0. Not surprisingly, the launch team developed an almost personal relationship with each orbiter. Lavishing their hands-on work into the preparations for flight, the launch team members were adamant to ensure that every part was working properly before committing to flight. No other organization—not the designers nor the flight controllers nor the astronauts—knew the intimate details of the Space Shuttle as well as the launch control team. Led by their own "general," the Launch Director, they ran the countdown with military cadence and superb professionalism. Many organizational experts have questioned the wisdom of separating the launch team from the flight team, but this worked superbly well for the shuttle.

After all the intense training, and worrying over the weather, the mission control team found that the actual shuttle ascent phase became much less stressful than the training. Only rarely did problems crop up in the ascent phase of shuttle launches. Even after the harrying launch phase, the actual execution on many flights became somewhat boring for the mission controllers in Houston; everything went according to plan, with no significant changes required.

On the other hand, there were just as many flights where events did not proceed smoothly. Scenarios that had been deemed "unrealistic" during training sometimes turned out to be very realistic. Tethered satellites broke free, deployed satellites malfunctioned, experiments went awry, robotic operations encountered difficulties, trajectories needed changing, and previously defined mission priorities were revised.

But after all the preparation, all the simulations, all the preflight practice, the astronauts and the mission control teams were able to handle those problems, not because they had trained for the specific problem that occurred, but because they had trained into a skill set of problem solving in a generic sense. The astronaut crews, if not perfect, were nearly so, and minor bobbles became the grist for minor changes during the "overnight" planning sessions.

Every flight day when the crew went to bed, the planning team would come into the Mission Control Center at JSC. Carefully directed, they would examine every minute of each operation for the next day, and to a lesser extent for the rest of the mission, and develop detailed plan changes. These changes were vetted through the shuttle design engineers in the Mission Evaluation Room and approved by the senior NASA managers at the Mission Management Team meeting. Every "morning" when the crew awoke, a crisp new plan that was ready for execution was uplinked to them. Woe to the planning team if an error showed up during the execution phase of the flight! Flight Directors are not known for their kind and gentle personalities. The Lead Flight Director, who had sweated for months planning and training for a specific flight and who was responsible for daily execution, did not tolerate any error in a plan.

Flight operations were a near-military operation. The stakes were high, in terms of both cost to the taxpayers and lives at risk. The near mythical founders of the mission control organization—the inimitable Eugene F. Kranz and the near-omniscient Christopher C. Kraft—had built an organization that was dedicated to the success of the mission and the safety of the crews, no matter what was required. The Flight Control credo was "Tough and Competent": tough enough to tell a senior management the truth, no matter how unpalatable that might be, and competent to know what to do in all circumstances.

The astronauts' dedication to their craft was legendary as well. They were selected from the top individuals from a vast pool of qualified applicants, and they had honed their interpersonal skills and technical knowledge. After years of training, they were ready for anything that might come, and they never admitted any fear except the fear that they might make a mistake. It was a cliché that one should never get into a competition with an astronaut, but it was also true that astronauts were the ultimate team players. In the future, when space flight might lose its broad-based appeal, it may be difficult to recruit and retain a similar cadre of dedicated and extremely competent individuals. Not only during flight but also during the flight preparation, the astronauts' review and participation were keystone elements to the success of each flight.

CONCLUSION

As each shuttle flight wound down to its close, with the mission objectives accomplished, attention returned to the weather forecasters. An extraordinary cadre from the National Weather Service known as the Spaceflight

Meteorology Group (SMG) spent years refining techniques to micro-forecast weather conditions at specific landing sites at specific times. They practiced every day, and their statistical accuracy for forecasts 90 minutes in advance was over 95%. This was required for a safe operation because the orbiter, operating as the world's largest and heaviest glider, would have only one opportunity to make a successful landing. NASA had prepared three landing sites for the normal end of shuttle missions: the Shuttle Landing Facility at KSC in Florida, the Dryden Flight Research Center at Edwards Air Force Base in California, and White Sands Space Harbor, also known as Northrop Strip, on the White Sands Missile Range in New Mexico. Each site had its peculiar weather patterns. Meteorological instruments at each site were critical for proper forecasting. NASA weather reconnaissance flew above each runway to assess the weather conditions from the pilot's point of view. With the shuttle normally operating flawlessly, the Entry Flight Director became the final authority to provide a GO for the deorbit burn and to commit the shuttle to landing. After that, it was in the hands of the skilled commander to land safely, even if the weather predictions turned out to be wrong. Every landing was tense because mission control provided last-minute advice to the commander. Following safe touchdown, the vehicle turnaround crew from KSC would take control of the orbiter and would take the first steps in preparing it for the next mission.

Following each successful flight were the nearly interminable debriefings—not the short, simple debriefings that followed the training sessions before flight, but detailed, inquisitive, painful retrospection of what went well, what did not go so well, and what mistakes should never be repeated. Detailed notes taken at every debriefing were rolled up into the crew report and the flight control team's post flight report. Lessons learned were documented and reviewed in very public forums. It was futile to hide any shortcomings or errors. Confession is supposed to be good for the soul; if so, many souls were improved during the shuttle post flight debriefings. Action items were taken to add new lessons to the planning documents, changes to the flight rules, and improvements to the constraints books. Over time, a great database of how best to plan shuttle flights—and what to avoid—was built up. In fact, NASA has captured this knowledge, and it exists inside the computerized firewall so that future spacecraft planners and operators can learn from it.

In addition, after the flight, there was always another flight waiting in the wings; there was another set of SRBs to stack, another set of SSMEs to green run, another orbiter to turn around, and a new payload complement to integrate into the payload bay. There were always new procedures to be written, priorities to be assessed, and training sessions to be exercised. For over 30 years, this process continued. Even during the hiatus following each accident, preparations were always going forward for the next mission, and NASA was working hard to make each flight as safe and successful as humanly possible.

This dedication to detail and the processes that were developed should be a model for future space operations to learn from as well.

REFERENCES

[1] "*Flight Production Generic Templates: Shuttle Templates*," NASA Lyndon B. Johnson Space Center, Houston, Texas, document JSC 25187 Appendix A Final Revision (28), December 2007.

[2] Schwartz, R. "Overview of STS Ground Operations/Orbiter Turnaround STS-1 through STS-7," *Space Shuttle Technical Conference*, edited by N. Chaffee, NASA Conference Publication 2342, 1985.

[3] Edson, W. F. Jr., "Transition to the Space Shuttle Operations Era," *Space Shuttle Technical Conference*, edited by N. Chaffee, NASA Conference Publication 2342, 1985.

[4] Augustine, N. R., *Augustine's Laws*, American Institute of Aeronautics and Astronautics, New York, 1983.

Chapter 8

Using the Shuttle: Operations on Orbit

Matthew H. Hersch

Introduction

In November 1985, an article in *Popular Mechanics* entitled "Space Vacation 1995" predicted that in the near future, America's latest spacecraft would accommodate 24 paying passengers at a time, in a commercially designed compartment containing seats, a bar, an observation deck, and an exercise area [1]. Paying $1 million each for the privilege, tourists would be medically vetted, rudimentarily trained, and offered a three-day orbital voyage—the first day of which would likely be consumed by almost incessant vomiting. Simultaneously grandiose and mundane, the image of this space hotel still fascinates us nearly 30 years later. These predictions were built upon the most optimistic assumptions about the capabilities of the Space Transportation System (STS) and the frequency with which NASA could launch its Space Shuttle fleet (an unreachable 24 flights per year) [2]. The predictions represented one of the last examples of the earliest wave of enthusiasm about the possibilities presented by a reusable orbital space plane: that it would restructure the economics of space exploration, making it more akin to airline travel than a polar expedition. The loss of shuttle orbiter *Challenger* two months after the publication of the *Popular Mechanics* article (and a damning exposé by Alex Roland in *Discover* magazine [3]) eviscerated this dream, but the accident helped to refocus the shuttle's purpose as a scientific and engineering craft. The STS would never quite bring holiday space travel to the masses, but it would become something far more important: a versatile and, at times, essential infrastructure for accomplishing less-glamorous scientific and technological goals [4].

When the Space Shuttle began flying in 1981, it was to become all things to all people, especially an orbital scientific laboratory and a low-cost, reusable spaceplane to ferry humans, cargo, and various technologies to and from LEO. Although the shuttle never met all of the expectations placed upon it, it met many of these challenges, occasionally surprising its crews and the larger public with both its technological capabilities and its ability to inspire. In the Space Shuttle's 30-year flight history, it demonstrated ample capabilities as a tool of scientific research, a platform for satellite deployment and recovery, and an element in larger infrastructure of exploration. Although the United States had other options for both satellite delivery and human spaceflight in the 1980s, the Space Shuttle provided both capabilities at a time of decreased spending and reduced urgency from the peak of the 1960s space race. In its 135 launches, the shuttle orbited more humans than any vehicle before or since, and it underwent continual improvement in thermal protection, escape equipment, computing, cockpit instrumentation, and other systems [5]. Even considering the Space Shuttle program's two fatal accidents, it is questionable whether any combination of expendable launch vehicles could have accomplished the shuttle's diverse achievements at a lower cost. Some of these—like the on-orbit repair of the Hubble Space Telescope (HST)—would have likely been impossible.

The creation of the Space Shuttle seemed to promise the fulfillment of NASA's mandate when Congress established it in 1958: to acquire and disseminate knowledge about space. It was designed and constructed specifically to carry larger numbers of scientists than any earlier vehicle could accommodate. Changes "in modes of flight and reentry" associated with the Space Shuttle, President Richard Nixon announced in 1972, would "make the ride safer, and less demanding for the passengers, so that men and women with work to do in space can 'commute' aloft, without having to spend years in training for the skills and rigors of old-style space flight" [6]. NASA redesignated its existing scientist-astronauts as "mission specialists" and recruited additional personnel to serve in a scientific role aboard the shuttle. Joining them would be "Payload Specialists": noncareer space passengers who were drawn from science and engineering communities and who flew in support of scientific, commercial, or military payloads and mission objectives. The shuttle would also offer unprecedented opportunities to fly individuals who had fared poorly in previous pilot-astronaut selections, including women, minorities, and international astronaut-hopefuls.

The shuttle, though, would not merely put scientists in space; it would allow them to bring into orbit—in the shuttle orbiters' crew spaces and voluminous cargo bays—everything from laboratories to standalone satellites previously hauled into space on expensive, expendable vehicles. Some missions would fly a specialized laboratory compartment (the European Space Agency's Spacelab) that was nestled in the shuttle's cavernous cargo bay to provide laboratory space and equipment for specialist crews conducting scientific research. The same vehicle that supported exploration and scientific goals would pay for these efforts by hauling into orbit commercial and military satellites formerly carried on other launch vehicles. By 1972, communications, weather, and other unmanned satellites had become, as President Nixon described them, "irreplaceable." As long as expendable launch vehicles remained the singular means for hurling them into space, though, launching such craft would require "special effort and staggering expense." In the same way that the shuttle would bring scientists aboard a craft traditionally ill suited to their work, the Space Shuttle would "take the astronomical costs out of astronautics" by piggybacking satellite launches on top of the human spaceflight program.

These goals, though, competed with other pressing requirements, especially that the shuttle serve as the principal means by which the DoD would accomplish its own objectives in space. That NASA was permitted to build the shuttle at all was principally because of promises that the craft would not only fulfill civilian space goals but also provide an economical means of supporting military space operations, including the launch of reconnaissance satellites. Unwilling to simultaneously support civilian and military space programs, President Nixon approved shuttle development only on the condition that NASA lend its vehicles periodically to military crews for national security missions, especially the deployment of military satellites, which were projected to constitute nearly one-third of America's space traffic in the decades that would follow [7]. This additional role demanded changes to early shuttle proposals and compromised the craft's utility as a scientific and commercial vehicle for years to come.

The shuttle would need to be reengineered for military compatibility, because military satellites needed to be deposited into orbits different than those required for most civilian spacecraft. Launched from eastward, shuttles enjoyed the advantage of the Earth's rotation to send them into orbit near the equator. Launched southward, though, satellites would overfly the entire globe in a single day, a trajectory requiring great lifting power but offering such satellites a daily view of the entire globe, essential for military photographic reconnaissance platforms. The shuttle would be capable of both launch modes, utilizing the facility at NASA's KSC in Florida for equatorial launches and a new California site for polar launches. Each site was chosen to minimize the exposure of populated areas to potentially deadly vehicle malfunctions.

Some military missions were to last only a single orbit, with the shuttle returning to the vicinity of the launch site after completing a satellite deployment. The Earth, revolving beneath the shuttle, though, would not cooperate;

by the time the shuttle returned to its launch latitude, the landing site would have moved 1500 miles away. This military design requirement eventually found its way into a large-winged vehicle with adequate "cross-range" to accommodate military flights, with the added weight and attendant structural and thermodynamic problems [8]. Although the loss of *Challenger* in 1986 and *Columbia* in 2003 undermined confidence in the shuttle as a military and civilian delivery vehicle, the shuttle succeeded in demonstrating most of the on-orbit capabilities its advocates had predicted.

The proposed STS achieved a major milestone upon its lift-off on April 12, 1981, flying piloted on its very first launch, a feat never before attempted in the U.S. space program (but one that was accompanied by significant delays). In keeping with NASA tradition, command of Space Shuttle *Columbia*'s 1981 first flight, STS-1, went to its most senior astronaut, Apollo veteran John Young, joined by pilot Robert Crippen, who transferred to NASA after participating in a DoD space program. Veteran spacecraft commanders had flown the lead mission of every new vehicle since 1965: Gemini (Grissom), Apollo (Schirra), Apollo lunar landing (Armstrong), Skylab (Conrad), and ASTP (Stafford). Young, the veteran of four previous flights, was the most experienced NASA astronaut on active duty, having flown first in 1965 and having commanded Apollo 16 in 1972. This skeleton crew would verify the flying characteristics of the orbiter and, critically, open its cargo bay doors, a procedure that was essential for the craft to be of practical value and necessary to ensure the vehicle's thermodynamic stability. Although undeniably successful, the flight provided an omen of future problems.

The discovery of damaged and missing thermal tiles on *Columbia*'s maneuvering pods provided the crew with the first surprise of the mission; without communications with Earth for up to one-quarter of their time in space, the crew waited until they were within range of a ground station to share the news [9]. Young and Crippen resigned themselves to their inability to repair them, or, indeed, to replace any tiles that might have fallen off the more critical spaces beneath the airframe. "...I know there was a lot of consternation going on the ground about it about...are the tiles really there...but there wasn't much that we could do about it if they were gone," Crippen recounted [10]. In fact, NASA had worried so much about *Columbia*'s underside that a DoD photographic reconnaissance satellite photographed it in orbit to verify that *Columbia*'s TPS was intact [11].

A similar test flight, STS-2, occurred in November 1981, establishing *Columbia*'s reusability (albeit with substantial post-flight maintenance) and the operation of the shuttle robot arm, the Canadian-built Remote Manipulator System (RMS). Two further test flights followed in 1982, including STS-4, which flew a classified military payload that failed to operate properly. Despite the shuttle's large seating capacity, the first four shuttle missions flew only the two-person flight crew necessary to pilot the vehicle [12]. NASA had expected that the first years of the shuttle fleet's operation would carry below-capacity crews and a launch schedule allowing only a few flights per year. Once it had proven itself, though, routine operations would pack the shuttle's flight decks to capacity with crews that might launch every two weeks. Although *Columbia* demonstrated its ability to fly repeatedly during 1982, minor anomalies throughout these flights indicated that shuttle orbiters would require modification to ensure safety and that post-flight repairs might be substantial. Yet following these largely successful missions, NASA declared the Space Shuttle operational, beginning an extensive series of flights by *Columbia*, *Challenger*, *Discovery*, and *Atlantis* that deployed and recovered several satellites and tested a variety of new technologies that emphasized the shuttle's exotic capabilities.

1982–1986: IMAGE AND REALITY FOR NASA'S "SPACE TRUCK"

The STS's early operational period constituted a critical era in spaceflight despite its brevity. Having seemingly demonstrated its value as both a scientific tool and a satellite delivery vehicle, the shuttle flew with increasing frequency amid expectations that routine, inexpensive access to space—and even a kind of democratization of the NASA astronaut corps—was just over the horizon. A 1983 NASA film documentary titled "We Deliver" chronicled the achievements of this fleeting period, describing successful satellite deployments, EVA, pharmaceutical research, materials processing experiments, and the orbiting of America's first woman and African-American astronauts.

Anticipating a rapid increase in the number of shuttle flights, NASA Administrator James Beggs issued a release on October 22, 1982. It announced that NASA would soon offer crew seats "concerning future flight crews on a reimbursable basis to all classes of Space Shuttle major payload customers, including foreign and domestic commercial customers, international cooperative partners, the scientific and applications community, and the Department of Defense" [13]. Beggs had explained this decision earlier that month in a letter to Representative Don Fuqua, chairman of the Committee on Science and Technology. Beggs wrote that the new policy would be accompanied by invitations to fly more scientists and military personnel after they participated in a "reduced training program" of "six months or less," half of what had previously considered the normal time. Flights "by U.S. citizens as passengers," Beggs hinted, were already under discussion [14]. In time, several nonastronauts, including foreign citizens and two U.S. legislators, joined early shuttle crews.

In the meantime, launch and deployment activities began in earnest, with scientific operations occurring alongside anxiously anticipated engineering goals. On STS-5, the first operational shuttle flight, *Columbia* flew the first NASA scientist-astronauts since 1974, Joseph Allen and Bill Lenoir, and deployed two Hughes communication satellites. At a press conference describing the flight, Allen was unable to describe his duties aboard the mission and joked that he was a mere "passenger" whom NASA had placed "in charge of religious activities" aboard the flight, a quip he immediately regretted and one that provoked NASA displeasure and a variety of concerned letters from the public [15]. The success of the flight, though, paved the way for similar missions, including STS-6's deployment of the first Tracking and Data Relay Satellite (TDRS), a system that enabled continuous contact between shuttle crews and ground controllers, [16] and the first launch of Spacelab, on STS-9.

Often, NASA's mission specialists assumed the principal role in EVA, an area in which the shuttle excelled. Previous U.S. spacecraft (Gemini and Apollo) had required a complete depressurization of the vehicle (with other crewmembers donning spacesuits) to facilitate a spacewalk. The shuttle's relatively spacious interior, airlock, and "back porch" (in the form of its cargo bay) presented a nearly ideal environment for EVA preparation and work outside of the shuttle, a task that ultimately became routine. In February 1984, Mission Specialist Bruce McCandless II achieved the first untethered spacewalk in Earth orbit using a cold nitrogen-powered maneuvering unit affixed to his suit (Fig. 8.1). That mission, STS-41B, also saw the deployment of two communications satellites whose upper stages failed, leaving them intact but stranded in orbits too low to be useful. (In 1984, NASA replaced

Fig. 8.1 Mission Specialist Bruce McCandless II hovers over *Challenger* in February 1984. (NASA Photo GPN-2000-001156)

the simple numerical naming scheme it had used for the missions with a number-and-letter code so confusing almost no one outside of NASA could keep track of the flights. NASA restored the old system after the loss of *Challenger*.) Later that year, astronauts James van Hoften and George Nelson attempted to use the new EVA backpack propulsion technology to recover and repair a malfunctioning satellite designed to study solar flares. (The satellite was instead recovered using the RMS.) In November 1984, STS-51A astronauts Allen and Dale Gardner recovered the two stranded communications satellites launched earlier that year and returned them to Earth in the shuttle's cargo bay for refurbishment and later relaunch.

NASA's insistence that the shuttle demonstrate its utility produced an operational tempo filled with both dramatic successes and increasingly unreasonable expectations. STS-51A Commander Rick Hauck, upon hearing that his crew would recover two derelict commercial communications satellites on their mission, was incensed, describing the optimistic prediction as a "f----g miracle," though the crew ultimately succeeded [17]. Subsequent flights might carry a rocket motor attached to the satellite, employing the same liquid hydrogen and oxygen propellants used by the shuttle's main engines at take off. Shuttle designer Max Faget had been concerned about the use of this super cold, highly explosive mixture in the shuttle's payload bay since the 1970s. According to Allen, astronaut David Walker secretly referred to the first such mission as "Death Star One" [18].

In describing the kinds of scientific work the shuttle would perform, NASA spokespeople often described basic research as well as made elaborate claims of potential commercial applications of space laboratories. However, NASA did not explain either in much detail. Early experiments focused on the effect of the space environment on living tissues: essential knowledge for long-duration spaceflight but of limited application on Earth. The weightlessness of Earth orbit, proponents argued, would eventually enable the growth of high-quality crystals and the production of extremely pure pharmaceuticals, at first a scientific curiosity but later the nucleus of a vast space manufacturing industry [19]. Like the "Space Truck," the "space factory" was an early element of shuttle mythology that offered promise for the future but that could not be realized in the near term [20].

NASA's promises of the shuttle's scientific return were often grandiose, but its flights at first yielded few significant achievements. NASA continued to fulfill a series of engineering challenges. However, scientific activities always piggybacked on flights, sometimes competing for space with commercial and DoD payloads, but occasionally operating alongside them within tight space and mass allowances. Scientific activities aboard early shuttle flights included both the launch of complex apparatus and "Getaway Specials," small experiment canisters nestled in the shuttle's cargo bay and filled with simple experiments often created by high school and college students. Some of these Getaway Specials cost their sponsors as little as $3000 to fly, but their experiments tended toward the inspirational rather than the transformational, studying the effects of the space environment on organisms and materials. Apparatus failures were frequent [21]. One ill-fated experiment involving an ant farm was mocked mercilessly on a 1996 episode of the long-running animated series *The Simpsons*.

For astronaut crews, the shuttle increasingly became a claustrophobic workplace where commercial, scientific, and military space activities coexisted, often uneasily, under the constant monitoring of ground controllers. Crammed with seven people with their own job responsibilities (eight people, on two occasions), all sharing a single bedroom, galley, bathroom, and office, shuttle orbiters of the early 1980s presented their crews with a confining, cacophonous workplace made all the more uncomfortable by the presence of nonprofessionals. The eight crewmembers of 1985's STS-61A mission shared four bunks (with one bunk reserved for the commander); one payload specialist slept in the airlock, conjuring up a murder scenario common in popular film [22]. Irritation, though, not murder, was the major problem [23].

The few Apollo scientist-astronauts remaining in NASA at the start of the Space Shuttle program were valued crewmembers, but although newer mission specialists were often assigned critical mission roles (especially EVA), in some cases, scientists joined crews almost as an afterthought and were not permitted free access to all of the shuttle's crew spaces. The shuttle's commander and pilot monopolized the flight deck, which housed most of the vehicle's control systems but also offered the best views. Down below, mission specialists floated in a windowless white room and conducted experiments or attended to housekeeping activities. With certain exceptions, pilot-astronauts responded to scientists aboard the craft with skepticism; never having faced life-threatening situations or achieving proficiency as jet pilots, mission and payload specialists could not be relied upon to handle contingencies in orbit [24]. On Gordon Fullerton's 1985 *Challenger* flight, NASA scientist-astronaut Story Musgrave provided Fullerton with a respected third member of the flight crew who could assist in monitoring systems during launch and reentry. However, the other four crewmembers (two mission specialists and two payload specialists) were regarded as hired hands working in staggered 12-hour schedules to monitor experiments and other in-flight operations [25].

To Fullerton, seven crewmembers meant "as many opportunities for somebody to blow it," mishandling equipment or throwing the wrong switches [26]. The presence of relatively inexperienced payload specialists was especially aggravating to the flight crew, who regarded them as unfit for spaceflight. Occasionally, payload specialists impressed their commanders, but mostly by keeping quiet. "I worked hard to integrate [payload specialist] Charlie [Walker] into the crew," Hartsfield recounted. "He was a good student. He was going into a strange environment and he wanted to learn, and he gave it his full attention." Another payload specialist, though, obsessed that the hatch on the mid-deck opened with the turn of a lever. "'You mean all I got to do is turn that handle and the hatch opens and all the air goes out'?" the man fretted. "It was kind of scary," Hartsfield recalled. "Why did he keep asking about that?" The concerned crew eventually locked the hatch to allay his fears [27].

The plethora of payload specialists aboard shuttle flights and constantly changing launch manifests and schedules confused career astronauts; at one point, crowded mission patches began to place payload specialists' names on ribbons attached to the crest (so they might be, imagined one astronaut, more easily removed at a later time) [28]. In the meantime, though, the shuttle's seemingly flawless safety record, and increasing flight of nonprofessional astronauts, suggested that the unique status enjoyed by NASA's career astronauts had ended. In December 1985, Deputy Assistant for National Security Affairs John Poindexter, writing on behalf of President Reagan, announced that President Johnson's policy of automatically promoting all military and civilian astronauts upon their first space flight would end. "Since space flight to low earth orbit is becoming less experimental, more frequent and more normalized, the policy to promote military and civilian astronauts for space flight is rescinded" [29]. Space travel had finally become safe.

More-frequent flights would resolve the competing priorities of NASA's constituencies, but by 1984, NASA was managing only five shuttle flights a year, one-fifth of its target rate and one-twelfth the rate shuttle advocates of the 1970s had promised [30]. Commercial obligations dominated shuttle manifests, placing a strain on crewmembers and ground scientists eager to fly delicate apparatus on the soonest available flight. Moreover, as an operational vehicle, the shuttle proved increasingly expensive to maintain and fly [31]. In 2003, the CAIB determined that the decision to certify the shuttle for operational use had been hasty given the small number of flights to date, the physical differences between orbiters, and their continual modification to suit safety and mission needs [32]. Yet as late as 1985, many outside of NASA still clung to the hope that the shuttle could meet most of the expectations placed upon it.

The landing strip at KSC (and an additional launch facility at Vandenberg Air Force Base) would, in theory, ensure rapid rotation of newly returned vehicles and a schedule that might permit every member of the astronaut corps (of between 130 and 190 men and women) the opportunity to fly at least once a year [33]. Yet the landing strip at Kennedy (a cost-saving measure that reduced turnaround time and transportation costs in flying newly returned orbiters cross-country to Florida from the primary landing strip in California) initially proved unpopular with shuttle astronauts. The runway at Edwards Air Force Base had been built upon a dry lakebed that could accommodate emergency landings; though of ample length to support shuttle landings, the strip at KSC was surrounded by swamps that would doom the shuttle if landing problems arose. "I often joke," Crippen noted, "that they've got a fifteen-thousand-foot runway, but they built this moat around it and filled it full of alligators to give you an incentive to stay on the runway" [34]. Manifold issues like these slowed vehicle turnaround and forced an accelerated maintenance schedule for the few flights NASA could launch in time. "[T]he whole system was starting to crater," Hartsfield later recounted, with scheduled flights hurriedly prepared so that NASA could meet its self-imposed guidelines [35].

DoD requirements added to these challenges. Early shuttle missions flew a variety of classified payloads, both onboard experiments (and, likely, a variety of communications and reconnaissance satellites) aboard six civilian flights and three dedicated military flights before 1986 [36]. Occasionally, the Shuttle program's uneasy relationship with the DoD led to operational problems. Missions earmarked for the deployment of classified military satellites interrupted NASA's flight schedule and replaced its regular crews with active-duty military personnel. NASA had tolerated DoD involvement in the Shuttle program out of necessity but still regarded with suspicion any flight personnel that were not career astronauts, regardless of their military pedigree. DoD-sponsored payload specialists dubbed "Manned Spaceflight Engineers" (MSEs) received little enthusiasm from NASA; of the 13 MSEs selected, only 1 flew before the MSE program was cancelled [37].

For shuttle crews, the additional security of military missions was a "burden," restricting the normally open communication flows that had long characterized the space program. Inside, the shuttle was reconfigured for defense work, with checklists encoded and documents maintained in a locked storage box on the shuttle. (Once safely in orbit and convinced no one would likely steal the sensitive documents, the flight crew removed the lock.) [38] Security rules prevented communications between shuttle crewmembers and principal investigators on the ground, for fear of tipping them off to the fact that their experiments were being flown on the mission [39]. The shuttle's proposed polar launch plans produced additional technical hiccups. As the shuttle's gross weight begin to climb, modifications to the shuttle's SRBs were required to enable the spacecraft to achieve the necessary velocity.

Despite the growing technical difficulties and the professional frustrations of its astronauts, NASA, through the mid-1980s, proceeded with its plans to expand the human spaceflight program to include relatively untrained civilian crewmembers. The Space Shuttle's large internal volume, comfortable environment, and comparatively gentle reentry profile would expose its passengers to pressures and forces any reasonably healthy person could endure without ill effect, although many would succumb to space adaptation syndrome (space sickness) during the first day or so in orbit. Like promises that the shuttle would visit an as-yet-unbuilt space station or retrieve and repair future satellites, the claim that the shuttle would carry everyday people was never fully clarified. The odds than any average citizen would fly one of NASA's four shuttle orbiters were astronomical, because even a "passenger" would require months of full-time training in basic spaceflight mechanics and emergency procedures before even being allowed near the vehicle. Although NASA hoped for as many at 20 or more launches per year, few crew positions were likely to be earmarked for essentially useless personnel when nearly 100 trained astronauts waited for a flight.

By the end of 1985, unable to achieve the economies of scale sought, the shuttle looked less and less like an efficient commercial satellite launcher than a contested workplace filled with complex compromises, despite heavy efforts by NASA to promote it [40]. By the end of 1985, NASA was still struggling to make 12 flights a year, with no indication that the tempo of operations would increase anytime soon [41]. Rather than ushering the routinization of spaceflight, *Challenger*'s first flight of 1986 ended in tragedy, claiming the lives of six NASA career astronauts and the first private citizen-astronaut, teacher Christa McAuliffe. Convinced of the safety of their vehicle, the crew had launched without wearing pressure suits; although the suits would not have saved their lives, this fact brought into sharp relief the discrepancy between what the shuttle was and what many wished it could be. The loss of *Challenger* ended this brief practice; it also brought an end to most commercial satellite delivery by the Space Shuttle, the launch of cryogenic upper stages aboard it, and, eventually, most DoD participation.

1988–1998: From "Space Truck" to Space Station

The loss of *Challenger* 73 seconds after liftoff on January 28, 1986, confirmed what shuttle critics inside and outside of NASA had feared: that the agency could not maintain its tempo of operations while maintaining acceptable margins of safety. More than that, though, it caused a re-examination of tradeoffs that had been made in the design and fabrication of the STS and many of the rationales that had sustained its use. In an instant, the notion of filling crew seats with private citizens (and even payload specialists) became unthinkable. Already concerned about the shuttle's reliability, the DoD continued an effort begun in 1983 to shift military payloads to a new generation of expendable vehicles [42]. (At first, though, military missions occupied several post-*Challenger* shuttle flights [43].) NASA, meanwhile, banned from the shuttle commercial payloads that could be carried on other launch vehicles, undermining one of the shuttle's principal benefits but in the process clearing the program's cargo manifest for scientific activities [44].

The loss of *Challenger* may have brought the story of NASA's space truck to a close, but it brought a reconfiguration of using the STS that ultimately led to the shuttle's most productive years in service. Flights before 1986 constitute only a minor fraction of missions the orbiters flew in their 30-year service life; though the loss of *Challenger* may have punctured the Space Shuttle's reason for its existence (as well as NASA's decades-old "aura of competence" [45]), it was only after the tragedy that STS flights actually became routine. Within a decade, flights by the three remaining orbiters (*Columbia*, *Discovery*, and *Atlantis*, eventually supplemented by *Endeavour*) were not only carrying Spacelab (and its successor, the privately developed SpaceHab modules) into orbit but also deploying and retrieving satellites, and visiting and servicing a space station in a manner close to that which its designers had envisioned. Shuttle flights increased in frequency and became increasingly ambitious [46]; public support for the space program, which had been in decline before the loss of *Challenger*, briefly rose [47]. The phase-out of noncareer astronaut crewmembers after *Challenger* resolved prior labor issues aboard the craft, and although astronauts would never again launch into space unprotected by pressure suits, the vehicles were considered safe enough that 77-year-old former astronaut and Senator John Glenn was permitted to fly aboard a 1998 *Discovery* flight, ostensibly as a biomedical test subject [48].

As a tool of laboratory science, the Space Shuttle continued to be an imperfect vehicle after *Challenger*, though it offered robust capabilities not found on other spacecraft. Routine flights by mission specialists resumed, and pressurized Spacelab and SpaceHab modules greatly increased the internal volume of shuttle crew spaces and enabled the carriage of palletized experiments for microgravity research, astronomy, and other scientific fields. (By 2007, the number of life sciences experiments flown had exceeded 2000 [49].) Yet unlike with research on the ground, shuttle experiments were limited to the craft's typical 14-day mission length and had to be prepared well in advance of flight. Damage or errors that might result in an experiment being repeated on Earth brought complete failure in space. Constrained by an externally driven launch schedule planned years in advance, one shuttle astronaut noted that even successful experiments were seldom repeated, despite the low cost involved, to make room for other payloads [50]. Despite its multibillion dollar cost, Spacelab pressurized modules, for example, flew relatively infrequently, though experiment pallets received somewhat more use [51].

With launch schedules planned well in advance and shuttle orbiters called upon to satisfy engineering objectives, laboratory scientists still often found their interests accorded a status secondary to satellite deployment missions. Carrying and deploying heavy scientific payloads, though, provided the STS program with its most dramatic scientific moments and ample opportunities for international cooperation. These brought tangible success to NASA and enabled the completion of big-budget programs that required launch capabilities only the shuttle (and a small number of expendable launch vehicles) could provide. Upon the Space Shuttle's return to flight in 1988, it began a successful series of flights that demonstrated its orbital delivery and retrieval

Fig. 8.2 Mission Specialist Story Musgrave, anchored to *Endeavour*'s RMS, prepares to service the HST in December 1993. (NASA Photo GPN-2000-001085)

capabilities in a scientific context; foremost among these achievements was the launch of the HST in 1990 and its subsequent retrieval, repair, and servicing during five additional flights from 1993 to 2009 (Fig. 8.2).

Among the oldest concepts for an orbiting scientific platform, the craft that became the HST was the subject of aggressive lobbying from the start of the Space Age; by positioning telescope optics outside Earth's atmosphere, astronomers hoped to achieve multispectral imagery of unprecedented sensitivity and clarity. This information would enable the study of well-known objects, the investigation of distant and dim deep-sky objects, and the study of light from the deep past, which would provide information on the origins of the universe. Including a range of cameras in the vehicle's design (including a planetary camera some thought useful only for public relations purposes) also enabled spectacular images of planets, nebulae, and other interstellar objects, and these pictures periodically renewed public interest in space exploration [52].

Developing alongside the Space Shuttle, the Hubble was designed to be launched and serviced by it, a controversial decision that enabled flexibility and longer operating life but tethered the satellite's future to continued shuttle operations, an issue that proved critical as the shuttle program wound down [53]. Later shuttle flights not only enabled the installation of corrective optics to compensate for deficiencies in the telescope's fabrication, but also upgraded or replaced key sensory and navigational equipment, extending both the operation life of the platform and its capabilities. It is doubtful that any craft other than the shuttle orbiters could have successfully captured the Hubble in orbit and conveyed both the equipment and personnel necessary to repair it.

Although many HSTs could have been funded with the shuttle's budget, it is unclear whether Congress would have allocated funds for the flights.

During this period, additional shuttle flights launched the Compton Gamma Ray (1991) and the Chandra X-Ray (1999) observatories into Earth orbit, as well as a number of deep-space probes. Magellan, launched in 1989, mapped the surface of Venus before undertaking a controlled entry into the Venusian atmosphere. Galileo, launched the same year, became the first spacecraft to orbit Jupiter, and it dispatched a probe into the planet's dense cloud layers. Ulysses, launched in 1990, utilized a Jupiter-fly-by to catapult itself into solar orbit out-of-plane with the Solar System, enabling examination of the sun from all latitudes. These missions were widely heralded as having made tremendous contributions to planetary science.

Increasingly regular shuttle flights in this period also enabled the launch and recovery of spacecraft intended to provide data about long-term exposure to the space environment. *Challenger*'s deployment of the Long Duration Exposure Facility (LDEF) in 1984 and *Columbia*'s retrieval of the satellite in 1990 provided unique information regarding the orbital debris and micrometeoroid environment of LEO as well as over 50 additional experiments. Never before had scientists been able to observe in detail a craft that had spent nearly six years in orbit and returned intact. Further experiments in 1996 involved the launch and release of the Orbiting and Retrievable Far and Extreme Ultraviolet Spectrometer-Shuttle Pallet Satellite II (ORFEUS-SPAS II) pallet, which contained ultraviolet spectrometers to study stars and the interstellar environment.

Although the return of this craft and the experiments routinely carried aboard shuttle flights generated a large quantity of scientific data, shuttle operations remained expensive through the end of the program. Each flight cost an estimated $500 million (comparable to the cost of Apollo-era launches), and payload rates into LEO hovered at approximately $10,000 per pound, more than 10 times the program's most optimistic forecasts had predicted (though still comparable to the cost of expendable launch vehicles). Although every shuttle crew conducted scientific research, and several missions were entirely devoted to scientific objectives, the peak years of shuttle activity produced no research that the mainstream scientific community regarded as transformative or Nobel Prize–worthy. Shuttle defenders, though, were always quick to point out that because sustained weightlessness is impossible to achieve on Earth, the shuttle represented a unique platform for research whose value would inevitably be demonstrated [54].

1998–2011: Toward the ISS

Throughout the late 1990s and early 2000s, the shuttle continued to demonstrate its ability to orbit large scientific payloads, completing a high-resolution radar mapping survey in 2000 and orbiting several other noteworthy instruments. At the same time, political transformations enabled the use of the shuttle in yet another capacity. In 1995, *Atlantis* fulfilled a key hope of shuttle proponents when it docked with the Russian Mir space station (Fig. 8.3). The first of nine American flights to the station, it not only demonstrated the shuttle's potential as a means by which to convey astronauts from Earth to orbiting space laboratories (one of the shuttle's original purposes) but the first joint U.S.-Russian space mission in two decades.

Shuttle visits to Mir—including crew exchanges and long-term stays—provided useful lessons in the design and operations of large space stations, but astronauts who visited it recalled a workplace fraught with frequent maintenance and safety issues, including incidents in 1997 involving uncontrolled combustion and a hull breach. The principal accomplishment of the joint program was political rather than technical: demonstrating the feasibility of international cooperation and ensuring the survival of the Russian space program through an influx of American funding [55]. This support discouraged the departure of Russian space engineers, potentially to developing nations seeking missile technology. On balance, though, the experiment provided NASA with its only experience in conducting long-duration space flights with an international partner, and provided an environment for American astronauts to stay in space significantly longer than the two-week missions customary on shuttle flights. Such missions yielded valuable data regarding the adaptation of human physiology to the space environment, a critical concern for interplanetary space flight [56].

Fig. 8.3 ***Atlantis*** **docked for the first time with Russia's Mir space station during 1995's STS-71 mission. The assembled spacecraft was the largest ever flown at the time. (NASA Photo GPN-2000-001315)**

Successful crew exchanges paved the way for a reconfiguration of previous American and Russian space station programs into a joint station, created in coordination with a variety of international partners (including Canada and Japan). This station—the ISS—would be visited and serviced by American, Russian, European, and, eventually, privately owned spacecraft. The construction of the ISS dominated the shuttle's final decade because the shuttle was called upon to haul into orbit its constituent modules, as well as crewmembers and supplies. The loss of *Columbia* in February at the conclusion of the STS-107 flight jeopardized the completion of the station, but the shuttle's central role in assembling and servicing it ensured the continued operation of the STS until the completion of its 135th flight in July 2011—a mission to deliver essential equipment to the ISS.

USING THE SHUTTLE: LESSONS LEARNED

The loss of *Challenger* in January 1986 ended the optimism that accompanied the STS's development and earliest flights. Before the ill-fated launch, Commander Dick Scobee confronted Teacher-in-Space Christa McAuliffe with this comment: "'[N]o matter what happens on the mission, it's going to be known as the teacher mission,'" a statement that intended to comfort her. However, it instead reflected Scobee's recognition that the public cared little for the difficult work of the flight, which would include astronomical experiments and a satellite deployment. It was in the accomplishment of seemingly mundane (but deceptively challenging) duties that the Space Shuttle excelled, though for crews of the vehicle, such work often brought professional frustration.

More a technology demonstrator than a craft capable of routine operation, the shuttle proved the theoretical capabilities of a large, reusable spacecraft, but its cargo bay often became an albatross to a vision of spaceflight to which the United States was unwilling to fully commit. NASA spent the better part of the next three decades attempting to fill the hole it had created, with commercial satellites (endangering the nation's space heroes with delivery duties more safely undertaken by expendable rockets); foreigners (in the form of dignitaries and payload specialists); scientists (in the Spacelab and SpaceHab modules nestled in the shuttle's cargo bay); reconnaissance satellites (stewarded into space by all-military crews that compromised NASA's peaceful façade); celebrities; journalists; and private citizens.

Flying the Space Shuttle, NASA's newest astronauts learned, was unlike the experience of astronauts flying two decades earlier, when short flights with modest goals made celebrities out of people very much like themselves. As complications slowed shuttle launch schedules, NASA continued to add astronauts almost every other year. For astronauts, particularly the dozens of anonymous new mission specialists, the wait for flight opportunities again became oppressive, as rosters filled and astronauts found their enthusiasm for spaceflight never completely satiated. With the shuttle able to hold a maximum of eight crewmembers, NASA made a point of filling seven seats on virtually every mission, crowding the vehicle and, to the chagrin of astronauts, overtaxing each shuttle's lone toilet [57].

Ultimately, through experimentation aboard shuttle orbiters and Spacelab by international scientific crews, the deployment and maintenance of critical payloads like the HST, and the support of the scientific work aboard Mir and the ISS, the Space Shuttle demonstrated the potential of a reusable orbital spaceplane as a tool for producing scientific knowledge, as an orbital cargo vehicle, and as a platform for assembling large structures in space. Although commentators have weighed the merits of the research produced by shuttle and ISS crews, little doubt exists as to the scientific value produced by a series of satellites launched into space by the shuttle, or its capacity to service the ISS. Although unable to meet all of its operational goals, the shuttle proved a hearty "first step" in the creation of a reusable space vehicle: a development still regarded as inevitable in the creation of future space systems.

USING THE SHUTTLE: ALTERNATIVES

Evaluations of the Space Shuttle inevitably turn toward analyses of what alternative paths NASA might have taken in pursuit of its spaceflight goals. In 1972, Project Apollo's enthusiasts still saw NASA's moon landings as the first step in an accelerating sequence of explorations that would combine moon and Mars bases, interplanetary tugs, and Earth orbital shuttles. Others within NASA saw the entire human spaceflight program so threatened financially in 1972 that only a radical restructuring—aimed at reducing the cost of access to space beyond that which Apollo could achieve—could save it.

As the most versatile element in an elaborate interplanetary space infrastructure NASA planned for the 1970s and 1980s, the Space Shuttle proved the element least vulnerable to budget cuts, even if that left it as only a small portion of a larger program of human space exploration. Comparing the Space Shuttle program to the most optimistic predictions of interplanetary travel, therefore, is inappropriate; the craft needed for such flights would have never been funded in the austere budget climate of the 1970s. Instead, as Administrator Fletcher explained in 1972, "the Space Shuttle is important and is the right step in manned space flight and the U.S. space program" [62], because it was "the only meaningful new manned space program which can be accomplished on a modest budget," and the only one likely to accomplish America's diverse space goals.

Envisioning a world without the Space Shuttle requires imagining not merely the expendable launch vehicles or piloted craft that might have operated in its place, but a range of possibilities, each with its own engineering or cost advantages but carrying substantial disadvantages as well. One commonly articulated position throughout the shuttle's program life (advanced by pioneering space scientist James Van Allen, among others) held that scientific benefits were insufficient to justify its development [58]. With the Moon Race won, human spaceflight had fulfilled its founding mandate and could be ended without loss of national prestige or technological capability; ending the human spaceflight program after Apollo would have eliminated NASA's most expensive budgetary category and enabled the construction of fleets of robotic craft and expendable launch vehicles, potentially duplicating the shuttle's scientific achievements at greatly reduced cost.

Yet although the shuttle proved approximately 20 times more expensive to operate than originally predicted, it was not more expensive than the expendable craft that might have replaced it, and when the cost is adjusted for inflation, NASA largely met its projections for the cost of its development. Historian and former NASA Chief Historian Roger Launius notes that cost per pound of payload for the Apollo-era Saturn launch vehicles actually exceeded that for the Space Shuttle, contrary to popular belief and published estimates [59]. Proponents of human spaceflight have always argued, in addition, that when it comes to space exploration, only "dramatic results" (that is, human exploration) served to advance the nation's political interests, and that the shuttle served an undeniable prestige function, particularly during its earliest years of flight [60]. Ultimately, the bipartisan political will that was required to terminate human spaceflight proved more difficult to conjure than that required to allow it merely to continue with more modest funding [61].

In the absence of the STS, American human spaceflight might have continued much as the Soviet (and later, Russian) program did, using 1960s flight hardware and capsule-based flight modes. For the United States, this would have meant extending the Saturn launch vehicle production line and continuing to rely on the three-person Apollo Command and Service Module (CSM)—a robust craft but one that afforded little room for additional crewmembers and placed great demands on their piloting abilities. Like the Soviet Union, the United States would have leveraged its existing arsenal of launch vehicles—principally the Saturn V—to orbit a second Skylab Orbital Workshop and possible follow-ons, servicing the space stations with Apollo launches using either the increasingly obsolete Saturn IB boosters or a similar vehicle based around the second and third stages of the Saturn V [62].

Such an infrastructure would have offered capabilities for delivering payloads into orbit and for rendezvous with other spacecraft, but virtually none for on-orbit repair and none for retrieval. In theory, Apollo CSMs could have supported deep space operations but in a cramped environment with minimum allowance for consumables and minimal protection against micrometeor and radiation hazards. One cannot assume, as Launius notes, that funding earmarked for shuttle development would have found its way into interplanetary spaceflight projects lacking even the basic military or commercial utility of shuttle missions [63]. Launch frequency would have been limited by the budgetary allotments for fabricating disposable launch vehicles and spacecraft, as well as the cost of launch, orbital operations, and recovery at sea, using a large and expensive naval armada. Judging by the Apollo launch schedule, two launches per year might have been typical.

Skylab's scientific crew was limited by the operational constraints of the Apollo ferry vehicles and NASA's aversion to flying professional scientists in crew positions normally held by test pilots. Without the Space Shuttle, NASA's astronaut corps would have remained small and comprised entirely of pilots, though many (like NASA's scientist-astronauts of the late-1960s) would have had substantial scientific training before joining NASA. As predicted, the Space Shuttle—a multipurpose vehicle with seats to spare on every flight—drove an unprecedented expansion in the astronaut ranks; the sheer number of vehicles, crew positions, and flights expected suggested a need for both pilots and researchers who could make use of the shuttles' surplus capacity. In the absence of the shuttle and its accommodations for a predominantly nonpilot crew, scientists would have found relatively fewer flying opportunities aboard American spacecraft. Flights by tourists or other nonprofessional astronauts would have been close to unthinkable.

Abandoning human spaceflight, developing the shuttle, and continuing Apollo were not, of course, the only options available to NASA in 1972. Throughout 1971, NASA, the DoD, and OMB considered alternative shuttle proposals, both for larger, more reusable, and more complex craft and for simpler vehicles that, as OMB suggested, would offer the United States the same benefits as a larger shuttle—including a public relations victory and the maintenance of a vibrant national space capability—at greatly reduced cost [64]. More elaborate shuttle designs ideas included air-breathing engines for the shuttle orbiters and a plan to loft the vehicle into the upper atmosphere on top of a larger, piloted, reusable rocket plane [65]. Of course, a massive, more expensive, and more sophisticated craft—a craft with rocket and jet engines; a piloted, fully reusable, liquid-fueled flyback booster—might have proven too difficult or expensive to construct. Previous space projects—the X-20 Dyna-Soar, the Manned Orbiting Laboratory, the National Aerospace Plane, and the X-33 Venture Star—met a similar fate, suffering cancellation before their first flight. Although likely robust, the larger and more sophisticated Space Shuttle alternatives may not have produced lower lifecycle costs than the more modest craft ultimately selected. They would have strained a spaceflight program that, even using a more modest craft, proved overambitious within a few short years, according to former NASA Administrator Mike Griffin [66]. In the case of the Space Shuttle, with the Apollo/Saturn production lines already long terminated by 1981, an early failure of an expensive, reusable, double-flyback launch vehicle would have left the United States to navigate the final years of the Cold War without any viable platform for human space exploration, and no desire to fund one.

Space historian and CAIB Member John M. Logsdon suggested in a 2011 evaluation of the shuttle program that an even simpler or less capable shuttle design than the one chosen might have provided an "evolutionary"—rather than a revolutionary—approach, with attendant reduction in expectations for the system. On balance, Logsdon suggests, the trade-off would have been a wise one, encouraging NASA to move more quickly beyond the original Space Shuttle architecture once its high costs became apparent [67]. Instead, NASA remained committed to its only launch vehicle large enough to support construction of the ISS, accomplishing that goal but delaying the development of a suitable replacement craft. A smaller shuttle, though, would have been unable to orbit the same number of astronauts or lift the most important scientific payloads into orbit, especially the Hubble, which closely matched in length the reconnaissance satellites the Air Force hoped the shuttle would launch into space, and for which it had sought a 60-foot payload bay.

Further questions arise in connection with NASA's decision to continue using the STS even after its costs became apparent. During the 1980s and 1990s, NASA examined a variety of options to leverage STS hardware to ameliorate the cost and safety issues associated with using the shuttle as an all-purpose launch vehicle. One such proposal—"Shuttle-C"—would have stripped the shuttle orbiters of wings and crew compartments to create cargo vessels capable of delivering heavy payloads to orbit without human participation [68]. Though resurrected frequently (most recently in 2009), similar shuttle-derived heavy lift vehicles never received funding from NASA. Instead, the agency continued to employ the shuttle for heavy payload delivery for two decades following the *Challenger* disaster, supplementing it with expendable Atlas and Delta launch vehicles. Ironically, the success of a shuttle-derived heavy lift vehicle would have invalidated the shuttle design concept, demonstrating that unpiloted craft were better suited to an orbital delivery role than a piloted, multimission craft and curtailing America's human spaceflight program. With the retirement of the STS in 2011, a shuttle-derived expendable launch vehicle is expected to take its place, though one that replaces the shuttle's side-by-side configuration with a more conservative staged design and capsule-shaped spacecraft.

CONCLUSION

After 1969, American human spaceflight lacked a clear mission and the kind of legislative mandate that would have funded the crash program required to produce a spaceflight architecture as robust as that contemplated by space program advocates [69]. Even a brief period during the mid-1980s, when Soviet space activities triggered brief concerns among some

Congressional representatives that cosmonauts were aiming for Mars, produced no significant human spaceflight initiative. Though able, effectively, to do nothing as well as its designers had intended, the shuttle was, oddly, the perfect craft for the era that produced it. Although the shuttle could not leave LEO, its literal emptiness meant that it could be a vessel in which to contain whatever national aspirations proved pressing: scientific, military, commercial, or popular.

In the Space Shuttle, NASA had vowed to continue a tendentious human spaceflight program but had failed to satisfy both the professional scientific community and the growing chorus of "fringe" thinkers who expected the moon to be a practice run for interplanetary voyages [70]. As the disparity between NASA's actual plans and the public's increasingly unreasonable fantasies grew, a space counterculture emerged in permanent opposition to the government policymakers it regarded as stuffy, unimaginative, and inflexible. Amid such criticisms, though, one may easily overlook the fact that before the flight of the Space Shuttle, no space vehicle had ever launched seven people into orbit for two weeks before gliding to a precision runway landing, let alone over and over again. That the shuttle could do all of these things and still face withering criticism indicates the exaggerated expectations that accompanied the arrival of the space vehicle, and the huge new responsibility NASA had imposed upon itself in promising routine, low-cost access to space. In the end, the Space Shuttle proved itself able to accomplish the most mundane tasks of space operations better than any other platform before or since: It was simultaneously too flawed to excel and too capable to disappear.

ACKNOWLEDGMENTS

The author would like to thank his fellow contributors to this volume for reviewing early drafts of this manuscript and for providing useful and insightful feedback. Suggestions from Wayne Hale, Dennis Jenkins, John Krige, Roger Launius, and John Logsdon were particularly helpful, as were comments from David DeVorkin.

REFERENCES

[1] Eskow, D., "Space Vacation 1995," *Popular Mechanics*, Vol. 162, Issue 11 (November 1985), pp. 59–60.

[2] Williamson, R. A., "Developing the Space Shuttle," *Exploring the Unknown: Selected Documents in the History of the U.S. Civil Space Program, Volume IV: Accessing Space*, edited by J. M. Logsdon, NASA, Washington, DC, 1999, p. 179.

[3] Roland, A., "Triumph or Turkey?" *Discover*, Vol. 6, Issue 11 (November 1985), pp. 14–24.

[4] Jenkins, D. R., *Space Shuttle: The History of the National Space Transportation System, the First 100 Missions,* Specialty Press, North Branch, MN, 2001; Hale, N. W. (ed.),

Wing in Orbit: Scientific and Engineering Legacies of the Space Shuttle, 1971–2010, NASA SP-2010-3409, Washington, DC, 2010; several astronaut memoirs have explored the vehicles' operational history, e.g., Jones, T. D., *Sky Walking: An Astronaut's Memoir,* Collins, New York, 2006.

[5] Rumerman, J. A., *U.S. Human Space Flight: A Record of Achievement, 1961–2006,* NASA, Washington, DC, 2007; also Jenkins, *Space Shuttle*, pp. 374, 375.

[6] NASA History Office, "President Nixon's 1972 Announcement on the Space Shuttle," http://history.nasa.gov/printFriendly/stsnixon.htm (updated: March 30, 2009; accessed: March 26, 2013).

[7] Logsdon, J. M., "The Space Shuttle Program: A Policy Failure?" *Science*, Vol. 132, No. 4754 (1986), pp. 1099–105, 1100.

[8] Allen, J. P., "Oral History Transcript #3 (J. Ross-Nazzal, Interviewer)," *NASA Johnson Space Center Oral History Project*, March 18, 2004), pp. 14–15; Logsdon, J. M., "The Space Shuttle Program: A Policy Failure?" pp. 1100–1101; Mattingly, T. K., II, "Oral History Transcript #2 (K. M. Rusnak, Interviewer)," *NASA Johnson Space Center Oral History Project*, April 22, 2002, p. 10.

[9] Williamson, R. A., "Developing the Space Shuttle," *Exploring the Unknown: Selected Documents in the History of the U.S. Civil Space Program, Volume IV: Accessing Space*, edited by J. M. Logsdon, NASA, Washington, DC, 1999, pp. 177–178.

[10] Crippen, R. L., "Oral History Transcript (R. Wright, Interviewer)," *NASA Johnson Space Center Oral History Project*, May 26, 2006, pp. 26–27.

[11] Allen, J. P., "Oral History Transcript #3 (J. Ross-Nazzal, Interviewer)," *NASA Johnson Space Center Oral History Project*, March 18, 2004, p. 11.

[12] Allen, J. P., "Oral History Transcript #3," p. 32.

[13] Cywanowicz, L., "NASA Expands Payload Specialist Opportunities," *NASA News (Marshall Space Flight Center)*, October 22, 1982, p. 1.

[14] "Letter from James M. Beggs to Don Fuqua," October 7, 1982, Folder 008960, NASA Historical Reference Collection, NASA Headquarters, Washington, DC.

[15] Allen, J. P., "Oral History Transcript #3," p. 32.

[16] Peterson, D. H., "Oral History Transcript (J. Ross-Nazzal, Interviewer)," *NASA Johnson Space Center Oral History Project*, November 14, 2002, p. 63.

[17] Allen, J. P., "Oral History Transcript #4 (J. Ross-Nazzal, Interviewer)," *NASA Johnson Space Center Oral History Project*, November 18, 2004, p. 12.

[18] Allen, J. P., "Oral History Transcript #3," pp. 8–9.

[19] Burrows, W. E., *This New Ocean: The Story of the First Space Age*. Random House, New York, 1998, p. 525.

[20] Wilford, J. N., "Big Business in Space: NASA's Next Mission," *New York Times*, September 18, 1983, p. SM46 [ProQuest Historical Newspapers].

[21] Boffey, P. M.,"'Getaway Specials' Seen as Less Than a Great Success," *New York Times*, October 11, 1983, p. C3 [ProQuest Historical Newspapers].

[22] Hartsfield, H. W., Jr., "Oral History Transcript #2 (C. Butler, Interviewer)," *NASA Johnson Space Center Oral History Project*, June 15, 2001, p. 19.

[23] Hartsfield, H. W., Jr., "Oral History Transcript #2, p. 10.

[24] Hartsfield, H. W., Jr., "Oral History Transcript #2, p. 41.

[25] Jones, *Sky Walking*, Mullane, M., *Riding Rockets: The Outrageous Tales of a Space Shuttle Astronaut*, Scribner's, New York, 2006.

[26] Fullerton, C. G., "Oral History Transcript (R. Wright, Interviewer)," *NASA Johnson Space Center Oral History Project*, May 6, 2002, pp. 32–33.

[27] Hartsfield, "Oral History Transcript (Carol Butler, Interviewer)," pp. 40–42.

[28] Hartsfield, "Oral History Transcript (Carol Butler, Interviewer)," p. 44.

[29] Poindexter, J. M., "Memorandum to Caspar W. Weinberger and William R. Graham," December 16, 1985, Folder 008949, NASA Historical Reference Collection, NASA Headquarters, Washington, D.C.
[30] Burrows, W. E., *This New Ocean: The Story of the First Space Age*, New York, 1999, p. 519; Presidential Commission on the Space Shuttle Challenger Accident, *Report to the President, Vol 1*. Washington: The Commission, 1986, p. 164.
[31] Williamson, R. A., "Developing the Space Shuttle," p. 180.
[32] *Columbia Accident Investigation Board: Report,* Volume I, NASA, Washington, DC, 2003, pp. 23–25.
[33] Shayler, D., *NASA's Scientist-Astronauts*, Springer, New York, 2007, p. 336.
[34] Crippen, R. L. "Oral History Transcript," p. 74.
[35] Hartsfield, H. W., Jr., "Oral History Transcript #2," p. 26.
[36] Jenkins, D. R., *Space Shuttle: The History of the National Space Transportation System, the First 100 Missions*, 3rd ed., Speciality Press, North Branch, MN, 2001, p. 331.
[37] Cassutt, M., "Secret Space Shuttles," A*ir & Space*, August 2009, pp. 45–46, http://www.airspacemag.com/space-exploration/Secret-Space-Shuttles.html
[38] Hartsfield, H. W., Jr., "Oral History Transcript," pp. 37–38.
[39] Bobko, K. J., "Oral History Transcript (S. C. Bergen, Interviewer)," *NASA Johnson Space Center Oral History Project*, February 12, 2002, p. 41.
[40] Roland, A., "Triumph or Turkey?"
[41] Hartsfield, H. W., Jr., "Oral History Transcript #2," p. 26.
[42] Williamson, R. A., "Developing the Space Shuttle," p. 182.
[43] Jenkins, D. R., *Space Shuttle*, p. 331.
[44] Logsdon, J. M., and Reed, C., "Commercializing Space Transportation," *Exploring the Unknown: Selected Documents in the History of the U.S. Civil Space Program, Volume IV: Accessing Space*, edited by Logsdon, J. M., NASA, Washington, DC, 1999, pp. 414–415.
[45] McCurdy, H. E., *Space and the American Imagination, Smithsonian History of Aviation Series*. Smithsonian Institution Press, Washington, DC, 1997, p. 84.
[46] Launius, R. D., "The Space Shuttle—Twenty-five Years On: What Does it Mean to Have Reusable Access to Space?" *Quest*, Vol. 13, Issue 2, 2006, pp. 4–20.
[47] Burrows, W. E., *This New Ocean: The Story of the First Space Age*, Modern Library, New York, 1999, p. 561.
[48] Clines, F. X., "John Glenn to Go Back Into Orbit, at Age 77," *New York Times*, January 16, 1998, p. A1 [ProQuest Historical Newspapers].
[49] Clines, F. X., "John Glenn to Go Back Into Orbit, at Age 77."
[50] Shayler, D., *NASA's Scientist-Astronauts*, p. 440.
[51] Allen, J. P., "Oral History Transcript #3," p. 7.
[52] Smith, R. W., *The Space Telescope: A Study of NASA, Science, Technology, and Politics*, Cambridge University Press, New York, 1989, p. 243.
[53] Smith, R. W., *The Space Telescope: A Study of NASA, Science, Technology, and Politics*, p. 75.
[54] Neal, R., "Value of Space Science?" *CBS News*, February 11, 2009, http://www.cbsnews.com/2100-500258_162-539214.html; Leary, W. E., "Loss of the Shuttle: History: Debate Over the Shuttle Fleet's Value to Science Has Been Raging From the Beginning," *New York Times*, February 10, 2003, p. A21 [ProQuest Historical Newspapers]; Lewis, P., "Europe's 'Factories In Space': Industrialists Are Doubtful," *New York Times*, May 14, 1984, p. D10 [ProQuest Historical Newspapers].
[55] Launius, R. D., "The Space Shuttle—Twenty-five Years On: What Does it Mean to Have Reusable Access to Space?" *Quest*, Vol. 13, Issue 2, 2006, p. 10.

[56] Burrough, B. *Dragonfly: NASA and the Crisis aboard Mir*, HarperCollins Publishers, New York, 1998.
[57] Slayton, D. K., and Cassutt, M., *Deke! U.S. Manned Space: From Mercury to the Shuttle*. St. Martin's Press, New York, 1994, p. 314.
[58] Launius, "The Space Shuttle – Twenty-five Years On: What does it Mean to Have Reusable Access to Space?"
[59] Logsdon, J. M., "Was the Space Shuttle a Mistake?" *Technology Review*, July 6, 2011, http://www.technologyreview.com/article/424586/was-the-space-shuttle-a-mistake/page/2/; Launius, R. D., "Assessing the Legacy of the Space Shuttle," *Space Policy*, Vol. 22, Issue 4, 2006, pp. 226–234, 227–228.
[60] Kennedy, J. F., "Memorandum for Vice President, April 20, 1961," *Exploring the Unknown: Selected Documents in the History of the U.S. Civil Space Program*, edited by J. M. Logsdon, NASA, Washington, DC, 1995, p. 424; Launius, R. D., "Assessing the Legacy of the Space Shuttle," p. 232.
[61] Weinberger, C., W., "Memorandum for the President," *Exploring the Unknown: Selected Documents in the History of the U.S. Civil Space Program*, edited by J. M. Logsdon, NASA, Washington, DC, 1995, p. 546.
[62] Shayler, D., *Apollo: The Lost and Forgotten Missions*. Springer, Chichester, UK, 2002; Shayler, D., *Skylab: America's Space Station*. Praxis, Chichester, UK, 2001.
[63] Launius, R. D., "Assessing the Legacy of the Space Shuttle," p. 229.
[64] Logsdon, J. M., "The Space Shuttle Program: A Policy Failure?" pp. 1102–1103. E.g., "Memorandum from George M. Low to Donald B. Rice, November 22, 1971," in *Exploring the Unknown, Volume IV*, pp. 231–238; Jenkins, D. R., *Space Shuttle*, pp. 140–141.
[65] Jenkins, D. R., *Space Shuttle*, p. 191.
[66] Launius, R. D., "Assessing the Legacy of the Space Shuttle," p. 226.
[67] Logsdon, "Was the Space Shuttle a Mistake?"
[68] Harsh, M. G., "Shuttle-C, Evolution to a Heavy Lift Launch Vehicle (Aim-89-2521)" (NASA/Marshall Space Flight Center, 1989).
[69] Portree, D. S. F., *Humans to Mars: Fifty Years of Mission Planning, 1950–2000, Monographs in Aerospace History*, NASA, Washington, DC, 2001, p. 61.
[70] Lapp, R. E., "Send Computers, Not Men, into Deep Space," *New York Times*, February 2, 1969, pp. SM32–33 [ProQuest Historical Newspapers].

Chapter 9

LOSING THE SHUTTLE (OR NEARLY): ACCIDENTS AND ANOMALIES

STEPHEN P. WARING

INTRODUCTION

"I'm hearing an echo here," Sally Ride remarked at a press conference in 2003 after the *Columbia* accident [1]. The physicist and former astronaut would know, having served on the investigation boards for the *Challenger* and *Columbia* accidents. Indeed, Ride was one of many echoes from the January 1986 tragedy to the February 2003 disaster. Technical causes originated in flaws with propulsion technology. Social causes stemmed from poor technical decision-making over a period of years. Early tests did not reveal design limitations. Later on, top NASA and contractor managers and expert engineers made analytical mistakes and tolerated anomalies. Officials had, and lost, last chances to reconsider and perhaps forestall tragedies. Similar political and managerial environments added risk. Independent boards investigated the accidents, with the perspectives of the first, and criticism of those perspectives, informing the second. Commentators on the first accident sometimes reprised their analysis for the second.

This chapter discusses the history of technical decision-making before the accidents, not the entire history of safety in the Shuttle program. To study accidents by definition leads to looking at bad news and could lead to rather jaundiced views of NASA or the shuttle. Critical perspectives existed before *Challenger* and spread after. This author has learned from such criticisms but finds blanket denigration of a broken technical culture at NASA to be overstated. Even from the perspective of flight safety, the shuttle's record has positive elements. Some of the vehicle's most risky and complex systems, such as the main engines or avionics, worked successfully for each mission. Out of 135 shuttle missions, 134 had successful launches, and 133 had successful launches and landings. Despite two fatal accidents, the failed *Challenger* launch and *Columbia* reentry, the shuttle had a reliability rate of 98.5%. This rate, although imperfect, was remarkable for a vehicle so complex and long-lived, indeed for any human space flight system [2].

Leaving out the launches after *Columbia*, the shuttle still had a reliability rate of 98% (111 missions of 113). The Aerospace Corporation, one of the organizations whose safety practices the CAIB upheld as a model for NASA, had a 2.9 percent "probability-of-failure rate" for expendable launch vehicles.

Through the lens of horrific accidents, this chapter looks at NASA and the shuttle's dark shadows and seeks the light of lessons learned. The method here is historical and comparative. First, the chapter looks at the accidents and then exposes the vehicle's architecture. It then turns to design, test, flight, and analysis of the fatal joints and foam. It pauses for lost last chances to save vehicles and crew. It then expands to wider contexts of the shuttle organization, economy, budgets, and schedules. The goal of exploring parallels and patterns is to yield lessons.

ACCIDENT AND ARCHITECTURE

After the accidents, investigation teams pieced together the chronology of technical failures. The sequences revealed the dangers of liftoff and reentry, and the vulnerability of the entire vehicle—and their crews—to single points of failure in spite of NASA's "fail-operational/fail-safe" design philosophy.

Independent panels directed the investigations and wrote the accident reports. Because of disarray at NASA headquarters at the time of the *Challenger* accident, President Reagan appointed a Presidential Commission, usually called the Rogers Commission after its chair, lawyer and former Secretary of State William P. Rogers. The technical analysis and recommendations of the Rogers Commission were generally sound. However, its panelists had little experience in accident investigations, and its report emphasized managerial malfeasance and violations of communications requirements [3]. A report by the House of Representatives Committee on Science and Technology radically downplayed such communications flaws and blamed weak technical decision-making [4]. Meanwhile, a sea change in academic interpretations of accidents—especially normal accident theory in the 1980s and high reliability organization theory in the 1990s—was occurring. These theories de-emphasized narrow examination of technical failures or operator error and stressed social processes and organizational breakdowns. In 1996, sociologist Diane Vaughan combined these academic approaches in an award-winning book, *The Challenger Launch Decision*. Vaughan's interpretation also dismissed communications errors and rules violations, and underscored normal routines of bureaucracy [5]. Although her book had no immediate impact on NASA, after the second shuttle accident in 2003, panelists of the CAIB had experience in accident analysis, and some had been influenced by social sciences. The head of the board, Admiral Harold Gehman, respected social science research. He added John Logsdon, a political scientist who facilitated use of social scientists like Vaughan in the investigation, to the CAIB [6].

Both investigations rested on hardware studies. The Rogers Commission concluded that the *Challenger* accident, Shuttle mission 51-L, stemmed from a leak in the joints of steel cases of the solid rocket motors (SRMs), the reusable strap-on boosters that provided most of the lift for the shuttle orbiter. The joints had to contain the gases of the solid fuel that burned at 5600°F and exerted internal pressures of more than 900psi. The aft field joint on the right motor leaked immediately after ignition, but charred materials created a glass-like composite that temporarily plugged the gap. Routine launch stresses and extreme wind shear broke the delicate plug, and at 58 seconds after liftoff, a flame emerged (Fig. 9.1). The plume severed the lower attachment strut, and the motor rotated around its forward strut. The hydrogen tank breached, creating forward thrust and destroying the ET, which not only stored fuel but also held the vehicle together. At 73 seconds after launch, aerodynamic, acceleration, and inertial forces tore the vehicle apart. The breakup fragmented the orbiter, and the crew cabin sheared away. The cabin

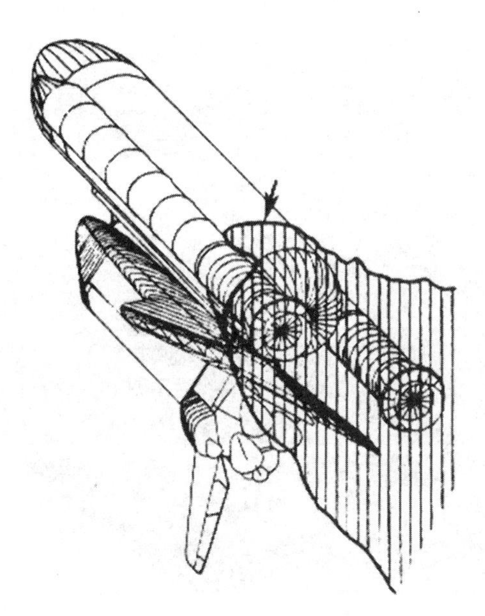

Fig. 9.1 During the launch of Challenger, 51-L, a flame escaped from the right SRM (arrow) and impinged on the ET. (*PCR*, Vol. I, 26)

lost pressure but remained more or less intact until it disintegrated on water impact [7]. The *Challenger* crew likely lost consciousness during the orbiter's breakup but probably did not die until two and a half minutes later, when the cabin hit the Atlantic Ocean [8].

Columbia also disintegrated because of a breach during liftoff, though the breakup did not happen until reentry. *Columbia's* breach was in the orbiter's TPS (see Chapter 5), which, like the motor's field joints, was designed to be reusable and to protect the orbiter from intense heat. Part of the solution for the leading edge of the wing were RCC panels, sophisticated carbon fiber reinforced panels molded to various contours. During *Columbia*'s launch, a large piece of insulating foam from the ET struck the panels of the left wing. The ET carried cryogenic fuel and oxidizer for the main engines and had Styrofoam-like insulation sprayed on to keep ice—ice that could fly off and damage the vehicle—from forming on the outside. A "bipod" strut connected the tank to the underside of the orbiter's nose. In front of where this strut attached to the ET was a bipod ramp of angled, hand-molded foam about a foot high and at least twice as long. At approximately 82 seconds after liftoff, a foam piece the size of a briefcase and weighing about 1.67 pounds, hit the leading edge of the left wing (Fig. 9.2). The foam had a velocity relative to the orbiter of about 545 miles per hour and struck with more than a ton of force. The collision created a hole about 6 to 10 inches wide in the carbon panels.

The breach caused *Columbia* to come apart days later during its reentry. As the shuttle sped downward into the atmosphere, friction generated superheated

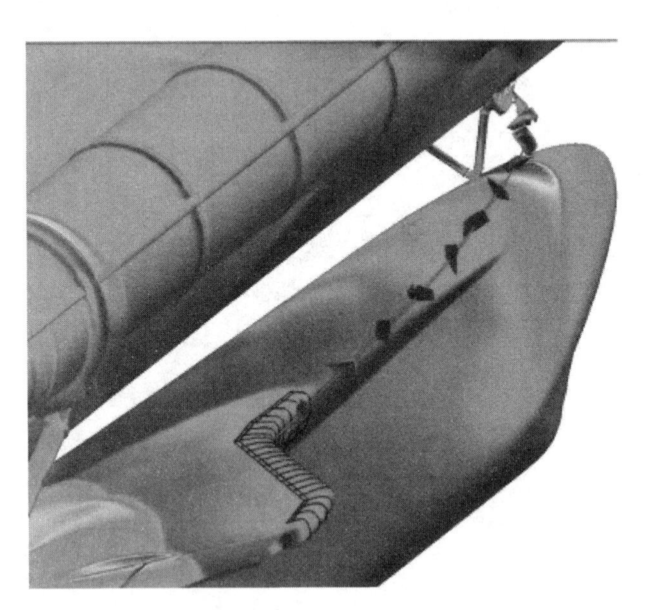

Fig. 9.2 The CAIB published this trajectory model of the bipod foam as it flew from *Columbia's* left bipod strut during liftoff and struck the left wing. (*CAIB*, Vol. I, 61)

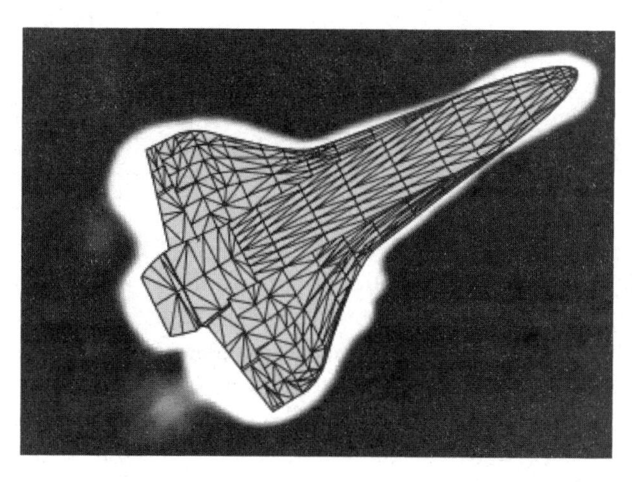

Fig. 9.3 Super-imposed on a diagram of the Orbiter's underside, this infrared image of *Columbia*, taken by scientists using commercial equipment as it passed over Kirtland Air Force Base, near Albuquerque, New Mexico, showed unusual distortions along the left wing, indicating a damaged leading edge. (*CAIB*, Vol. I, 71)

plasma, but a shock wave that was formed by compressed air also insulated it. However, the hole in *Columbia*'s wing allowed superheated air of over 2000°F to enter the hole in the wing; over California, structural supports melted, debris began to shed away, and the wing deformed (Fig. 9.3). The orbiter lost its aerodynamic shape, lost control, and began to pitch end over end. About 15 minutes after entering the atmosphere, it disintegrated southeast of Dallas, over east Texas and western Louisiana [9]. During the violent tumult, the crew module lost pressure but maintained structural integrity for about 38 seconds. *Columbia*'s crew likely lost consciousness during depressurization and probably died from blunt force trauma during the violent gyrations of the cabin's tumbling and breaking apart [10].

Besides technical flaws, the accidents also implicated the Space Shuttle's architecture. The configuration had a long, heavy spacecraft piggybacking on an external liquid fuel tank and strap-on boosters [11]. The parallel design put the spacecraft in harm's way; one failure during liftoff precipitated a cascade of failures that led to the loss of the vehicle and crew. In contrast, the Apollo-Saturn family of launch vehicles had spacecraft and launch vehicle in series, with one on top of the other. With such a configuration, a scenario like *Columbia* would have the spacecraft in relative safety above flying foam. Surviving a leaky joint in a strap-on booster during the first minutes of launch would be more problematic, but if the spacecraft had an Apollo-style escape rocket, the odds of surviving the scenario would improve.

Moreover, the patterns of the accidents show the relevance of academic "normal accident" theory, developed in the 1980s and given classic expression in 1984 by Charles Perrow. The theory did not influence the Rogers

Commission, but the CAIB endorsed it. A "normal" or "system" accident occurs when a complex technological system has interactive, tightly coupled weak points inherent in its design. A failure of one part can cascade uncontrollably and unpredictably. Such a failure was inevitable eventually, despite best efforts to prevent it. Because system accidents originated from interactions of a complex design rather than from solitary widget failure or operator error, the thesis had a strong element of technological determinism. To reduce catastrophic accidents, only so much could be done through redesigning parts or improving operations. Some overly complex systems should be abandoned and replaced by technology with greater engineering margins and redundancies and more loosely coupled subsystems [12]. The CAIB, of course, recommended that the shuttle be retired and replaced [13].

The evolution of NASA's configurations for post-shuttle space vehicles seemed to show how the agency learned from the accidents. Subject to fits and starts and political vagaries, NASA's plans for a replacement responded more to changing budgets and mission strategies than to safety concerns. However, designs evolved from those characterized by ambition and complexity to those of conservatism and utility reminiscent of Apollo-Saturn. Reusable spaceplane plans of the 1980s and early 1990s like the X-30 and X-33 integrated propulsion and spacecraft in ways similar to the shuttle. After *Columbia*, several configurations of the Constellation program, the Space Launch System, and the Commercial Crew Integrated Capability Initiative had the spacecraft on top with escape rockets like Apollo's [14].

ACCIDENTS IN HISTORY

Sadly, the technical problems that caused the accidents—flawed joints on motors and foam strikes on the orbiter—had long histories stretching back to the origins of the shuttle. Documents reveal not only the technical troubles, but also warnings of danger and sometimes controversial resolutions to continue flying without redesign. With 20-20 hindsight, an observer can select documents that show foolish errors, prophetic warnings, willing ignorance, and ridiculous arrogance. James Chiles, an expert on the history of disasters, observed that after any technological catastrophe, investigators and journalists typically presume that participants had full information, and therefore assume recklessness [15]. However, shuttle engineers and managers did not have the hindsight and learning that comes after an accident. Consequently, tracing the history of joints and foam, fairness and completeness requires consideration of not only mistakes but also the limitations of knowledge and the rationales for engineering decisions.

Two organizations, Thiokol and NASA-Marshall, developed the SRMs from 1973 to 1980. Thiokol Chemical Corporation, later Morton-Thiokol of Wasatch, Utah, won the contract, having had long experience with solid rocket propulsion since 1948. SRM segments had joints of two types: factory and field. The Thiokol plant connected factory joints, and contractors at the KSC connected four field joints. The field joints were of two types: one that connected case segments and one that connected the case and nozzle segment.

Though very experienced in liquid propulsion and contract management, MSFC in Alabama had little background in solid rockets. Thiokol worked out most of the design, imitating the strap-on solids of the Titan III-C, the biggest unmanned space launcher used by the Air Force between 1965 and 1982. At 125 feet long when assembled, the shuttle's strap-on motors were too big to transport by rail from Utah to Cape Kennedy, Florida, and back again for refurbishment. Accordingly, the motors had 4 segments approximately 26 feet long by 12 feet wide and each weighing 150 tons. The challenge for the engineers was to design joints for the steel segments that would be leak-free and reusable; the joint requirements called for redundant and verifiable seals. The design had a metal tongue on the upper segment, and this fit into a metal groove on the lower segment. In theory, ignition pressure would squeeze the parts together. Two rubber O-rings, one-quarter inch in diameter and the thickness and color of a twist of licorice, would function like gaskets in a garden hose and prevent gases from flowing through the joint. Asbestos-filled putty would provide thermal protection. The second O-ring would provide redundancy, and a leak-check would compress nitrogen (much like inflating a tire) into the gap and verify the primary O-ring [16].

NASA believed in the maturity of solid rocket technology, and this led to shortcuts in the test program. It minimized component testing and proceeded to full-scale prototypes. Unfortunately, early tests revealed problems. A large SRM is expensive and difficult to test; the solids had only 7 firings for fewer than 1000 seconds before the first flight, whereas the SSMEs had 700 firings totaling 110,000 seconds. The SRM static firings were horizontal, but this caused the heavy propellant to "slump," distorting the cases and opening gaps in the joints. Engineers crawled up the hollow bore of the motor to pack in extra putty wherever they saw gaps, an option not available for a launch. Rather than interpreting these problems as masking design flaws and invalidating tests, they reasoned slumping would not occur in vertical launch configuration, and they thought the seven firings without leaks confirmed the design. Leaks did occur during hydro-burst tests where motor segments were filled with fluid and pressurized to 1.5 times the peak operating pressure at launch. After more than a dozen pressurization cycles, one teaspoon of fluid leaked out. Engineers attributed this to "joint rotation," meaning that the metal parts of the joint slightly opened rather than closed under pressure (Fig. 9.4). Rotation, some Marshall engineers worried, could mean that the secondary O-ring would not always seal, and one memo in January 1978 recommended a redesign of the joints. However, Thiokol and most Marshall engineers attributed the leaks to the repeated pressure cycles, which jammed

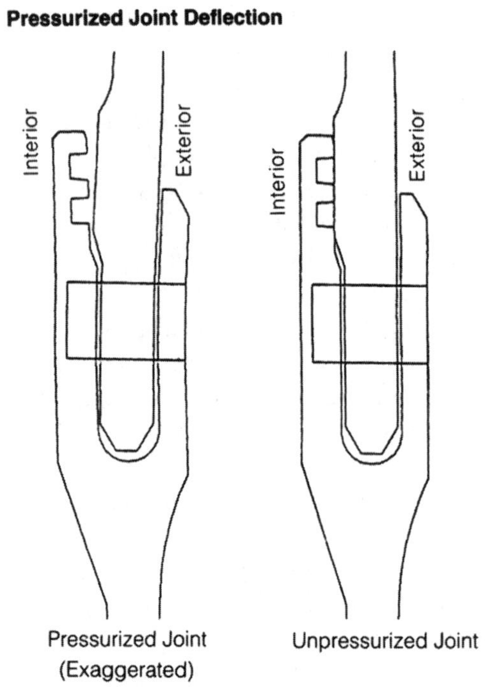

Fig. 9.4 The drawings show how during pressurization the metal segments of the SRM joint "rotated" and deflected, opening a gap at the O-ring slots. (*PCR*, Vol. I, 60)

the gaskets into the grooves. During launch, O-rings would pressurize only once, and so Thiokol and Marshall remained confident [17].

Besides tests and analysis that later seemed unsound, tests at temperature extremes were not done. Design requirements called for all components to function at 31°F to 99°F in the natural environment, and at 25°F to 120°F in the induced environment (the temperature around the shuttle with the ET loaded with cryogenic fuel). No hardware tests proved that the motor met the requirements; the coldest static firing was at 40°F. Temperature tests of the O-ring material in its applications also never happened. Thiokol apparently misunderstood the requirements, and a dozen formal reviews by Marshall, NASA, and independent certification boards before the first flight did not correct the mistake. The shuttle team and independent boards believed tests had adequately verified motor safety [18].

Equally complex requirements shaped the design of the TPSs of the orbiter and ET. SRMs were regarded as "off-the-shelf," but designers chose new technology for the orbiter's TPS. The orbiter's aluminum air frame could withstand temperatures of only about 347°F, but superheated air develops as the vehicle reenters the atmosphere, subjecting the nose and leading edge of the wings to 2800°F to 3000°F, and the underside of the wing and fuselage to

2300°F. Apollo spacecraft had used an ablative heat shield of epoxy resin, but this was heavy, single use, and expensive. The shuttle system would have several components, including thousands of very light ceramic tiles for the orbiter's underside and dozens of carbon panels for the nose cap and leading edge of the wing. The tiles and panels were reusable and replaceable. The ET, modeled in part on the Saturn S-II stage, was disposable and carried the cryogenic fuel and oxidizer for the SSMEs in the orbiter. Its contents were liquid oxygen at −297°F and liquid hydrogen at −423°F. To keep the inside from boiling and the outside from icing, the structure was insulated. To save costs and weight, the foam insulation was on the outside and mainly machine sprayed. Because of the complex shape and structure of the bipod struts and their heating during launch, workers manually shaped the bipod foam ramp. Design requirements for the shuttle intended that no ice or foam debris strike it during ground preparation or launch [19].

Tests for the TPSs had significant flaws concerning impacts. Tests showed both the orbiter and ET systems were very resistant to heat and cold. Orbiter tiles worked only when they adhered, and teams worked on many structural and bonding tests. The early struggles with poor tile adhesion became a national story in the spring of 1979 and led to improvements in adhesive techniques and protective materials that continued throughout the shuttle lifetime (as described in Chapter 5). Engineers subjected the tank foam to subscale wind tunnel testing, environmental tests of rain and temperature, bend tests, and full-scale ground vibration tests. However, because of the size of the ET, full-scale tests could not simultaneously simulate the various and changing launch stresses, including changing temperatures, air pressures, and aerodynamic forces. Moreover, no wind tunnel tests were done for manually molded bipod foam applied over another layer of foam insulation. Recognizing that small pieces of debris could hit the orbiter, engineers conducted impact tests of cigarette butt-sized pieces of foam and ice on individual tiles. Because no one envisioned strikes by large objects, tests for large object impact were not conducted. No test studied the impact of large foam chunks on the RCC panels [20].

During launch, the SRM joints and TPS produced anomalies. From the second shuttle flight, post-flight inspections of recovered boosters revealed anomalies in the motor joints; nevertheless, engineering teams accepted them. In about four percent of the field case joints, they found O-ring "erosion" that resulted from combustion gases melting small sections of primary O-rings. On less than three percent of the field case joints, "blow-by" occurred, in which the primary O-ring failed to seal and so soot collected between O-rings. Blow-by meant that combustion gases reached the secondary O-ring, and this condition was more dangerous than erosion because joint rotation might prevent the secondary O-ring from sealing. Even so, the expert consensus considered the design safe, because the SRMs worked successfully despite violating requirements. A successful launch with erosion showed that the O-rings worked despite damage; blow-by showed that even when the primary O-ring failed, the secondary O-ring provided redundancy. Hence, the SRM team created a standard of "acceptable damage"—erosion of 0.050 inch of the 0.25-inch diameter—which dismissed any lesser melting "as within the experience base." Tests showed that an O-ring with more than 0.090 inch of damage would seal. The engineers also interpreted data to mean that blow-by past the primary was "self-limiting" because once the small amount of gas in the joint burned away, the secondary one would seal [21].

NASA and Thiokol took steps to solve the problems while maintaining flights. They sought to ensure careful assembly of the joints, and they raised O-ring leak check pressures for field joints in November 1983 and January 1984. They raised pressure despite concern that the check could create gaps in the putty—"blowholes"—which could direct jets of combustion gas and produce O-ring erosion. Before the January 1984 increase in the pressure for field case joints, post-flight inspection had found only one anomaly in nine flights. After the increase, over half the missions had blow-by or erosion on field case joints. Concerns escalated during the summer, and in the fall of 1985, after the April 51-B launch, inspection revealed that a case-to-nozzle joint's primary O-ring had erosion and blow-by, and that its secondary O-ring had erosion. This led to the creation of an O-ring Task Force at Thiokol, which would begin studying a hardware redesign. In a major briefing to NASA headquarters in August 1985, Thiokol and Marshall described problems and outlined initial plans for a redesign. They also concluded that it was "safe to continue flying [the] existing design." After the *Challenger* accident, the Rogers Commission lamented the lost opportunity to stop flights and fix the problem, concluding that the August presentation was "sufficiently detailed to require corrective action prior to the next flight." During the summer and fall, memos from the Task Force circulated internally in Thiokol; they warned of potential "catastrophe of the highest order" and waved "a red flag" about how resource obstacles delayed progress. Meanwhile, launches continued in the fall, but continued anomalies were reported as "within experience base" or "No Major Problems or Issues" [22].

As with the motor joints, the shuttle had persistent anomalies with TPSs. Every ET lost foam pieces, and some hit the orbiter's TPS. *Columbia*, on the first shuttle flight in 1981, had more than 300 scarred tiles; the missions on average had 101 divots under the wings and fuselage. Photo evidence showed foam loss on more than 80 percent of the 79 missions for which imagery was available. The foam pieces were usually marshmallow size and caused shallow scrapes, but shuttle teams knew of 12 missions that had broader or deeper gouges that threatened the aluminum structure. Of these 12, 5 came from foam shedding from the left bipod ramp. After the *Columbia* accident, investigators learned that approximately 1 of every 11 missions lost foam from the left bipod. Before the accident, engineers developed various explanations for foam loss, including variability in the foam, application defects and voids, and gasses trapped in the voids that changing stresses opened [23]. Post-accident studies confirmed that the foam stayed on the ET except where voids existed in the foam, and that thermal cycles of filling the tanks with cold liquids had cracked the foam, especially where foam was applied over foam like the bipod ramp [24].

Before the *Columbia* accident, however, the shuttle organization tracked the foam strikes but mainly classified them as acceptable and not threatening. In dozens of engineering and flight management meetings, engineers and managers determined the hits were "in-family"—known, understood, inevitable, and not a safety-of-flight issue. They saw hits as a maintenance issue that affected "re-use" and "turnaround" for the next flight. One sign of confidence was that the launch pad and the shuttle itself had inadequate cameras and equipment to monitor damage. Engineers applied various palliatives, including improving foam application, reducing the foam's thickness, and poking tiny holes to vent gases from voids. Extreme events received greater attention but still a confused and ultimately complacent response. Managers usually labeled a serious event as an "In-Flight Anomaly," which meant the condition had to be corrected or proved to be safe before the next flight. After a bipod foam loss in June 1992 for STS-50, the Marshall Center's External Tank Project designated it as "not a safety-of-flight issue," meaning it was not threatening, whereas the Johnson Center's Integration Office called it "an acceptable risk," which implied that the threat was real but unlikely to recur. Ten years later and two flights before STS-107, another bipod foam loss occurred on STS-112; this caused a 4x3-inch divot in SRM foam during launch. The Shuttle program, however, called only for an "action," which required that the External Tank Project find the cause and propose a fix. This classification, rather than the usual "In-Flight Anomaly," allowed two flights to launch before the problem was resolved, something the CAIB called "an unprecedented step." The CAIB called the choice a way for mission managers to classify the foam strike as a "maintenance and turnaround concern rather than a safety-of-flight issue" and found it "a serious error" [25].

Clearly, the experience of flight success despite problems—even because of problems—bred almost unshakable confidence. Commentary on *Challenger* exposed the issue of accepting anomalies as signs of success. Richard Feynman, a Nobel Prize–winning physicist and a member of the Rogers Commission, said, "When playing Russian roulette the fact that the first shot got off safely is little comfort for the next. Erosion was not something from which safety can be inferred," he said, but "was a clue that something was wrong" [26]. Diane Vaughan, the sociologist whose book on *Challenger* won several prizes, labeled it "the normalization of deviance."

The shuttle team mistakenly departed from safety rules by accepting the first anomaly and then using their initial acceptance to justify the safety of later ones. This process became routine, she argued, and led the team to predict problems and rationalize them [27].

Even formal analytical methods failed to reveal danger. Thiokol engineers analyzed flight and test data in a computer model called ORING, correlating pristine joints with joints that had erosion or blow-by. The April 1985 model predicted that catastrophic joint failure was "improbable." ORING was reductionist and could not warn of potential danger because its database included only historic information of past success. Feynman mocked this as "empirical curve fitting." The Rogers Commission criticized the motor team for failing to compute statistical correlations, or even elementary bi-variate plots, of O-ring performance with any other factor, such as temperature or leak check pressure; such correlations could have promoted an understanding of risk. Statistical studies after the accident demonstrated that if this had been done by the time of the 17th mission, STS-51-B, in April 1985, these studies could have shown the risks of low temperatures [28]. Similarly, Boeing, working from an Apollo-era program, developed an algorithm called CRATER, which predicted the depth of tile penetration from small pieces of foam and ice. A CRATER-like model also characterized damage for ice pellets on the carbon panels. Both were called "conservative" because they predicted worse damage than was actually observed. As with ORING, the CRATER tools had limitations; CRATER was based on a database of "in-family" strikes, and so its utility for predicting the damage of a large foam object was unknown [29].

The test programs for the SRM joints and for the TPS had substantial limitations, as did the analytical processes developed in the wake of the flight anomalies. This left unresolved weaknesses in design and limitations of knowledge. Even so, in trials by fire, experts voiced concerns and created opportunities to reconsider.

LAST CHANCES?

The accident investigation revealed in detail how managers and engineers in the Shuttle program assessed risk. At the 11th hour, they met to discuss potential dangers and possible solutions.

For *Challenger*, the final discussions concerned the impact of predicted 31°F weather on O-rings. On the evening before launch, groups of Thiokol and NASA-Marshall personnel conversed via teleconference from Utah, Alabama, and Florida. The Thiokol engineering presentation argued that the temperature was "outside of the database"—much colder than the previous coldest launch (at 53°F on mission 51-C in January 1985), which had experienced O-ring blow-by. They argued cold made O-rings lose resiliency and the rubber became like a brick rather than a sponge. Given the unknowns, the recommendation was to wait until temperatures reached 53°F. Rather than accepting Thiokol's recommendation or requesting input from Marshall experts, NASA-Marshall managers questioned this rationale and offered sarcasm; one comment was "My God, Thiokol, when do you want me to launch, next April?" They doubted that cooler temperatures caused O-ring problems and observed that O-ring blow-by had occurred at 75°F [30].

Thiokol had raised the possible risks of cold temperature in formal reviews twice before, after the 51-C launch and before the August 1985 briefing at headquarters, and Marshall had dismissed their concerns. Thiokol engineers had not followed up, assuming that cold weather was unlikely to repeat. They were unready to deal with temperature risks. Immediately before the ill-fated *Challenger* mission in January 1986, mission 61-C had several launch scrubs for technical reasons; several of these aborted launches had temperatures fall below 53°F, including one scheduled launch that was in the 40s, and Thiokol had not protested. NASA initiated the pre-*Challenger* discussions with Thiokol about the cold; without NASA's request, company engineers gave no sign of calling a launch delay or even requesting a discussion [31].

Responding to Marshall's challenges, Thiokol engineers admitted they could not quantify their concerns but argued that a cold launch was "away from goodness" in the motor's flight history. After inconclusive discussion, Thiokol management went offline for an internal caucus, during which Marshall managers discussed preparations to postpone the launch. Meanwhile, Thiokol managers, who unbeknownst to Marshall included two vice presidents and the space programs general manager, decided that Marshall's reasoning had merit. Thiokol managers therefore overruled protestations of some engineers and came back online to repeal the no-launch recommendation and to accept that a launch in the cold was no more risky than any other. Because the teleconference had not reversed earlier go decisions, Marshall officials did not report the teleconference, let alone the initial no-go decision, to NASA-Johnson or headquarters [32].

The *Columbia* deliberations were eerily similar but more complex, prolonged, and shambolic. On Flight Day Two, NASA personnel reviewed blurry video and noticed a large piece of debris hit the left wing of the orbiter. This blurry video was the only real evidence, and the level of risk had to be assessed before the orbiter's planned reentry 12 days later. Formal and informal discussions occurred throughout the Space Shuttle program; emails circulated imagery and information. Deliberations took place within a very complex shuttle organization. Participating were NASA experts from the Kennedy, Johnson, Marshall, and Langley Centers. Contractors were also involved. In 1996, in an effort to streamline the shuttle organization, NASA had consolidated 86 separate shuttle contracts with 56 firms into one operations contract with USA created by Boeing and Lockheed Martin. The contract made NASA dependent on contractors for technical expertise, and its success depended on close communication and cooperation [33].

An initial photo analysis of the debris strike on Flight Day Two proved amazingly accurate about the bipod origins, large size, angle of incidence, high speed, and possible impact on the tiles or RCC panels; this led to an initial classification as "out-of-family" and hence a threat to flight safety. Over the weekend of Flight Day Three and Four, NASA, USA, and Boeing experts in a Debris Assessment Team used the CRATER model to predict damage; some scenarios were optimistic, but others predicted penetration deeper than the depth of TPS tiles and, more presciently, through an RCC panel. But because CRATER was "conservative" and predicted damage greater than what was actually observed in post-flight inspection, considerable uncertainty remained. To resolve the uncertainty, experts wanted more imagery of the left wing, either from crew video or from military spy cameras. The CAIB investigation found three unfilled requests for military imagery and counted eight missed opportunities for actions that might have led to the discovery of wing damage. Mission evaluation and shuttle management teams were informed; they knew of the potential safety-of-flight issue and requests for imagery, and they discussed the concerns. These managers, however, did not approve imagery requests and thus forestalled deeper investigation. From the beginning, mission managers treated the event as "in-family" and not a threat to flight safety. Managers paid attention to optimistic scenarios and ignored pessimistic ones. Their discussions showed an understanding of the debris strike primarily as a schedule and maintenance issue rather than as a safety concern, and thus they saw an attempt to get imagery for the wing as a waste of time [34].

Several commonalities stand out. Managers made the final decisions. In the teleconference and debris team presentation, managers ran key technical meetings as bureaucratic reviews with formal presentations. Although they asked for dissenting views, they did not conduct informal polls among the experts; the formal mode and managerial statements that no safety-of-flight issue existed suppressed discussion [35]. The process disempowered experts who had concerns about flight risks, and engineers silenced themselves.

In both cases, the experts who warned of risks had correct conclusions but less than optimal presentations of technical information. Edward Tufte, a graphics and presentation guru, later described the engineering presentations as examples of how not to display information. He coined the term *chartjunk* to characterize distracting, irrelevant information in Thiokol's charts. At no point did engineers juxtapose paired data on temperature and anomalies. Rather than using available data from 22 previous launches and including the temperature data for flights without O-ring problems and those with O-ring erosion, they truncated the number set down to two data points where O-ring blow-by had occurred on the launches at the warmest and

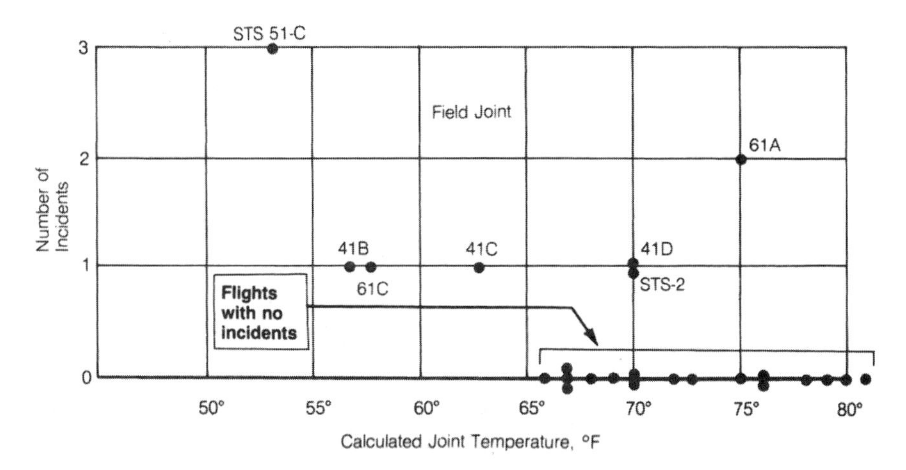

Fig. 9.5 The Rogers Commission published a plot to demonstrate the relationship of ambient launch temperatures to flights with and without incidents of O-ring distress, with distress defined by O-ring erosion, blow-by, or excessive heating. (*PCR*, Vol. I, 146)

coldest temperatures (Fig. 9.5). The instances cancelled each other out and yielded no clear warning. Thiokol's flawed presentation, Tufte bluntly concluded, signified "statistical stupidity" [36].

Tufte also criticized the ultimate presentation of CRATER results by the Debris Assessment Team to the *Columbia*'s mission evaluation manager on Flight Day Nine, which the CAIB incorporated into its report. To be fair, the team felt demoralized because it believed its imagery request had already been rejected; consequently, the team did not include a formal imagery request in its presentation, which would have indicated its concern. Its PowerPoint slides contained inappropriate scenarios, used single words to mean different things, and used outlines with excessive hierarchy and fine print. Large fonts and upper-level bullets emphasized CRATER's "conservatism," which seemed to deny any safety-of-flight issue but which actually indicated a choice among several analytical models. Managers interpreted the presentation to be a rigorous confirmation of their optimism. However, a confusing fourth-level bullet in fine print indicated that the model was being used outside of experience, and to predict damage for a foam object 640 times larger than normal (Fig. 9.6). The team did not effectively communicate the model's uncertainty, the possibility of damage to carbon panels, and the significant possibility of a life-threatening situation [37].

Whatever the problems with the presentations, the engineering teams sought to learn something new, and many managers did not. Indeed, managers set an almost impossible standard, asking experts to prove a negative, to show that a safety-of-flight issue existed. Normally, NASA flight reviews asked experts to demonstrate safety. Thiokol engineer Roger Boisjoly said, "This was a meeting where the determination was to launch, and it was up to

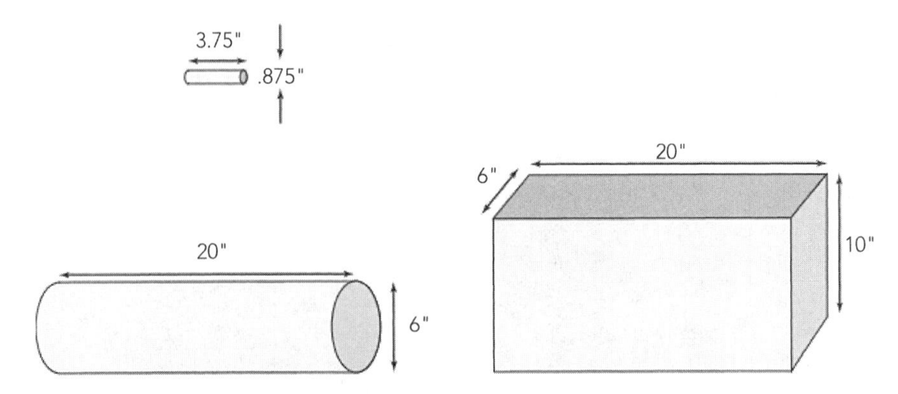

Fig. 9.6 The small figure at top illustrates the size of debris CRATER was intended to analyze. The Debris Assessment Team used the larger cylinder for STS-107 in-flight analysis. The block at right is the estimated size of the foam that hit *Columbia*. (*CAIB*, Vol. 1, 143)

us to prove beyond a shadow of a doubt that it was not safe to do so." Normally, Thiokol simply had to "prove we had enough data to launch" [38]. Flight managers similarly asked debris team analysts to show a "mandatory need" for the imagery request, in effect asking them to prove the wing was unsafe [39]. Rodney Rocha, chief engineer for the TPS for STS-107, said later [40]

> It's like someone saying, "I want you to tell me how bad that car accident is that you just heard out the window...." And you say, "Well, I'll go look out the window" and...someone says, "No, you may not look out the window. You do your analysis first."

Although managers applied strict standards to formal engineering teams, they drew reassuring information from an "informal chain of command." For *Columbia*, shuttle and STS-107 mission managers turned to a long-time NASA engineer specializing in TPS tiles. Highly respected as the tile expert, he had regular contact with the managers and provided an encouraging perspective. Although he was a member of the Debris Assessment Team, his optimism and certainty were outside the consensus of the team, and he had no expertise on RCC panels [41]. For *Challenger*, engineering managers at Marshall before the teleconference also turned for out-of-channel advice to materials engineers, including the highly respected director of their Material and Processes Laboratory. They called a former colleague who had retired in 1981 to help find military specifications for material properties of O-rings. The materials engineers believed that cold did not affect the ability of the rings to seal, an over-simplification given the dynamics of the joint. In both cases, managers appeared to be gathering from favorites data that conformed to a predetermined point of view rather than organizing open discussions and empowering experts [42].

Hierarchical structures in both instances worked to stifle the flow of information upward to managers. Some engineers involved in the teleconference did not speak up even if they believed that the cold weather was an unnecessary risk. They were accustomed to letting managers speak for the organization, and they silenced themselves after managers gave their opinions [43]. A 1988 NASA "Lessons Learned" report blamed the agency for having no strong "forgiveness policy" that would protect dissenters, especially contractor engineers, from punishment or job loss. Without such a policy, engineers would likely lay low and defer to managers. NASA safety officials did not like the report and excised its bad news [44]. Post-*Challenger* reforms intended to protect dissent did not work to foster the upward flow of information during the *Columbia* mission. Engineers circulated emails that discussed risks from the foam strike, but the emails did not go up to top shuttle and NASA managers. Perhaps the most remarkable example came from Langley and Johnson engineers who explored possible scenarios of damage to the main landing gear door; they conducted studies outside channels, but the Mission Management Team never learned of their concerns. The CAIB contended that managers had unintentionally created a "cultural fence" that cut themselves off from bad news [45].

The tangled politics of the NASA-contractor relationships also figured prominently in shaping managerial behavior. For *Challenger*, NASA depended on Thiokol for risk assessment, but Thiokol depended on NASA for business. During the teleconference, Thiokol's corporate managers had just heard Marshall management, a primary customer, sabotage the presentation of corporate engineers. Although not intended to force risky behavior, Marshall's criticism pressured Thiokol. In December 1985, NASA had opened bids for a potential second source of SRMs, and discussions for the renewal of Thiokol's contract would begin in January 1986. During the caucus, a senior corporate vice president told the managers present to "Take off your engineering hat and put on your management hat." Afterwards, Thiokol managers reversed their no-launch recommendation. The Rogers Commission concluded that Thiokol, dependent on NASA, changed its position "contrary to the views of its engineers in order to accommodate a major customer." The contractor relationship seemed to create a disincentive for safety [46].

For *Columbia*, NASA was even more dependent on contractors for risk assessment. The Debris Assessment Team was primarily composed of experts from Boeing and USA. It also had NASA personnel, a change from the first shuttle accident designed in part to eliminate adversarial contractor relationships, and so experts worked together for the same purpose. The team had co-chairs, one from USA and one from NASA. However, neither NASA mission nor mission evaluation room managers designated it as a "tiger team," which would have given it formal authority in the command chain. Nor did they take ownership of its direction. The debris team, the CAIB observed, worked in "a kind of organizational limbo." Lacking status, it routed an indirect request for imagery through JSC's Engineering Directorate. To mission managers, the indirect route was a sign that the request was not urgent, so they rejected it. Again, the status of a contractor did not work to ensure a vigorous risk assessment [47].

If the 11th-hour discussions had determined that freezing temperatures and flying foam had threatened flight safety, would that have saved *Challenger*, *Columbia*, and their crews? Counter-factual analysis is problematic at best, but it offers perspective. Noting many comments from mission managers that "nothing could be done" after a TPS breach, the CAIB asked NASA to study options for saving STS-107's crew. The scenarios assumed that NASA confirmed the breach by Flight Day Four. The studies found that an in-flight wing repair was "high risk" and unlikely to lead to a safe reentry. More likely to work was an orbital rescue by the Shuttle *Atlantis*. *Columbia*'s crew would modify its activities to conserve life support resources while ground crews accelerated *Atlantis*'s preparation. Presuming preparations succeeded, an admittedly high hurdle, the shuttle could launch between Flight Day 25 and 30, and *Columbia* crew members could transfer via spacewalks to *Atlantis*. Therefore, an affirmative response to an early request for imagery might have saved the crew [48].

The Rogers Commission conducted no formal study, but part of its report presumed rather dubious counter-factuals. The Commission Report concluded that Marshall had a general propensity to withhold bad news from shuttle officials at NASA headquarters and Johnson, and specifically withheld information about the teleconference, the initial no-launch recommendation, and the continued opposition of Thiokol engineers. The Commission conjectured that if this information had been reported, it was "likely that the launch of [*Challenger*] 51-L might not have occurred when it did." The House of Representatives' Committee on Science and Technology report on the accident found "no evidence" of the Commission's general contention that Marshall withheld bad news about the joint and "no evidence to support a suggestion that the outcome would have been any different had they [top shuttle officials] been told [49]." After the teleconference, Thiokol provided a written recommendation to launch, and, if top managers had read it, they likely would have accepted it. Under similar circumstances on the eve of *Challenger*'s launch, shuttle managers from Johnson had information that Rockwell, the orbiter contractor, had not wanted to launch because of ice on the launch pad, but that top NASA shuttle managers had accepted a recommendation to do so by Johnson technical managers. Top shuttle managers in the past followed the advice of technical managers, who spoke for a whole team, rather than individual engineers. And so even with better communication, the House Committee report contended that the cold weather launch, and the accident, would likely have happened anyway.

The last-chance discussions reveal the difficulty of overturning entrenched safety rationales in the hothouse atmosphere of prelaunch and in-flight decision-making. Managers showed a tendency to dismiss as inconsequential what seemed to be another familiar event and to apply lesser standards to their own processes than those of support teams. Engineers found it difficult to demonstrate what was new and dangerous without access to preexisting pattern data and trend analysis. Presentations suffered from poor graphics, statistics, logic, and rhetoric. The processes of formal reviews stifled discussion and silenced dissent.

To transcend the history of poor technical decision-making, both the Rogers Commission and the CAIB recommended a range of sensible changes, with the CAIB seeking help from academic social scientists and endorsing "high reliability organization" theory. This academic theory, developed in the 1990s, emerged from study of accidents such as *Challenger* and of organizations that safely managed complex technology [50]. The CAIB described "high reliability" theory and reported on best practices of high-tech independent safety programs in the Navy and Air Force. As a guide to creating strong safety culture, the theory assumed that safety could be built into each person, team, and process, thus making the entire system reliable. The CAIB's recommendations included written peer-reviewed technical reports, training based on accident studies, empowering technical experts, improved databases, communication of technical information in redundant paths, as well as broader organizational changes [51].

ENGINEERING ENVIRONMENTS

Shuttle decisions did not happen in a vacuum but rather in a complex political and organizational environment. Engineers and managers did not make decisions in open fields but instead within structures of budgets and schedules, politics of patrons and customers, management directives, and organizational cultures. These structures limited resources, constrained choices, and added risk to the inevitably risky business of space technology.

The Rogers Commission and the CAIB identified a basic mistake, the arrogance of NASA and the shuttle organization. The Rogers Commission saw NASA's overconfidence in its "can-do" attitude, a feeling that had developed during the Apollo program. The agency's optimism that "we can do anything," the Commission argued, meant that it often tried to do things without proper resources and pushed to meet unrealistic schedules. The CAIB believed that this can-do attitude recurred after the *Challenger* accident. The CAIB drew on the work of organizational theorist Garry Brewer, who argued that the NASA community had an idealized vision of itself as a "perfect place." Brewer concluded that NASA's very successes led to bad habits, including "righteousness, flawed decision making, self-deception, introversion, and a diminished curiosity about the world outside the perfect place" [52].

Excess confidence worked its way in from the beginning of the shuttle when NASA decided to depend less on in-house engineers and rely more on contractors. For Apollo, NASA had usually maintained close supervision over contractors. Marshall had used an "arsenal system" that sometimes built prototypes in-house and then turned the plans over to contractors, and Marshall practiced "penetration" of contractors by maintaining resident offices with large staffs at firms' plants. The system meant two teams performed engineering, checking and balancing one another. In part, the old ways were simply no longer possible as post-Apollo budget cuts had ended the arsenal system and had reduced funding. But NASA made a strategic shift, directing its centers to imitate Air Force styles of technology management that shifted engineering to contractors. Political scientist Howard McCurdy has argued that "no single factor affected NASA's technical culture more than the increased use of contractors" [53].

The approach reflected NASA's belief in "success-oriented" engineering. A "success-oriented" approach meant that NASA expected to design and build shuttle equipment so well that it would pass tests the first time. The approach came from efforts to save costs and worked on the assumption that engineers had learned enough from Apollo to make aerospace "mature" technology. The aerospace industry had built lightweight, leak-proof fuel tanks with insulation on the outside, like the Saturn S-II stage, and large solid fuel rockets, like the Titan III-C. NASA planned that, in comparison to Apollo, "mature" pieces of the shuttle would have less parallel development of technology, fewer component tests, and fewer tests overall [54].

The relationship of Thiokol and Marshall showed belief in aerospace maturity. Thiokol had made parts of the solid fuel Minuteman missiles for the military since the mid-1960s. Marshall, with unparalleled expertise in liquid propulsion, lacked expertise in solids and tended to defer to Thiokol. In the words of Marshall's booster project manager at the time, the center trusted Thiokol because solid propulsion was mature technology and the shuttle's SRMs were "within the state-of-the-art" or at least "the state-of-experience" [55]. The *Challenger* accident revealed mistaken assumptions, and afterwards, during the redesign of the joints, NASA returned to contractor penetration and added levels of engineering review. New test facilities at Marshall and Thiokol included test-stands for full-scale but short-stack motors that allowed engineers to study various configurations and joint failure scenarios [56].

After the shuttle's return to flight, NASA Administrator Daniel Goldin (1992–2001) introduced a policy of "faster, better, cheaper." Goldin, embracing efforts to reinvent government and coping with diminished funding, directed the agency to "do more with less" [57]. Following a temporary increase in funding after the accident, NASA budgets declined and flattened, losing about 13 percent of purchasing power during the decade. As the NASA budget declined, the shuttle got less, whereas new technology projects like the space station got more [58]. NASA pursued efficiency, and a key milestone was the 1995 Report of the Space Shuttle Management Independent Review Team. This report, called the Kraft Report, contended the shuttle was "a mature and reliable system…about as safe as today's technology will provide," and that safety could be maintained by delegating "greater responsibility to the contractor" with "only the necessary checks and balances." Some space policy experts warned that the Kraft Report's recommendations were a "recipe for disaster," and NASA's Aerospace Safety Advisory Panel complained that "the assumption that the Space Shuttle systems are now 'mature' smacks of a complacency which may lead to serious mishaps" [59]. Even so, by the end of the year, NASA privatized much of the shuttle operations by creating the USA super-contract. NASA continued cutting shuttle operating costs, with reductions in contractor and NASA personnel from 30,000 in 1992 to almost half by 2002, with proportionally more cuts falling on agency employees until they were only 10 percent of shuttle workers. A 1999 review by the Shuttle Independent Assessment Team worried that the Space Shuttle program had become less able to assess risk "by the desire to reduce costs" and failures of communication within the complex network of contractors and government. The CAIB's report showed how NASA's efficiency drive had helped leave the TPS issues unresolved and produced tangled organizations with communication problems [60].

The erosion of risk assessment was most evident in the "silent safety program," which both investigations found had degraded crosschecks. The Rogers Commission found that NASA's internal quality and safety offices lacked independence and reported to the engineering bosses whose teams they checked. In 1965, NASA-Marshall's Quality Laboratory had 629 people who conducted tests and studies. After abandoning (U.S. Army Redstone) Arsenal practices in the 70s, which had served well for Apollo, NASA relied more on hardware contractors to monitor their own work, and it used DoD inspectors. In 1985, Marshall's quality office had only 88 people and no longer conducted tests. They lacked the personnel and expertise to duplicate analyses. Instead they tracked problems reported in formal documents. They did not challenge the preflight engineers' assessments of flight safety, and no one invited safety personnel to the last-minute teleconference [61].

After Challenger

After *Challenger*, NASA beefed up its safety functions, adding personnel and crosschecks. But the agency never followed the Rogers Commission recommendation that safety teams become centrally managed and independent;

they continued to get funding from the programs they monitored. The Kraft Report derided the post-accident "safety shield" as creating "a safety environment that is duplicative and expensive." As NASA sought to streamline safety, it transferred more safety functions to contractors. The CAIB found the changes "rendered NASA's already problematic safety system simultaneously weaker and more complex." The layered organization "almost defies explanation," and contracting created "a relationship in which programs sustain the very livelihoods of the safety experts hired to oversee them." In the events leading up to *Columbia*'s fatal reentry, safety personnel attended meetings but were passive and did not channel dissenting views or challenge managerial consensus [62].

The interaction of budgets and schedules also constrained choices and shaped outlooks. NASA had justified the shuttle as being cost effective compared to expendable launch vehicles. But it proved very expensive, reaching more than half a billion dollars per flight and requiring several months of repair and refurbishment before reflight. Slips in schedule meant budget repercussions and scrutiny from Congress and the White House. After the first four flights, NASA declared the shuttle "operational," meaning that the vehicle was no longer developmental and had achieved its designed safety objectives. The shuttle's customers were the U.S. military and commercial satellite companies, and meeting schedules (with airline-like regularity) would reduce operating costs. The number of annual missions increased from 4 in 1983 to 5 in 1984 to 9 in 1985; 15 were planned for 1986 and 24 for 1990. After delays to the mission before *Challenger*'s last flight, the shuttle manager, on January 14, 1985, called on the Shuttle program to "safely and consistently launch on time." But the Rogers Commission concluded that the two goals conflicted, contributing to tolerance of anomalies and risky launch decisions. After *Challenger*, NASA ended use of the shuttle as a commercial launch vehicle, and the military phased out missions with strictly military payloads [63].

For a time in the 1990s, the Shuttle program served various science and technology missions, but by 1998, the schedule and budget paradox returned with flights to the ISS. The station reduced shuttle budgets and increased schedule stress. NASA accepted as a programmatic goal meeting the space station schedule for "U.S. core complete" at the risk of jeopardizing future station funding and the agency's political status, with the benchmark of lifting "Node 2" to orbit on February 19, 2004. NASA headquarters distributed a computer screensaver counting down days until that date. To meet the goal, the shuttle would have 10 launches in 16 months. STS-107 was not a mission to the space station, but its schedule would impact later station missions. The chair of the STS-107 Mission Management Team had safety responsibility but also had a schedule role as acting manager of shuttle launch integration for the next flight; the CAIB called this "a dual role promoting a conflict of interest." The board concluded that the countdown to Node 2 and the goal of staying on schedule "may have begun to influence managers' decisions," including their handling of foam strikes for STS-112 and STS-107 [64].

To improve safety, the CAIB recommended that NASA make organizational changes that would restructure some of the engineering environment. Again, these proposals drew from high reliability and normal accident theory. It called for the agency to create an independent Technical Engineering Authority that would be in charge of engineering standards, hazard analysis, and definition of anomalies. This organization would have independent funding from headquarters and would have no responsibility for schedule or program cost. Similarly, safety and mission assurance should be independent and funded from headquarters [65]. These changes would restore checks and engineering redundancy.

Because NASA's decisions about program planning, technology development, technical culture and contractors, schedules, and even vehicle design depended on budgets, the indirect responsibility of Congress and the presidency for the accidents was an open question. After *Challenger*, rumors circulated that President Reagan had directed NASA to launch so that he could talk to Christa McAuliffe, who was to be the first teacher-in-space, while in orbit during his State of the Union Address. Although the rumor was widely believed, neither the Rogers Commission nor the House Committee found direct evidence. The more difficult question was whether Congress and the presidency constrained NASA and the Shuttle program's budgets so as to jeopardize safe decisions. The Rogers Commission's answer was indirect. It noted that after NASA classified the shuttle as "operational"—a status publicized with great fanfare by the Reagan White House—and the flight rate escalated, the program lacked the resources to monitor and address flight anomalies before the next mission [66]. More frankly, the CAIB concluded that the cost controls during shuttle development had been a mistake and that the decade of belt-tightening and downsizing before the *Columbia* accident had led to "operating too close to too many margins" for the aging shuttle fleet [67].

Retrospection on the accidents and technical decision-making after the end of the Shuttle program yields a few lessons. By their nature, the lessons are simple, but simplicity does not (always) mean they arise from stupidity. To borrow from von Clauswitz, in war everything is very simple, but the simplest things are difficult—and the same holds for spaceflight. The CAIB report began with "building rockets is hard" [68]. The lessons are mixed because "rocket science" combines rocket politics, engineering, and management. A vehicle for human spaceflight should at all times position the spacecraft relative to the propulsion system so as to put safety first. Test what you fly, and fly what you test; in other words, tests should duplicate flight hardware and conditions. The longer the span of flight success, the more likely complacency will develop. Successful flights with anomalies are not proof of no risk.

Anomalies, even seemingly innocuous ones, should be analyzed and subjected to hardware tests. Dissent from a safety consensus is the highest form of engineering. Engineering and safety organizations should be independent from program and mission management. A research and development organization should not be envisioned as a quasi-business with customers and timetables. A fixed or fast schedule jeopardizes safety. And remember, humans, more than budgets, ride on each rocket.

REFERENCES

[1] Cabbage, M., and Harwood, W., *Comm Check...: The Final Flight of Shuttle Columbia*, Free Press, New York, 2004, p. 202.

[2] Roland, A., "The Shuttle: Triumph or Turkey?" *Discover*, Vol. 6, No. 6, November 1985, pp. 29–49; McConnell, M., *Challenger: A Major Malfunction, A True Story of Politics, Greed, and the Wrong Stuff*. Doubleday, Garden City, New York, 1987; Klerkx, G., *Lost in Space: The Fall of Nasa and the Dream of a New Space Age*, Pantheon Books, New York, 2004; *Columbia Accident Investigation Board: Report*, Vol. I of 5 vols, NASA, Washington, DC, 2003, p. 184; "List of Space Shuttle Missions," Wikipedia.org, August 21, 2012, http://en.wikipedia.org/wiki/List_of_space_shuttle_missions.

[3] *Report of the Presidential Commission on the Space Shuttle Challenger Accident*, Vol. I of 5 Vols., U. S. Government Printing Office, Washington, DC, 1986, pp. 206–214.

[4] House Committee on Science and Technology, *Investigation of the Challenger Accident*, 99th Cong., 2d sess., 1986.

[5] Vaughan, D., *The Challenger Launch Decision: Risky Technology, Culture, and Deviance at NASA*, University of Chicago Press, Chicago, 1996; Galison, P., "An Accident of History," *Atmospheric Flight in the Twentieth Century*, edited by Galison and A. Roland, Kluwer, Dordrecht, 2000, pp. 2–43.

[6] *Columbia Accident Investigation Board: Report,* Vol. I, pp. 231–234.

[7] *Report of the Presidential Commission on the Space Shuttle Challenger Accident*, Vol. I, pp. 19–39.

[8] Kerwin, J. P., [Bio-Medical Report] to Richard H. Truly, July 28, 1986, NASA Historical Reference Collection, NASA Headquarters History Office; Harwood, W., "The Fate of Challenger's Crew," *The CBS News Space Reporter's Handbook*, accessed April 14, 2001, pp. 23–25 at http://www.cbsnews.com/network/news/space/SRH_A5_Challenger.pdf.

[9] *Columbia Accident Investigation Board: Report,* Vol. I, pp. 34, 38–39, 49–84; NASA/CAIB Integrated Time Line Team, "Summary Timeline," March 10, 2003, http://history.nasa.gov/columbia/Troxell/Columbia Web Site/Timelines/summary_timeline.htm.

[10] NASA, *Columbia Crew Survival Investigation Report*. NASA SP-2008-565, http://www.nasa.gov/pdf/298870main_SP-2008-565.pdf.

[11] Launius, R. D., "NASA and the Decision to Build the Space Shuttle, 1969–72." *Historian*, Vol. 57, No. 1, 1994, pp. 17–34; Heppenheimer, T. A., *The Space Shuttle Decision: NASA's Search for a Reusable Space Vehicle*, NASA SP-4221, Washington, DC, 1999; Jenkins, D. R., *Space Shuttle: The History of the National Space Transportation System The First 100 Missions*, 3rd ed., Voyageur Press, Stillwater, MN, 2001; Harland, D. M., *The Story of the Space Shuttle*. Springer, Praxis, London, 2004.

[12] Perrow, C., *Normal Accidents: Living with High-Risk Technologies*, Princeton University Press, Princeton, NJ, 1999, originally Basic Books, 1984; *Columbia Accident Investigation Board: Report*, Vol. I, pp. 231–234.

[13] *Columbia Accident Investigation Board: Report*, Vol. I, pp. 210–211.

[14] Launius, R., "After Columbia: The Space Shuttle Program and the Crisis in Space Access," *Astropolitics*, Vol. 2, 2004, pp. 277–322; Hallion, R., The Hyp*ersonic Revolution: Case Studies in the History of Hypersonic Technology*, Vol. III, Air Force History and Museums Program, 1998; Bergin, C., "X-33/VentureStar–What Really Happened," *NASASpaceflight.com*, January 4, 2006, http://www.nasaspaceflight.com/2006/01/x-33venturestar-what-really-happened/; Reeves, J., "NASA is Borrowing Ideas from the Apollo," *USA Today*, August 14, 2006, http://usatoday30.usatoday.com/tech/science/space/2006-08-14-nasa-apollo_x.htm?csp = 34; Amos, J, "NASA Names Post-Shuttle Shortlist." *BBC*, August 3, 2012, http://www.bbc.co.uk/news/science-environment-19118764.

[15] Chiles, J. R., *Inviting Disaster: Lessons from the Edge of Technology*, Collins, New York, 2001.

[16] *Report of the Presidential Commission on the Space Shuttle Challenger Accident*, Vol. I, pp. 19–39; Bell, T. E., and Esch, K., "The Fatal Flaw in Flight 51-L," *IEEE Spectrum*, February 1987, pp. 40–42.

[17] Dunar, A. J., and Waring, S. P., *Power to Explore: A History of Marshall Space Flight Center, 1960–1990*, NASA SP-4313, Washington, DC, 1999, pp. 340–350.

[18] Sutter, J., and Lee, T. J., "NASA Development and Production Team Report," Appendix K, *Report of the Presidential Commission on the Space Shuttle Challenger Accident*, Vol. II, pp. 12–13, 30; Presidential Commission, "Design, Development, and Production Panel," meeting at Thiokol, Utah, March 17, 1986, pp. 32–64, 168–169, 181.

[19] *Columbia Accident Investigation Board: Report*, Vol. I, pp. 51–52; Cabbage and Harwood, *Comm Check...: The Final Flight of Shuttle Columbia*, pp. 55–58.

[20] Lewis, R., *The Voyages of Columbia: The First True Spaceship*, Columbia University Press, New York, 1984, Ch. 7, p. 168; Heppenheimer, T. A., *History of the Space Shuttle. Volume 2, Development of the Shuttle, 1972–1981*, Smithsonian Institution Press, Washington, 2002, pp. 183–190, Ch. 6; Cooper, P. A., and Holloway, P. F., Paul, "The Shuttle Tile Story," *Astronautics & Aeronautics*, January 1981, pp. 24–36; Cabbage and Harwood, *Comm Check*, pp. 101, 122; Pessin, M., NASA-MSFC, "Lessons Learned from Space Shuttle External Tank Development: A Technical History of the External Tank," October 2002, http://www.nasa.gov/pdf/2241main_shuttle_et_lesson_021030.pdf; *Columbia Accident Investigation Board: Report*, Vol. I, pp. 51–52, 122.

[21] Dunar and Waring, *Power to Explore : A History of Marshall Space Flight Center, 1960–1990,* pp. 350–362.

[22] Dunar and Waring, *Power to Explore : A History of Marshall Space Flight Center, 1960–1990,* pp. 362–370; *Report of the Presidential Commission on the Space Shuttle Challenger Accident*, Vol. I, pp. 124–125.

[23] *Columbia Accident Investigation Board: Report*, Vol. I, pp. 50–55.

[24] Hale, N. W., "How We Nearly Lost Discovery,"Space *Safety Magazine*, April 20, 2012, http://www.spacesafetymagazine.com/2012/04/20/lost-discovery/; Hale, N. W., "Shared Voyages: Lesson Learned from the External Tank," presentation, Project Management Challenge 2007, n.d, http://www.slideshare.net/NASAPMC/hale-wayne.

[25] *Columbia Accident Investigation Board: Report,* Vol. I, pp. 121–130.

[26] *Report of the Presidential Commission on the Space Shuttle Challenger Accident*, Vol. II, pp. F1–F2.

[27] Vaughan, D., *The Challenger Launch Decision: Risky Technology, Culture, and Deviance at NASA*, pp. 62–64, 75, 114, 140, 150.

[28] Salita, M., Morton-Thiokol, Inc, "Prediction of Pressurization and Erosion of SRB Primary O-rings during Motor Ignition," TWR-14952, April 23, 1985, PC Records, National Archives; *Report of the Presidential Commission on the Space Shuttle*

Challenger Accident, Vol. I, pp. 133, 146, F-2; Lighthall, F. F., "Launching the Space Shuttle Challenger: Disciplinary Deficiencies in the Analysis of Engineering Data," *IEEE Transactions on Engineering Management, 38*, February 1991, pp. 63–74.

[29] *Columbia Accident Investigation Board: Report*, Vol. I, pp. 143–144, 168.

[30] *Report of the Presidential Commission on the Space Shuttle Challenger Accident*, Vol. I, pp. 82–111; Dunar and Waring, *Power to Explore : A History of Marshall Space Flight Center, 1960–1990*, pp. 370–379.

[31] House Committee on Science and Technology, *Investigation of the Challenger Accident*, 99th Cong., 2d sess., 1986, pp. 148–151, 27, 71; McDonald, A., and Hansen, J. R., *Truth, Lies, and O-Rings: Inside the Space Shuttle Challenger Disaster*. University Press of Florida, Gainesville, 2009.

[32] *Report of the Presidential Commission on the Space Shuttle Challenger Accident*, Vol. I, pp. 96–104; Dunar and Waring, *Power to Explore : A History of Marshall Space Flight Center, 1960–1990*, pp. 374–376.

[33] *Columbia Accident Investigation Board: Report*, Vol. I, pp. 108–109.

[34] *Columbia Accident Investigation Board: Report*, Vol. I, pp. 143–145, 147, 150, 166–169.

[35] *Columbia Accident Investigation Board: Report*, Vol. I, p. 202; Vaughan, *The Challenger Launch Decision,* p. 367.

[36] Tufte, E. R., *Visual Explanations: Images and Quantities, Evidence and Narrative*, Graphics Press, Cheshire, CT, 1997, pp. 19–27. The Thiokol presentation is in *Report of the Presidential Commission on the Space Shuttle Challenger Accident*, Vol. IV, pp. 664–673; Robison, W., Boisjoly, R., Hoeker, D., and Young, S., "Representation and Misrepresentation: Tufte and the Morton Thiokol Engineers on the Challenger." *Science and Engineering Ethics*, Vol. 8, no. 1, 2002, pp. 59–81.

[37] Tufte, E. R., "Edward Tufte Forum: PowerPoint Does Rocket Science—and Better Techniques for Technical Reports," September 6, 2005 http://www.edwardtufte.com; *CAIB*, Vol. I, pp. 201, 191; NASA, Boeing Debris Impact Assessment Charts, "STS-News, Records Released under Freedom of Information Act" at www.nasa.gov/columbia/foia/index.html.

[38] *Report of the Presidential Commission on the Space Shuttle Challenger Accident.* Vol. IV, pp. 793–794, 811, 822.

[39] *Columbia Accident Investigation Board: Report*, Vol. I, pp. 156, 190, 201.

[40] PBS, "NOVA—Space Shuttle Disaster," October 14, 2008, http://www.pbs.org/wgbh/nova/space/space-shuttle-disaster.html.

[41] *Columbia Accident Investigation Board: Report*, Vol. I, pp. 149, 151, 154, 156, 160, 162.

[42] Presidential Commission interviews of Peoples, J, (March 12, 1986), Riehl, W. (March 13, 1986), Schwinghamer, B. (March 13, 1986), Schell, J. (March 25, 1986), PC Records, National Archives.

[43] Vaughan, *The Challenger Launch Decision: Risky Technology, Culture, and Deviance at NASA*, pp. 358–361; *Report of the Presidential Commission on the Space Shuttle Challenger Accident*, Vol. I, pp. 88–89, 93.

[44] NASA Headquarters, Safety Division, Office of Safety, Reliability, Maintainability and Quality Assurance, "Lessons Learned from Challenger [Dahlstrom Report]" February 1988, www.internationalspace.com/Challenger.htm; Klerkx, *Lost in Space*, pp. 241–247.

[45] *Columbia Accident Investigation Board: Report*, Vol. I, pp. 163–164, 169.

[46] NASA, "NASA Proposes Solid Rocket Motor Second Source," PR85-178, 26 December 1985; Bell, T., and Esch, K., "The Fatal Flaw in Flight 51-L," *IEEE Spectrum*, February 1987, pp. 46–46; Isikoff, M., "Thiokol was seeking New Contract when Official Approved Launch," *Washington Post*, 27 February 1986; *Report of the Presidential Commission on the Space Shuttle Challenger Accident*, Vol. I, pp. 19–39.

[47] *Columbia Accident Investigation Board: Report*, Vol. I, pp. 142, 151–152.

[48] *Columbia Accident Investigation Board: Report*, Vol. I, pp. 173–174.

[49] *Report of the Presidential Commission on the Space Shuttle Challenger Accident*, Vol. I, p. 105; House Committee on Science and Technology, *Investigation of the Challenger Accident*, 99th Cong., 2d sess., 1986, pp. 10–11, 27, 29–30.

[50] LaPorte, T. R., and Consolini, P. M., "Working in Practice But Not in Theory: Theoretical Challenges of 'High-Reliability Organizations'," *Journal of Public Administration Research and Theory*, Vol. 1, no. 1, January 1991, pp. 19–48; Roberts, K., "Some Characteristics of One Type of High Reliability Organization," *Organization Science*, Vol. 1, no. 2 (March 1, 1990), pp. 160–176; Rochlin, G., 'Defining "High Reliability" Organizations in Practice: A Taxonomic Prologue', *New Challenges to Understanding Organizations*,edited by K. H. Roberts, Macmillan, New York, 1993, pp. 11–32; Weick, K. E., and Roberts, K. H., "Collective Mind in Organizations: Heedful Interrelating on Flight Decks," *Administrative Science Quarterly*, Vol. 38, No. 3, September 1, 1993, pp. 357–381.

[51] *Columbia Accident Investigation Board: Report*, Vol. I, pp. 180–193; Starbuck, W. H., and Farjoun, M. (eds.), *Organization at the Limit: Lessons from the Columbia Disaster*, Wiley-Blackwell, Englewood Cliffs, NJ, 2005.

[52] *Report of the Presidential Commission on the Space Shuttle Challenger Accident*, Vol. I, pp. 172–73; *Columbia Accident Investigation Board: Report*, Vol. I, pp. 101, 102, 118, 138, 199; Brewer, G., "Perfect Places: NASA as an Idealized Institution," in R. Byerly, Jr., (ed.), *Space Policy Reconsidered,* Westview Press, Boulder, CO, 1989, 157–174.

[53] Dunar and Waring, *Power to Explore : A History of Marshall Space Flight Center, 1960–1990*, pp. 19, 43, 92, 165; McCurdy, H., *Inside NASA: High Technology and Organizational Change in the U.S. Space Program*, Johns Hopkins University Press, Baltimore, 1993, p. 134.

[54] "Budget Constraints Pare Test Efforts," *Aviation Week and Space Technology*, November 8, 1976, pp. 64–66; Lewis, R., *The Voyages of Columbia: The First True Spaceship*, Columbia University Press, New York, 1984, pp. 56–60, 67.

[55] Sutton, E. S., "From Polymers to Propellants to Rockets: A History of Thiokol," 35th AIAA/ASME/SAE/ASEE Joint Propulsion Conference, Los Angeles, CA, June 20–24, 1999; Hunley, J. D., "The Evolution of Large Solid Propellant Rocketry in the United States," *Quest*, Vol. 6, Spring 1998, pp. 22–38; Hunley, J. D., "Minuteman and the Development of Solid-Rocket Launch Technology," chapter draft of June 2002, pp. 94, 97–99; Andrepont, W. C., and Felix, R. M., "The History of Large Solid Rocket Motor Development in the United States," AIAA, Indianapolis, IN, June 27–29, 1994, p. 14; Covault, C., "Solid Rocket Motor Designed with Conservative Margins," *Aviation Week and Space Technology*, February 10, 1986, pp. 53–55.

[56] Dunar and Waring, *Power to Explore : A History of Marshall Space Flight Center, 1960–1990*, pp. 407–420; Stever, H. G., and National Research Council, *Collected Reports of the Panel on Technical Evaluation of NASA's Redesign of the Space Shuttle Solid Rocket Booster*, National Academy Press, Washington DC, 1988; Logsdon, J. M., "Return to Flight: Richard H. Truly and the Recovery from the Challenger Accident," *From Engineering to Big Science*, edited by P.E. Mack, NASA SP-4219, Washington, DC, 1998, Ch. 15.

[57] McCurdy, H. E., *Faster, Better, Cheaper: Low-Cost Innovation in the U.S. Space Program*, Johns Hopkins University Press, Baltimore, 2001; Lambright, W. H., *Transforming Government: Dan Goldin and the Remaking of NASA,* PricewaterhouseCoopers Endowment for the Business of Government, 2001.

[58] *Columbia Accident Investigation Board: Report*, Vol. I, pp. 102–110.

[59] "Report of the Space Shuttle Management Independent Review Team [Kraft Report]," February 1995, pp. 7, 18–19, http://www.fas.org/spp/kraft.htm; Carreau, M., and Clayton, W. E., Jr., "Shuttle's managers say go slow; Contractor plan stirs safety fears," *Houston Chronicle*, March 16, 1995, p. A17, http://www.chron.com/CDA/archives/archive.mpl?id=1995_1262378; *Columbia Accident Investigation Board: Report*, Vol. I, p. 108.

[60] *Columbia Accident Investigation Board: Report*, Vol. I, pp. 108–110; Mahler, J. G., and Casamayou, M. H., *Organizational Learning at NASA: The Challenger and Columbia Accidents*, Georgetown University Press, Washington, DC, 2009, pp. 89–91, 123–126.

[61] *Report of the Presidential Commission on the Space Shuttle Challenger Accident*, Vol. I, Ch. VII; Dunar and Waring, *Power to Explore: A History of Marshall Space Flight Center, 1960–1990*, pp. 397–398; *Columbia Accident Investigation Board: Report*, Vol. I, p. 25.

[62] Report of the Space Shuttle Management Independent Review Team [Kraft Report], February 1995, p. 8, http://www.fas.org/spp/kraft.htm; *Columbia Accident Investigation Board: Report*, Vol. I, pp. 231–234.

[63] *Report of the Presidential Commission on the Space Shuttle Challenger Accident*, Vol. I, Ch. VIII, p. 177; Dunar and Waring, *Power to Explore: A History of Marshall Space Flight Center, 1960–1990*, pp. 399–401.

[64] *Columbia Accident Investigation Board: Report*, Vol. I, pp. 116–117, 139, 200.

[65] *Columbia Accident Investigation Board: Report*, Vol. I, pp. 184, 192–193; "NASA Selects Admiral Cantrell for Independent Technical Authority," *SpaceRef*, June 4, 2004, http://www.spaceref.com/news/viewpr.html?pid=14350.

[66] *Report of the Presidential Commission on the Space Shuttle Challenger Accident*, Vol. I, pp. 176, 177; Cook, R., *Challenger Revealed: An Insider's Account of How the Reagan Administration Caused the Greatest Tragedy of the Space Age*. Basic Books, New York, 2007.

[67] *Columbia Accident Investigation Board: Report*, Vol. I, pp. 18, 210.

[68] *Columbia Accident Investigation Board: Report*, Vol. I, p. 19.

Chapter 10

Constructing a Port in Orbit: The Space Shuttle and Building the Space Station

Howard E. McCurdy

Introduction

In the minds of the people who conceived the U.S. space effort, the Space Shuttle and space station were unalterably intertwined. Shuttle designers viewed the Space Shuttle as the primary vehicle for moving parts, people, and supplies to a large space station. Station designers viewed the station as the primary destination for a large space shuttle. Existence of the space shuttle provided the primary rationale for establishing an orbital outpost once the shuttle flew. Building upon the flights of the NASA space shuttle, station advocates called the facility the "next logical step" in the stages of extraterrestrial flight [1]. Founders of the space station and space shuttle viewed each as halves of the same system.

Assembly of the ISS using the space shuttle proved especially challenging. Assembly and resupply during the 13-year period from 1998 to 2011, when the space station was under construction, required 36 flights of the NASA Space Shuttle. Works of imagination, which for more than 100 years anticipated the construction of an orbital outpost, generally failed to anticipate the scope of the difficulties involved. A station assembled in orbit had to be capable of controlled flight from the launch of the first element onward. It required extensive spacewalks and new robotic techniques. It necessitated extensive international cooperation. It tested the process for making national space policy in the United States.

The people who designed and assembled the ISS constructed the equivalent of a 35-story office building in space. They did so without the advantages of a solid foundation, relying instead upon Newtonian physics for the station's stability. The risks were considerable, the probabilities of success uncertain. Yet no catastrophes occurred. No major components launched into space failed to reach their destinations. None collided with the station or failed; no astronauts transferring components from the shuttle to the station suffered serious injuries or lost their lives [2]. NASA officials overcame the loss of the Space Shuttle *Columbia* on February 1, 2003 (flying a nonstation mission), and the 30-month suspension of flights that followed, readjusted the assembly schedule, and completed the job. Given the engineering challenges involved, assembly of the ISS can be judged an engineering achievement of historic magnitude. From a policy and budgetary perspective, however, the results were not as inspirational.

THE VISION OF A SPACE STATION

People imagining the construction of orbital stations gave little attention to the details of assembly. When Edward Hale in 1869 began serializing *Brick Moon*, his story imagined that the masonry outpost could be flung into orbit with the help of two giant Earth-based flywheels. In the March 22, 1952, issue of *Collier's*, Wernher von Braun proposed an inflatable space station, a 250-foot-wide rotating wheel constructed from sections delivered by a winged space shuttle. His shuttle could not dock with the inflated wheel, necessitating the use of open-cockpit space taxis and an extensive regimen of space walks.

Stanley Kubrick and Arthur C. Clarke envisioned the use of a winged space shuttle to deliver passengers to a 900-foot-wide rotating wheel in *2001: A Space Odyssey*. In the film, released in 1968, astronauts can be seen bolting metal sheets to the underlying framework of an unfinished portion of the large wheel. Presenting his vision for gigantic space colonies in orbits well removed from the Earth, Gerard O'Neill suggested the use of a lunar "mass driver." The mass driver consisted of a large bucket on a metal track powered by electronics that could achieve sufficient velocities to hurl materials taken from the moon to the appointed locations in space.

The Soviet Union resolved the assembly challenge in an ingenious way when it began orbiting a succession of small Salyut workshops in 1971. Instead of assembling structures in space, it replaced older models with newer ones. Soviet engineers launched the first workshops whole, on a large Proton rocket. NASA utilized a similar approach with the launch of the Skylab orbital workshop in 1973 (Fig. 10.1). It dispatched the entire workshop, complete with its supplies of water and fuel, with the single launch of a Saturn V rocket. Crews rocketed to and from Skylab in Apollo capsules atop Saturn IB rockets. (In an early effort to extend the use of Apollo for further applications, NASA established the Apollo Applications Program [AAP] in August 1965.) Skylab fell back to Earth in 1979, with pieces appearing in Australia. In a

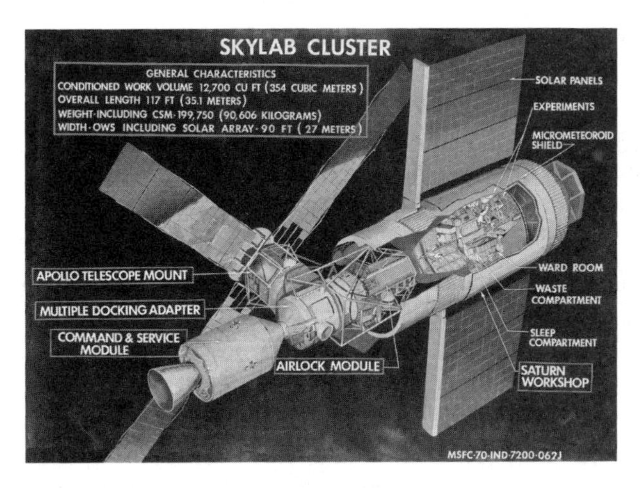

Fig. 10.1 General characteristics of the Skylab workshop with callouts of its major components. (NASA photo no. MSFC-0101537)

similar manner, the Soviet Union allowed used Salyut stations to reenter the Earth's atmosphere and disintegrate.

The Soviet Union relied upon automated rendezvous and docking techniques to assemble the Mir space station, beginning in 1986. Five of the seven station components flew to their appointed destinations under their own power. A fifth used a space tug. Robot arms guided the modules into place. The automated rendezvous and docking technique reduced the need for extensive spacewalks, which were confined to small assembly and maintenance tasks and averaged about six two-person walks per year. The Mir flight inaugurated the first use of the space shuttle for station assembly work when the Space Shuttle *Atlantis* attached the Russian-built docking module to the station complex in 1995.

NASA officials had originally planned to use a similar approach to assemble their very large space station. In the late 1960s, NASA proposed launching a limited number of 33-foot diameter canisters that could be fit together in space. Canisters would be launched using the Saturn V rocket, the giant launch vehicle that dispatched Americans to the moon. Each canister would take the place of the rocket's third stage. Station engineers planned to divide each crew canister into four decks, incorporating staterooms, lavatories, showers, a conference room, laboratory facilities, and a computer terminal. Two more decks, above and below the crew area, housed equipment, storage tanks, and large fans [3].

Planners anticipated the launch of 4 canisters for the crew, up to 4 additional canisters for utilities, a 359-foot-long axis, and a nuclear power generating plant. They planned to join the crew quarters and utility canisters to the axis pinwheel style, with the nuclear power plant isolated at the far end. The pinwheel-shaped base could rotate so as to produce artificial gravity or remain still for microgravity work. Plans called for the station or "space base" to house 100 astronauts or more.

Station advocates prominently featured the proposed facility in the 1969 report of the Space Task Group, a presidential commission appointed to assess the future direction of the nation's space effort. Under one of the group's more conservative options, members suggested that NASA could be ready to place the first canister, which would house 6 to 12 occupants, in orbit by 1977 [4].

NASA officials knew that a large space station, even one with limited assembly requirements, would require multiple missions to resupply. Occupants would shuttle to and from the large space station. They would need food, supplies, equipment, and spares. One NASA planner estimated that the agency would need to conduct 91 resupply missions to a large space station over a 10-year period [5]. The prediction was prophetic. The international partners dispatched 81 resupply and logistics flights to the ISS between the commencement of construction in 1998 and the official conclusion in mid-2011.

The need for assembly and resupply defined the relationship between the proposed space station and the proposed space shuttle. Planners sought to use the Saturn V rocket to deliver a limited number of components for a large space station, thereby simplifying assembly. They planned to use a shuttle designed primarily for crew transfer and resupply, thereby promoting economy. A large rocket was cost effective for a limited number of assembly flights, whereas a smaller shuttle would be more economical for a much larger number of resupply flights.

The two-vehicle justification for building and resupplying a large space station appeared throughout the early discussion of the undertaking. George Mueller, NASA's associate administrator for manned space flight, explained the relationship to members of the British Interplanetary Society in the summer of 1968. The Saturn V rocket, Mueller stated, would remain "the foundation of the U.S. manned space program." NASA used the Saturn V to launch the first orbital workshop, Skylab, the precursor to the large space station. An economical space shuttle, meanwhile, would produce "an order of magnitude reduction in operating costs." The expenses associated with maintaining and operating a large space station could be significant, Mueller warned. Without a cost-effective vehicle like the space shuttle, he predicted, the cost of resupplying a small orbital workshop for one year could equal the original cost of the whole facility [6].

Between 1966 and 1968, NASA officials provided more detail on the reusable space vehicle. The resulting shuttle plan was much more station focused than the one eventually approved; it was also a smaller vehicle. The basic mission of the new spacecraft, a 1968 Statement of Work explained, would be logistic missions for an orbiting space station. For this purpose, the spacecraft might be configured to hold 10 passengers in addition to its 2-person crew. Alternatively, the vehicle might be reconfigured to replace the passengers with a payload consisting of equipment and supplies. An earlier document identified "a nominal discretionary lofted cargo of 25,000 pounds" but encouraged contractors to investigate alternatives that ranged from 5,000 to 50,000 pounds. The vehicle needed to be able to bring cargo back from space as well [7].

Drawing on the optimism contained in these early studies, Clarke and Kubrick visually portrayed the anticipated shuttle-station relationship in *2001*. Clarke had access to the concepts guiding NASA's long-range planning efforts when he prepared the screenplay. The modern sequence opens with the image of a winged space shuttle approaching the 900-foot-wide wheel, both rotating in time to the music of "The Blue Danube." Clarke's spacecraft was essentially a passenger vehicle, ferrying people to and from the large orbiting station. His shuttle had room for 20 passengers in addition to a stewardess and 2 pilots. Following NASA's conceptual plan, Clarke explained that the shuttle would begin its journey attached to an also-winged reusable first stage, would disengage, would rocket to the space station, and then would glide back to Earth. Clarke predicted an operational cost of $1 million per flight, or $50,000 round trip for each of the 20 seats. The film version showed 36 seats [8].

As of 1969, the concept of the space shuttle relative to the anticipated space station seemed fairly well defined. The exact shuttle configuration—its wingspan, its booster stage, and its payload size—had not yet taken form. Neither had the exact scope of the vehicle's responsibilities beyond station tending, such as the satellites it might deliver. Yet in relation to the anticipated space station, the shuttle mission was clear. Government planners wanted to follow the Apollo program with the construction of a large space station and to use the space shuttle to resupply it with people and material. Members of the Space Task Group urged that both station and shuttle be developed simultaneously. Both could be ready to start flying, the members predicted, in 1977 [9].

THE VISION UNFOLDS

In a now-familiar story (see Chapter 2), station advocates were obliged to set aside their plans for a large orbital base and focus instead on developing the space shuttle. President Richard Nixon and his aides allotted funds sufficient for only one new human space flight initiative, not two. The new policy had a profound effect on the relationship between the now-lingering station initiative and its resupply vehicle.

First, the policy obliged station advocates to take an incremental approach toward their long-term goal. They did not abandon their hopes for a large space station; they merely deferred them. They would first develop the space shuttle and then request approval for a station to which it could fly.

This approach complicated the process of defining the purpose of the space shuttle considerably. Station advocates could not advance station resupply as the primary function of the proposed space shuttle because the station had been set aside. At the same time, they tried to retain that prospective capability for a future date when the government might approve the orbital base.

Second, lacking an approved space station—and facing severe budget cuts—NASA officials could not afford to continue producing Saturn V rockets. The last two models to come off the assembly line wound up in museums. By the end of 1969, it was obvious that Saturn V would be discontinued.

Henceforth, all considerations of the large space station fell on the shuttle design. NASA needed the shuttle to deliver and resupply the prospective space station, even though no official requirement existed for the shuttle to do so.

Contractor studies suggested how the assembly process for a shuttle-delivered space station might occur. Such studies continued into 1971 until newly appointed NASA Administrator James Fletcher told agency planners to forget the space station and focused on getting the space shuttle approved [10].

A contractor study from North American Rockwell summarized a typical design. Company officials drew on the popular pinwheel-shaped Space Base concept, substituting 14 × 40-foot shuttle delivered modules for the earlier 33-foot-wide cans. The initial configuration would require seven flights of the space shuttle. Shuttle astronauts would deliver a station core, an electric-power axis that held solar arrays, two crew and control modules, two laboratory modules, and one cargo closet. To compensate for the smaller volume of the shuttle-derived modules, officials suggested adding a second circle of modules. The resulting configuration resembled a pinwheel with two rotors flanked with solar arrays [11].

For reasons that had little to do with the prospective space station, NASA officials won White House approval for a space shuttle with a payload bay large enough to deliver station parts. Among other purposes, the large payload bay met military requirements for delivering large reconnaissance satellites, a purpose that White House officials did support.

NASA employees and their contractors worked to develop the space shuttle for nearly 10 years, from its approval by the president in early 1972 to the first test flight in the spring of 1981. Fourteen months later, in the summer of 1982, astronauts flew the fourth, and what NASA executives perceived to be the final, test flight of the new launch system.

Completion of the test flights allowed NASA Administrator James M. Beggs to resurrect hopes for the construction of an Earth-orbiting space station. Beggs called it "the next logical step" in the agency's long-range but incremental vision of space exploration [12].

Beggs had explained the shuttle-station relationship in a speech to the Economic Club and Engineering Society of Detroit one month before the final test flight. "The shuttle program originally was conceived to include a space station," he observed. "More than a decade ago a total system was envisioned in which the shuttle would transport payloads routinely to such a station." Now that the space shuttle had flown, advocates could address the next step. "What better way to make full and complete use of the shuttle than to develop a manned space station in orbit tended and supplied by the Columbia and her sister ships," he said [13].

During his Senate confirmation hearings in the summer of 1981, Beggs had indicated that he favored a space station as "the next step" in the national space program [14]. Three days before delivering his Detroit speech, Beggs had established an internal Space Station Task Force to prepare the conceptual studies necessary to win approval for the initiative.

President Ronald Reagan attended the landing of the fourth and final test flight at Edwards Air Force Base in California. In words suggested by NASA officials and accepted by White House speechwriters, Reagan announced that the *Columbia* and her sister ships were "fully operational" [15]. The announcement was premature, but it allowed Beggs to marshal agency efforts behind the station plan. Extensive agency and White House review activities occurred. President Reagan directed NASA to develop a space station in his 1984 State of the Union address [16].

Design work on NASA's permanently occupied space station began in mid-1984. By that time, the United States and the Soviet Union had experimented with a number of station assembly and resupply techniques, with one more to follow shortly. Designers of the first proto-stations—orbital workshops, really—had avoided the necessity for extensive orbital assembly and resupply by launching their workshops intact. The first astronauts to visit Skylab needed to repair the workshop's sun shield and fix a jammed solar panel, but otherwise little external work occurred. The Soviet Union had adopted a similar strategy with the launch of its Salyut workshops. Astronauts docked with intact orbital workshops to exchange crews, but little resupply occurred. When one Salyut workshop wore out, the Soviets simply launched another.

The Soviet Union experimented with resupply techniques on its later Salyut workshops. Soviet engineers installed on Salyut 6 and Salyut 7 docking ports that could receive cargo vessels. The vessels flew to and rendezvoused with the workshops in an automated mode, without cosmonaut crews in the vessels to guide them.

The Mir space station was much more ambitious. It consisted of seven modules linked together in space. Soviet engineers launched six of the modules on Proton rockets. The modules automatically rendezvoused with the core facility. A series of docking and robot arm maneuvers put the modules in their designated places. In one case, an automated space tug assisted; in another, the module arrived in the Space Shuttle payload bay. Extensive reliance on rendezvous and docking reduced the need for space walks considerably. Mir crews conducted 80 two-person spacewalks for station assembly and maintenance between 1987 and 2000, or an average of just 6 per year.

Availability of the NASA Space Shuttle took station assembly and resupply planning to a new level. Since the completion of the Space Shuttle test program provided the rationale for building a large space station, planners assumed that the shuttle would be used for both assembly and resupply. As of 1985, NASA officials still presumed that the Space Shuttle could be launched cheaply. The cost of launching the Space Shuttle was thought to be so low that Space Station Task Force members did not include transportation expenses in their cost estimate for the initial space station [17].

Availability of the Space Shuttle transformed the assembly process from a rendezvous and docking challenge into what could be characterized as a construction project. Astronaut crews would accompany station elements delivered to space and could assist with their assembly. Availability of the shuttle's crane-like Canadarm meant that shuttle astronauts could move station components without the need for those components to have their own flight and rendezvous capability. The strategy encouraged a larger number of deliveries than would be necessitated with a launch-and-rendezvous approach, which favored fewer launches of more massive components. The strategy also encouraged a larger number of spacewalks for assembly work. Plans for station assembly using shuttle-delivered modules began to resemble the process for constructing a large office building in space.

The assembly process would require shuttle astronauts to lift large components from the shuttle's cargo bay and place them in the appropriate position for attachment to existing sections. Astronauts in spacesuits would complete component assembly by linking cables and securing fittings. Depending upon the chosen configuration, truss sections, modules, nodes, airlocks, docking compartments, pressurized mating adapters (hallways to arriving spacecraft), a cupola, a central truss structure, solar panels, radiators, robotic arms and smaller components like rotary joints, nitrogen tanks, and thermal covers would require installation and maintenance (Fig. 10.2).

So long as the expense of flying the Space Shuttle fell as anticipated, NASA officials could also utilize the vehicle for station resupply and crew transfer. Its larger payload and crew capacity gave it an advantage over alternative spacecraft such as the Soviet Progress resupply spacecraft and Soyuz crew transfer vehicles. NASA used the fleet of space shuttles in this manner to assist the Mir space station between 1995 and 1998. Nine docking missions occurred. Some modest delivery and assembly work took place, but most of the effort involved crew transfer and cargo resupply.

During the planning process leading up to the president's directive, members of NASA's Space Station Task Force had not addressed assembly

Fig. 10.2 Astronauts Mike Fossum (left) and Ron Garan work outside the International Space Station to assemble components on one of the station's pressurized modules, part of the STS-124 shuttle mission. (NASA photo no. s124e006361)

techniques in detail. Instead, Task Force members concentrated their attention on the functions that a large space station would perform and the resulting technical requirements for items like electric power and crew space. John Hodge, the director of the Space Station Task Force, precluded the analysis of assembly techniques by refusing to let Task Force members draw configurations of the space station. He wanted station planners to concentrate on missions and requirements instead. "People got very mad at me because of course there were a lot of engineers on the Task Force, mostly engineers, and they want to start designing things." As an engineer himself, Hodge understood the tendency. "I wouldn't let anybody draw any pictures" [18].

A hint of one possible configuration had occurred when workers at NASA's Langley Research Center constructed a station model that Task Force members took to the White House for policy review. President Reagan subsequently took the model to the May 1984 London economic summit in an effort to encourage international participation. Task Force members used the model to demonstrate the missions a space station would perform, not how it would actually look. The model featured a pallet for experiments, attachments for satellite servicing, equipment for refueling satellites and other vehicles, a robot arm, two solar arrays, and four 14-foot-wide modules arranged pinwheel-style around a fifth modular core.

In the letter explaining their station cost estimate, agency officials listed the components that the appropriated funds would supply. Again, the listed elements suggested a modest station in modular form with not-extensive delivery requirements. The basic $8.8 billion space station consisted of crew quarters, two laboratories, a utility module, a docking module, a robotic arm, a space tug, and two independently-flying automated platforms.

One of the first assembly plans for the new space station appeared in 1986. During the planning phase that followed approval of the initiative, the station design expanded considerably. Planners pressed to construct an ambitious "dual keel" space station that departed significantly from earlier pinwheel designs. To minimize the gravity effects that the station itself would produce, station planners placed the crew and laboratory modules at the station's center of gravity. Planners transferred the solar power generating equipment to the ends of a 574-foot-long truss. To complement the central truss, they added two 361-foot-long keels to which very large structures (such as a Mars expedition spacecraft) could be attached. The resulting structure resembled a football field-sized structure with living quarters in the middle.

Station planners offered an assembly strategy for the large structure. Assembly would require 14 flights of the Space Shuttle, they said. According to this plan, assembly would begin with the orbiting of solar arrays and truss segments. Nodes, modules, and more solar arrays would follow. Planners allocated four flights for delivering the central truss and keels. The planners estimated that assembly would take three years, with the station ready for occupancy at the end of the second year. Once the dual keel configuration was approved, they predicted, its assembly could begin in early 1993 and end in late 1995 [19].

The proposed configuration faced a number of challenges. First, it cost too much. The expected cost of the core facility (excluding transportation charges) grew from $8.8 to $14.5 billion. Cost growth upset White House budget analysts, who forced NASA officials to forgo the dual keel design in favor of a less ambitious "revised baseline configuration" organized along a single truss.

Second, planners vastly underestimated the number of assembly and maintenance spacewalks that the facility would require. In 1990, NASA astronaut Bill Fisher and JSC engineer Charles Price released a study discussing the challenges of keeping a space station in good working order during assembly, as well as the requirements of maintaining it when complete. Fisher and Price predicted that what had by then evolved into Space Station Freedom would require over 11,000 hours of space walks during the anticipated four and one-half year construction period. This would require an average of 18 two-person spacewalks each month—exponentially more than the Russian practice of 6 spacewalks per year with Mir. Fisher and Price predicted that the high pace of maintenance work would continue undiminished after assembly. So much space work was dangerous and clearly impractical [20].

Third, the estimated cost of delivering and resupplying the space station soared with the cost of operating the Space Shuttle. Instead of falling to anticipated levels, the cost of shuttle operations remained persistently high. The average cost of flying the shuttle during consideration of the space station initiative in fiscal 1983 and 1984 was $343 million per mission. NASA officials pressed to cut the cost by flying more missions on a steady appropriations base. In fiscal 1985, NASA launched eight shuttle missions for an average obligation of $164 million per flight.

The January 1986 loss of the Space Shuttle *Challenger* followed. It destroyed the strategy for cutting cost by flying more. Operational costs in the post-*Challenger* era rose to nearly $400 million per flight. Station planners no longer measured assembly and resupply flights in millions of dollars. Assembly and resupply with the Space Shuttle would cost billions.

The combined effect of these events nearly caused Congress to cancel the space station initiative in 1993. To meet cost concerns, NASA planners shaved back the original space station configuration so much that some began to jokingly refer to the remaining parts as Space Station Fred. The size and scope of the orbital undertaking rebounded considerably with the post-Cold War agreement to broaden participation by including Russia as an international partner along with Canada, Japan, and the European Space Agency. What had been scrubbed back to Space Station Alpha became the plan for the ISS, a much more ambitious facility with far more elaborate assembly requirements.

That assembly process still depended primarily upon the NASA Space Shuttle. As the station configuration grew, so did the number of shuttle flights and spacewalks required to assemble it. The stubbornly high cost of flying the shuttle, however, precluded its extensive use as a resupply vehicle. For that purpose, the international partners would come to depend more extensively upon automated Progress-type vehicles for cargo resupply and eventually Soyuz spacecraft for crew transfer. The Space Shuttle held more passengers and cargo, and could return equipment to Earth, but NASA did not use it extensively for those purposes after the second accident restricted its use.

Assembly Occurs

Assembly of the ISS began on November 20, 1998, and ended more than 12 years later on May 19, 2011. In all, NASA flew the Space Shuttle 36 times to the emerging space station during the assembly period. It conducted 27 shuttle flights for what could be defined as station assembly. It conducted nine flights for what were primarily station resupply and crew transfer, and it added one post-assembly resupply mission.

A Russian Proton rocket delivered the first segment of the ISS. The Zarya module provided electrical power, propulsion, stabilization, and storage during construction. Launch of the Space Shuttle *Endeavour* followed two weeks later, delivering the first node, called Unity. Station nodes can be viewed as hallways, designed to provide connecting points between other components of the facility.

The technique used to join Zarya and Node 1 illustrated the challenges associated with station assembly. Astronaut Nancy Currie used the shuttle's robot arm to place Unity on top of the orbiter's docking mechanism. Astronauts Frederick Sturckow and Robert Cabana flew *Endeavour* to within 10 meters of the Zarya module, at which point Currie grabbed the Zarya module with the shuttle robot arm and placed Zarya on the node's mating adapter. Their work not yet done, astronauts Jerry Ross and James Newman donned spacesuits and left the orbiter to connect power and data cables. The shuttle crew, led by commander Cabana and cosmonaut Sergei Krikalev, entered the pressurized node, switched on the lights, and began the internal work of setting up the two modules. As is normal on a mission of this complexity, anomalies occurred. One of the battery discharge units in the Zarya module failed, a problem resolved with equipment on board.

Joining the two station elements required a combination of rendezvous techniques, docking procedures, robotic manipulation, spacewalks, and internal checkout.

The next assembly mission did not occur until July 2000, one and a half years later, when a second Proton rocket delivered the Zvevda service module. Zvevda contained life support equipment for the Russian section of the space station and living quarters for its crew. Flight controllers directed the docking of the Zvevda and Zarya modules from the ground. Seven weeks later, crew members from the Space Shuttle *Atlantis* arrived at the station, put on spacesuits, and connected power and communication cables between the two Russian modules. They also installed a six-foot-long compass to help the station maintain a proper orientation relative to the Earth. They transferred 6000 pounds of cargo to the station, including food, water, office supplies, and a vacuum cleaner. Completion of this assembly phase allowed astronauts to occupy the ISS. On November 2, 2000, two years after assembly began, three crew members arrived on a Russian Soyuz spacecraft, docked with the Zvevda service module, and began the permanent occupancy of the orbital facility.

During various periods between first launch and permanent occupancy, controllers flew the embryonic station from the ground. The absence of a permanent crew did not preclude the need for resupply and maintenance flights. NASA flew two resupply flights using the Space Shuttle during this period, one in May to June 1999 and a second in May 2000. Shuttle crews repaired ductwork. They replaced fire extinguishers, smoke detectors, air filters, and the still malfunctioning batteries. They delivered two small Russian Strela space cranes and boosted the station to a higher altitude. Russian ground crews flew one resupply flight during this time—an automated Progress cargo vehicle that docked with the Russian Zvevda module. No crew was on board when the Progress vehicle arrived.

To complete assembly, the international partners and their flight crews repeated this process for 10 years. More than 100 shuttle flights, Soyuz crew transfer vehicles, and automated resupply craft arrived and left (Fig. 10.3).

Fig. 10.3 The robotic arm on the Space Shuttle *Atlantis* appears to be saluting "goodbye" to the station. Earth's airglow is seen as a thin line above the horizon (photo taken on the final day of joint activities between the crew members for the STS-135 and Expedition 28 missions on July 18, 2011). The Raffaello multipurpose logistics module, full of items to be returned to Earth, is seen in the aft cargo bay. (NASA photo no. ISS028-E-017801)

Installation of the 357-foot-long station truss, the backbone of the orbital facility, required 11 assembly flights of the NASA Space Shuttle over nearly 9 years. The girder-like truss supported the station's radiators (which dissipated heat) and the electricity-generating solar arrays. It provided mounting points for various instruments and logistic carriers and the framework along which the station's mobile robots crawled. It contained rotary joints that maneuvered the solar arrays so that they pointed toward the sun. The truss segments arrived in sections. They were large, easily dwarfing the space-suited human beings working around them.

With assembly well underway, a safety task force assessed the risks that station directors might encounter as they assembled and maintained the orbital facility. Project managers commonly commission a formal risk analysis for projects large and small. This safety task force identified 53 risks that could result in what they termed an undesirable "end state." A visiting vehicle or arriving component could collide with the ISS. An astronaut could become separated from the space shuttle or station during an assembly spacewalk. A micrometeorite or piece of orbital debris could puncture an astronaut's spacesuit. Components or their associated software could fail. A fire could occur. An inadvertent but critical command from a ground controller could create an unstable situation. The suspension or termination of shuttle flights due to a flight accident could deprive the station of needed logistical support, spare parts, or elements still to be delivered. Engineers had designed most of the station's components for transport on the Space

Shuttle; suspension of shuttle flights for any reason could have a serious effect on the assembly process [21].

A frightening incident on the Mir space station foreshadowed the type of safety issues that worried station directors. On June 25, 1997, a Russian Progress resupply vehicle had collided forcefully with the Mir space station. The collision showed how dangerous a mishap on a supply or assembly mission could be. The collision knocked out electric power, the station's stabilization system, instruments, and computers and sent the Mir station into an uncontrollable spin. NASA Astronaut Mike Foale held his thumb up to the station's airlock window, used the rotating star field to calculate the station's spin rate, and communicated the results to flight controllers in Moscow, who fired stabilization engines for an appropriate period from the ground [22].

In the assembly phase of a project as complicated as that imposed by the ISS, something invariably goes wrong. In this case, the loss of the Space Shuttle *Columbia* (on a nonstation mission) created a two and one-half year suspension of flights during which the Space Shuttle was unavailable for assembly or resupply. NASA's Space Shuttle did not visit the ISS between the *Endeavour* assembly flight of late 2002 and the *Discovery* resupply flight of mid-2005. During this period, waste products accumulated. The international partners depended upon nearly two dozen flights of the Russian Soyuz spacecraft and Progress cargo carrier to resupply the station, transfer crew, and remove as much waste as they could. Shuttle-derived assembly operations did not resume until September 2006—a nearly four-year hiatus.

A serious assembly incident occurred during the late 2007 mission to complete the installation of the P6 truss and deploy its 120-foot solar arrays. During deployment, a guide wire snagged and tore the array, preventing it from extending fully. Astronauts retired to the station and, using aluminum strips, bolt connectors, and 66 feet of wire, fabricated a set of "cuff links" that might repair the tear and allow the panel to fully deploy.

The repair mission was dangerous. The solar arrays produced electricity, a danger to any astronaut who got too close to them. Astronauts had never rehearsed such a maneuver. To reach the solar array, they attached the boom used to inspect the shuttle to the end of the station's robotic arm. A photograph of Scott Parazynski shows a human in a spacesuit carefully balanced above a large electrified panel on the tip of a long pole. The makeshift repair effort succeeded, and astronauts unfurled the whole array.

Other problems occurred. Astronauts detected a jammed rotary joint in the mechanism that points the solar arrays toward the sun. A Soyuz imagery check revealed a damaged cooling panel on the station's radiator. During a 2007 shuttle assembly flight, a computer malfunction in the Russian segment left the station temporarily without thrusters, oxygen generators, or carbon dioxide scrubbers. The computer reboot undertaken to fix the problem set off fire alarms.

Assembly of the ISS and maintenance during assembly required 155 spacewalks totaling 974 hours—roughly 12 spacewalks per year [23]. This was far less than the 18 spacewalks per month predicted in the 1990 Fisher-Price report. Robotic technology, both on the Space Shuttle and on the station itself, significantly reduced the amount of human effort required to assemble and maintain the orbital facility.

During the 1984 debate over approval of the orbital facility, members of Congress insisted that the station program include robotic capabilities. Lawmakers directed NASA to conduct an automation and robotics study and to spend money on such systems. Additionally, they insisted that the station be automated in such a fashion that it could be flown from the ground, and they involved Congress in the assembly sequence in such a manner as to assure this outcome.

Station advocates viewed these early directives as an effort to recast the facility in the direction of a "man-tended" station that would not always have humans on board. Beggs objected. "A space station that is not manned full-time from the start would be far less capable than a permanently manned space station" [24].

The congressional view prevailed. The international partners began the assembly sequence with a station that could be flown in an automated mode and made use of numerous robotic entities during the assembly process. Astronauts utilized the Space Shuttle's Canadarm to lift components from the shuttle payload bay and maneuver them into position. Astronauts installed a second, more elaborate Canadarm2 on the outside of the station framework. With an overall reach of nearly 50 feet, Canadarm2 could move from place to place on the station and assist with the installation of station parts.

In 2008, shuttle astronauts delivered Dextre, a space robot with a 30-foot arm span fingertip to fingertip. Like Canadarm2, Dextre lives on the outside of the space station and can move around. After extensive testing, robotic flight controllers at the JSC instructed Dextre to unload items from a Japanese cargo vehicle, the robot's first official job. While Dextre worked, astronauts on the station slept.

Shuttle astronauts delivered four Russian-built Strela space cranes, telescopic poles mounted on the Russian orbital segment that can extend to 46 feet and assist with the movement of equipment and spacewalks. The Japanese aerospace agency attached a robotic arm to the end of its Kibo module, where the arm could help lift experiment packages from an open space pallet. The European Space Agency built a large robotic arm to help service the Russian section of the ISS. On the final shuttle mission, astronauts delivered a robotic refueling module that future station crews could use to refuel satellites. In 2011, shuttle astronauts delivered Robonaut2 to the crew of the ISS. Robonaut2 is a humanoid personal assistant robot that lives inside the space station, where it can assist the crew with vacuuming and cleaning filters.

After 12 and one-half years of assembly work, NASA officials declared the ISS assembly complete with the installation of the Alpha Magnetic Spectrometer on May 19, 2011. Scientists designed the instrument to search for exotic substances like dark matter and antimatter. The shuttle robotic arm lifted the spectrometer, with a mass of nearly 15,000 pounds, from the spacecraft's payload bay and handed it to Canadarm2, which placed the instrument on the station truss. Installation was handled autonomously, with no space walks involved. Like a homeowner contemplating remodeling, the participating nations will never complete work on the station so long as it flies. More modules and robotic equipment will arrive. Resupply flights and crew transfers will occur. NASA flew one final flight of Space Shuttle *Atlantis* to the space station after declaring assembly complete, a resupply mission in July 2011.

The Space Shuttle played a critical role in the assembly of the ISS. Yet it did not play an exclusive role. The United States and its international partners conducted more than 100 flights in support of the ISS during the 12 and a half year assembly phase—a more exact estimate totals 112. This number incorporates 27 missions of the Space Shuttle fleet for assembly work, 9 shuttle flights primarily for resupply and crew transfer, 42 resupply missions using the Russian Progress vehicle, 26 crew transfer missions using the Soyuz spacecraft, 2 Russian assembly flights using the Proton rocket, 2 Russian assembly and resupply flights using the Soyuz rocket, 2 European resupply missions, and 2 Japanese resupply missions (Fig. 10.4).

Fig. 10.4 Backdropped by a cloud-covered part of Earth, the ISS is seen from Space Shuttle *Discovery* as the two spacecraft begin their relative separation. Earlier, the STS-124 and Expedition 17 crews concluded almost nine days of cooperative work onboard the shuttle and station. Undocking of the two spacecraft occurred on June 11, 2008. (NASA photo no. s124e009968)

As predicted much earlier, resupply missions during the assembly phase of a large space station outnumbered missions devoted to delivering and installing parts by a substantial margin. In this case, the exact ratio was 81:31. In what was truly an international partnership, the United States assumed the primary responsibility for assembly using the NASA Space Shuttle, while the international partners conducted much of the work of crew transfer and resupply.

Assessment and Lessons Learned

In an odd reversal of the original concept, NASA needed the Space Shuttle for station assembly and relied upon other vehicles for station resupply. When first thinking about the shuttle-station relationship, station advocates planned to rely upon the shuttle for crew transfer and resupply but launch station components using large rockets like the Saturn V.

It is doubtful that the Space Shuttle as built could have served the purpose of resupply, especially in a cost-effective manner. During the initial assembly period beginning in late 1998, the median cost of a Space Shuttle flight exceeded $600 million. When flights resumed after the loss of the *Columbia*, the median cost jumped to $1 billion per flight. These figures include only operational expenses, not the amortized value of funds spent to build and upgrade the vehicles.

Early space station advocates envisioned a large facility with a crew of 50 to 100 individuals, necessitating a fleet of orbiters in constant motion to transport occupants to and from space and to keep the crew resupplied. NASA officials initially believed that they could do this work with a reusable spacecraft that cut transportation costs "by a factor of ten" [25]. The goal was set during the approval process for the Space Shuttle during 1971 and 1972. At that time, both the Titan IIIC and the Saturn IB rockets carried payloads into orbit for roughly $1000 per pound. The Saturn IB, a vehicle with a payload capacity roughly equal to that of the planned Space Shuttle, cost $55 million per flight as of 1972 [26]. The comparative cost figures took on mythical proportions during the debate over whether to approve the Space Shuttle, since they served as the official rationale for making the investment needed to produce the new vehicle.

In 1972, NASA officials estimated that a well-designed reusable space shuttle could be launched for about $10 million per flight. Such a vehicle would have provided an efficient method for resupplying any future space station. The estimate, NASA officials wrote, was "based on a careful assessment of NASA and contractor studies" [27].

In the first full years of shuttle flight (fiscal 1983 and 1984), the operational cost of each mission averaged $343 million. NASA officials flew four missions each year. Officials believed that they could depress flight costs by increasing the number of flights while maintaining a fixed annual budget. In 1985, they flew eight missions (instead of the previous four) on a fixed $1.3 billion operations budget. Per-flight cost fell to $164 million. NASA officials sought to triple the rate to 24 missions per year, aiming for an operational cost of $54 million per flight.

A $54 million per flight rate would have provided an effective mechanism for resupplying and even assembling a large space station. The inflation-adjusted expense of a Saturn IB launch by 1983, with space station planning underway, would have been $138 million. With experience, alas, the logic collapsed. The calculations misestimated the marginal costs of launching additional shuttle flights each year. When NASA and its contractors attempted to increase the flight rate in 1986, the *Challenger* exploded. When missions resumed, the average flight cost rose to $393 million.

NASA officials genuinely believed that they could make their cost models work. They relied heavily on airline analogies as their means to that end. Airline companies use robust engines, computerized checkout, and components that can be replaced without disassembling aircraft to keep them in the air earning revenue [28]. NASA officials planned to use similar techniques. For justification as well as transport, station planning between 1981 and 1986 depended heavily upon the logic of the Space Shuttle—both its capabilities and projected costs. It would not be unfair to say that the future course of the space station initiative rode on every flight of the NASA Space Shuttle.

The actual course of shuttle events undercut this plan and precluded it as the primary launch vehicle for station resupply. Even by consolidating the 81 resupply and crew transfer flights that took place during assembly on a lesser number of shuttle missions, it is doubtful that the Space Shuttle could have provided a cost-effective method for delivering crew and supplies to a growing orbital base. Moreover, the suspension of station flights following a shuttle accident without a primary backup vehicle would have left the station crew in a very precarious position. The Space Shuttle ceased to be the primary vehicle for crew transfer and station resupply.

That left the shuttle as the primary launch vehicle for station assembly. Space station planners from 1971 on never questioned its role in this regard. NASA officials presented the Space Shuttle as the nation's primary launch vehicle for military, scientific, and commercial payloads. Assembly of a large space station, should one be approved, was not exempt from this logic—even after the *Challenger* accident. In fact, after the second shuttle accident, station assembly became the primary justification for continuing flights.

Engineers designed most station components with the shuttle in mind. At 14 × 28 feet, the U.S. Destiny station laboratory neatly fit into the payload bay of the Space Shuttle *Atlantis*. The *Atlantis* also delivered the European space laboratory *Columbus*, which measured 15 × 23 feet. When a 780-pound pump powering the station cooling system failed in the summer of 2010, engineers realized that the spare could be transported only in the soon-to-be-retired Space Shuttle [29]. Astronauts replaced the unit and returned the failed module to Earth for analysis.

With the assistance of a few other launch vehicles, the shuttle fleet delivered to orbit most of the components of an ISS with a mass of one million pounds. It did so with 27 flights averaging $845 million per mission. The total expense, once thought to be so small it did not deserve inclusion in the station's overall cost estimate, exceeded $22 billion.

A heavy lift vehicle like the Saturn V rocket might have been more effective. The Saturn V rocket could place a payload with a mass approaching 285,000 pounds in LEO. The price of procuring and launching a Saturn V in 1969 was $225 million. The inflation-adjusted value of that price during the period of station assembly would be about $1.3 billion. The cost of four or five flights to deliver the components of a large space station would not have approached the $22.8 billion spent to fly the Space Shuttle for assembly work. The savings would have left a considerable sum available for any crewed flights needed to finish the assembly process.

Such comparisons are mostly hypothetical. The United States stopped making Saturn V rockets three decades before station assembly began. Two working models were left over, but they wound up in museums instead of space.

A number of lessons can be drawn from this experience. First, the actual course of events confirms the original supposition that the shuttle and station were deeply interdependent. The shape of the vehicle designed to serve the space station strongly influenced the shape of the resulting facility and its method of assembly. The shape of the shuttle dictated the size of station components. It forced a modular approach to station design—not Skylab-size modules like originally proposed, but smaller modules that fit in the shuttle payload bay. It encouraged a part-by-part construction technique. Small parts arrived on a growing number of shuttle flights. With astronauts in the delivery vehicle, it allowed the use of spacewalks for finishing work. It prompted the extensive use of space robotics, with astronauts using shuttle and station cranes to handle individual components. This approach in turn lessened the need for autonomous fly-and-dock techniques and eliminated the need for autonomous space tugs.

The modular design of the station, a consequence of the shape of the shuttle payload bay, invited assembly creep. Beggs, NASA administrator when the space station was approved, insisted that a station constructed from relatively small modules could be purchased like a bolt of cloth "by the yard." By that, he meant that the size and capabilities of a modular space station could be expanded as U.S. funding and international contributions enlarged [30]. True to his characterization, the number of shuttle assembly missions grew along with the number of component parts once the original configuration was approved.

Interdependence of the shuttle and station leads to a second observation. NASA works best when it designs launch vehicles with a clear vision of how they will be used. Design of the Space Shuttle proceeded without a simultaneous consideration of the space station design. This imposed a severe hardship on NASA engineers. It forced them to design the shuttle/station system in an incremental fashion—first the shuttle, then the station. As a consequence, the shuttle could not be designed with reference to the primary purpose for which it was conceived. Instead, it was designed in such a fashion as to accomplish a substantial number of activities once thought to be secondary to its station functions. The decision to set aside station design during the shuttle approval process resulted in a launch vehicle that, as members of the CAIB wrote, sought to be "all things to all people" [31]. The resulting vehicle was too large to play a useful resupply role and too small for cost-effective assembly work.

NASA officials sought to use the Space Shuttle for many activities. Indeed, more than 70 percent of the total number of shuttle flights wound up somewhere other than the ISS. Yet those capabilities drove the expense of working in space well beyond the level that anyone imagined when the shuttle was approved.

Third, the shuttle/station experience reaffirms the value of a mixed fleet for complex space activities. In fact, a mixed fleet built and resupplied the ISS. For assembly, the international partners utilized the U.S. Space Shuttle, the Russian Proton rocket, and the Russian Soyuz rocket. For crew transfer and resupply, they used the Soyuz spacecraft, the Russian Progress cargo vehicle, the European Automated Transfer Vehicle, the Japanese H-II Transfer Vehicle, and the NASA Space Shuttle. The idea that just one or two vehicles could do all of this work in a cost-effective manner was probably naive.

Finally, the long history of shuttle/station design and assembly demonstrated the value of international cooperation for undertakings of this scale. International participation rescued the U.S. component of the space station from a slow death by retrenchment. It created a station with far more capability than originally conceived. It proved that people from different political and engineering cultures could produce a phenomenally complex facility. It provided the station with a much-needed mixed fleet of launch and resupply vehicles.

Construction of the ISS was a great technical achievement. It ranks with the greatest engineering projects in the history of humankind. The U.S. Space Shuttle played a critical role in the station's design, shape, and assembly mode. From the perspective of the original shuttle/station advocates, however, the shuttle did not accomplish its primary purpose of crew transfer and station resupply, nor did it complete its work in a cost-effective manner.

ACKNOWLEDGMENTS

The author expresses his appreciation to Matthew Vanderschuere, a doctoral student at American University, who assisted with collecting data for this chapter.

REFERENCES

[1] Froehlich, W., *Space Station: The Next Logical Step*, NASA EP-213, Washington, DC, 1984.

[2] NASA, "Final Report of the International Space Station Independent Safety Task Force," NASA Historical Reference Collection, NASA History Office, NASA Headquarters, Washington, DC, Feb. 2007.

[3] NASA/Marshall Space Flight Center, "33-Foot-Diameter Space Station Leading to Space Base," 1 Jan 1969, available on-line at http://archive.org/details/MSFC-9801788 (accessed 26 July 2010); see also Heppenheimer, T. A., *The Space Shuttle Decision: NASA's Search for a Reusable Space Vehicle*, NASA SP-4221, Washington, DC, 1991, p. 227.

[4] Space Task Group, "The Post-Apollo Space Program: Directions for the Future, Report to the President," Washington, DC, Executive Office of the President, Sep. 1969.

[5] Heppenheimer, T. A., *The Space Shuttle Decision: NASA's Search for a Reusable Space Vehicle*, NASA SP-4221, Washington, DC 1999, pp. 228–229.

[6] Mueller, G. E., Associate Administrator for Manned Space Flight, NASA, "Honorary Fellowship Acceptance," address delivered to the British Interplanetary Society, University College, London, England, 10 Aug. 1968, pp. 2, 5, NASA Historical Reference Collection; see also Mueller, "The New Future For Manned Spacecraft Developments," *Astronautics & Aeronautics*, Vol. 7, Mar. 1969, pp. 25–27.

[7] Peterson, C. B., "Integral Launch and Reentry Vehicle Study Information Data Package," Marshall Space Flight Center, 28 Feb. 1969, JSC History Collection at the University of Houston—Clear Lake archival collection; See also Heppenheimer, *Space Shuttle Decision*, p. 117; and NASA Space Shuttle Task Group Report, Vol. 2, "Desired Systems Characteristics," in Logsdon, J. M. (ed.), *Exploring the Unknown: Selected Documents in the History of the U.S. Civil Space Program, Vol. IV: Accessing Space*, NASA SP-4407, Washington, DC, 1999, pp. 206–210.

[8] Clarke, A. C., *2001: A Space Odyssey*, New American Library, New York, 1968, Chap. 7; also Kubrick, S., *2001: A Space Odyssey,* Metro-Goldwyn-Mayer Pictures, Beverly Hills, CA, 1968.

[9] Space Task Group, "The Post-Apollo Space Program," pp. 14, 20.

[10] McCurdy, H. E., *The Space Station Decision: Incremental Politics and Technological Choice*, Johns Hopkins University Press, Baltimore, MD, 1990, p. 27.

[11] Cole, E. G., Program Manager, Space Station Program, *Modular Space Station Phase B Extension, Preliminary Performance Specification, Vol. II—Project*, North American Rockwell, Space Division, Downey, CA, December 1971, pp. 14–52; see also Hook, W. R., "Historical Review," *Transactions of the ASME,* Vol. 106, November 1984, pp. 276–286.

[12] Beggs, J. M., "Why the United States Needs a Space Station," remarks prepared for delivery at the Detroit Economic Club and Detroit Engineering Society, 23 June 1982, NASA Historical Reference Collection, also in *Vital Speeches*, Vol. 48, 1 Aug. 1982, pp. 615–617.

[13] Beggs, "Why the United States Needs a Space Station."

[14] U.S. Senate, Committee on Commerce, Science, and Transportation, "Nominations—NASA," 97th Cong. 1st sess., June 17, 1981, p. 22.

[15] Reagan, R., Text of remarks of the President on the landing of the Space Shuttle *Columbia*, Dryden Flight Research Facility, July 4, 1982, NASA Historical Reference Collection; also in *Weekly Compilation of Presidential Documents.*
[16] McCurdy, *The Space Station Decision,* p. 85.
[17] McCurdy, *The Space Station Decision*, pp. 171, 232.
[18] McCurdy, *The Space Station Decision*, p. 85.
[19] Office of Space Station, *The Space Station: A Description of the Configuration Established at the Systems Requirements Review (SSR)*, NASA Headquarters, Washington, DC, June 1986, p. 35.
[20] Fisher, W. F. and Price, C. R., "Space Station Freedom External Maintenance Task Team, Final Report," July 1990, NASA Historical Reference Collection.
[21] NASA, "Final Report of the International Space Station Independent Safety Task Force," p. 67.
[22] Morgan, C., *Shuttle-Mir: The U.S. and Russia Share History's Highest Stage,* Lyndon B. Johnson Space Center NASA SP-2001-4225, Houston, TX, 2001.
[23] Harwood, W. "ISS Spacewalk History," *CBS Space News*, 20 Aug. 2011, available online at www.cbsnews.com/network/news/space/home/spacecalc/evastats.html (accessed 7 November 2012).
[24] Beggs, J. M., to Senator Edwin (Jake) Garn, 5 June 1984, quoted in McCurdy, *The Space Station Decision*, p. 220.
[25] Office of the White House Press Secretary, "Press Conference of Dr. James Fletcher, Administrator, NASA, and Deputy Administrator George M. Low," 5 Jan. 1972, p. 8, NASA Historical Reference Collection.
[26] McCurdy, H. E., "The Cost of Space Flight," *Space Policy*, Vol. 10, Nov. 1994, p. 286.
[27] Fletcher, J. C., Administrator, to the Honorable Walter F. Mondale, 25 Apr. 1972, NASA Historical Reference Collection.
[28] Heppenheimer, *Space Shuttle Decision*, pp. 246–254.
[29] Klotz, I., "Space Station Problem Shows How Shuttle will be Missed," *Discovery News*, August 2, 2010.
[30] McCurdy, *Space Station Decision*, pp. 169–170.
[31] *Columbia Accident Investigation Board Report*, Vol. 1, NASA, Washington, DC, Aug. 2003, p. 23.

SUGGESTED READING

Burrough, B., *Dragonfly: An Epic Adventure of Survival in Outer Space*, HarperCollins, New York, 1998.
Harland, D. M. and J. E. Catchpole, *Creating the International Space Station*, Springer Praxis, New York, 2002.
Kitmacher, G., *Reference Guide to the International Space Station: Assembly Complete Edition*, NASA, Washington, D.C., November 2010.

Chapter 11

A Victory for Clean Interfaces: Europe's Participation in the Space Shuttle Program

John Krige

Introduction

When President Nixon announced in January 1972 that the United States should proceed at once to develop the Space Shuttle, he made only passing reference to international cooperation. However, just before the president spoke, he mentioned to NASA Administrator Jim Fletcher and his deputy, George Low, who were with him at the San Clemente White House, just how important international collaboration was, particularly flying astronauts from all nations, East and West (he specifically suggested putting a Bulgarian into orbit) [1]. Nixon also said that it would be valuable to encourage "meaningful participation" in experiments "and even in space hardware development" [2].

Standing beside the president, Fletcher picked up on these hints at once. He gave "greater international participation in space flight" as one of four main justifications for developing the shuttle as the core of the American post-Apollo program. The State Department was also pleased. When U. Alexis Johnson, the under secretary for political affairs, phoned Fletcher a couple of days later to congratulate him, he remarked that the shuttle "would be damn useful and valuable from a foreign policy and public-relations point of view" [3]. Thus was NASA's new STS officially born: as a vector for American leadership in the new place being carved out for humans in space, and as a platform for international cooperation. In so doing, it ably exemplified the two main aspects of NASA's mandate as laid down in the National Aeronautics and Space Act of 1958.

For the Nixon administration, the post-Apollo program was always going to have an international component. The matter was broached while the president was flying to the splashdown of the crew of the Apollo 11 mission along with Secretary of State William P. Rogers, National Security Adviser Henry Kissinger, and NASA Administrator Thomas Paine [4]. Paine was emphatic that the United States should now emphasize international cooperation rather than costly and duplicative competition. Those on board Air Force One agreed. With the space race won, détente was in the air, new space powers were emerging, and America had to position itself in a multipolar world.

As originally conceived, the shuttle was just one element of a far more ambitious post-Apollo program that included an orbiting space station that was to serve as a base for further exploration of the cosmos, including the moon and Mars. Paine actively promoted this agenda in several countries—Australia, Canada, Japan, and Western Europe. Japan was hesitant, eventually deciding that it lacked the indigenous technological capability to take advantage of the American offer [5]. In 1974, the Canadian National Research Council decided to finance the construction of the highly successful Canadarm—a 15-meter-long space crane that was attached to the edge of the orbiter vehicle's cargo bay [6]. The Europeans, for their part, were immediately intrigued by the offer. For two years, from mid-1970 to mid-1972, they had active and intense discussions with the agency over what they could contribute to the post-Apollo program.

NASA actively supported European involvement and proposed a wide variety of collaborative options with varying degrees of hardware integration. At one extreme, the Europeans were invited to build parts of the orbiter itself. At the other, they were encouraged to build modules that could be plugged into the shuttle's cargo bay: a Sortie Can or shirtsleeve experimental laboratory and/or a Research and Applications Module (RAM), an instrument-carrying pallet flown on the shuttle or attached to the station. The tug was a third, intermediate option. It was a cryogenically fueled propulsion system that snuggled in the shuttle's cargo bay and that was used to carry (and recover) satellites to (and from) orbits other than the shuttle's LEO—notably communications satellites in geosynchronous orbit.

During 1970 and 1971, notwithstanding hesitations on both sides of the Atlantic, both partners became increasingly invested in this joint venture, above all in the tug. Then, in June 1972, Herman Pollack of the State Department announced to a European delegation in Washington that only the Sortie Can remained as a suitable project: Cooperation in building parts of the orbiter itself, and the development of the tug, was off the table. The immediate reaction was one of shock and dismay. Only Germany stayed on board, in part to gain access to NASA's project management skills, in part for its own foreign policy reasons. It took the lead in building Spacelab, a version of the Sortie Can. Spacelab carried a European astronaut, Ulf Merbold, into space on its first mission in 1983, planting the seed for Germany to host the European Space Agency's astronaut training center, which was established in Cologne in 1990.

The international legacy of the Space Shuttle resides in Spacelab, Canadarm, and the use of the shuttle by foreign astronauts (see Chapters 8 and 12). The other ambitious projects originally embarked on with the Europeans—participation in building parts of the orbiter and building the tug—are often overlooked because they never materialized. Yet from the point of view of the "lessons of history," they provide a valuable supplement to the more visible and well-known international achievements. International technological collaboration in a strategic domain like space triggers concerns regarding its potential threat to national security and economic competitiveness. It is an interagency issue that engages the highest levels of the state apparatus. By studying why NASA eventually walked away from these modes of technological collaboration with Western Europe, this chapter throws light on the *limits* of what the agency could achieve as it navigated through this complex field of force in the early 1970s. NASA was well aware that it was taking on a challenge when it decided to engage others technologically in the orbiter. In the event, the measures that it took to avoid jeopardizing American national security and economic advantage were overwhelmed by the forces aligned against it and by the agency's own doubts about the wisdom of the path it had taken. International collaboration is one of NASA's success stories. A case study of the evolution of the U.S.-European technological collaboration on the orbiter shows just how difficult it can be, and it provides insights into the pitfalls to be avoided.

THE BIG PICTURE

In February 1971, Arnold Frutkin, who was responsible for NASA's international affairs for 20 years (1959–1979), explained the benefits of including Europe in the core of the post-Apollo program to Robert Behr of the National Security Council [7]. Europeans would contribute $1 billion to the STS, 10 percent of its estimated development cost of $10 billion. Participation would encourage "cooperative dependency on the United States" so hampering parallel programs to develop European independence of American technology. European participation would also impede the proliferation of missile technology and quench a challenge to U.S. control over the telecommunications satellite market: Europe could not afford both to develop an autonomous launch capability and to engage $1 billion in post-Apollo. Including international partners would also enable NASA to carry out a "personal and repeated directive of the president." Behr extended the list by adding the industrial angle in a long memorandum that he prepared for Kissinger in March 1971, in which he spelled out both the pros and cons of European participation. Many large U.S. firms, he said, had discovered that their international business depended on their having access to pertinent foreign capabilities and skills. For them, even a 10 percent participation in the STS program would build the kind of "framework" that would narrow the vast technological gap between the United States and Western Europe, and would facilitate transatlantic collaboration at the level of the firm [8].

Several factors drove the whittling down of the scope of international collaboration in the shuttle from joint technological development to building a separate module or Sortie Can:

1. Fears of accelerating Europe's rising technological strength in the space sector, ironically the very fact that also made it a desirable partner
2. Hostility among senior members of the Nixon administration to Paine himself, and to his international vision for the post-Apollo program
3. Reluctance to fight for that vision by Paine's successor, Fletcher, and his deputy administrator, Low
4. Ambiguity about the precise scope of the president's commitment to internationalizing the shuttle
5. The elevation of the Soviet Union and China, at the expense of Western Europe, to Nixon's foreign policy priorities in the early 1970s.

This immensely rich and complex story is described in greater detail elsewhere [9]. I will concentrate here on one main problem that loomed over the entire project from its inception: technology transfer [9]. The orbiter that was eventually built exemplified the conservative solution to international collaboration that Frutkin had put in place in the mid-1960s: no exchange of funds between partners, and the preservation of clean technological and managerial interfaces [10]. Hostile forces in the White House including the presidential science adviser, Ed David, strongly urged the agency to accept this solution. Fletcher and Low adopted it as their own in 1972. Only the State Department could oblige them to be more flexible in the interests of diplomacy, but it too yielded to their wishes. By June 1972, no one could justify an intimate collaborative relationship with Western Europe that engaged sensitive technologies and that involved the risk of significant knowledge flows between American and European aerospace firms.

DEFINING THE SCOPE OF COLLABORATION

Within months of Armstrong and Aldrin landing on the moon, Paine was touring European capitals whipping up enthusiasm for the post-Apollo program and stressing how much international partnership mattered to NASA. The Europeans were entering a prolonged crisis of their own at the time, redefining the mission of their first space research organization (ESRO) beyond science to include applications, notably telecommunications satellites, and grappling with political divisions over the need to pursue their deeply flawed launcher program, Europa. Participation in post-Apollo was at once flattering and intimidating. On the one hand, NASA, which had just performed an awesome space first, deemed them worthy partners. On the other, given their limited resources, participation in post-Apollo seemed possible only at the expense of developing an indigenous launch capability. The one certainty that had given focus to their program in the late 1960s—acquire independent access to space to secure a slice of the lucrative and politically important communications satellite market—crumbled before them. To clarify their position, they needed an indication from NASA of where they might contribute.

In response to European requests, NASA arranged for briefings on the station and the shuttle in Europe in summer 1970. There they were told that participation could take four forms. Two of them—conceptual and R&D studies, and using the shuttle for foreign experiments and astronauts—will not detain us here. There were other, more substantial alternatives.

First, there was the development of major modules of a total system, where each side could, with relative independence, contract with its own industry for the required technology. The orbit-to-orbit space tug to be carried aloft in the shuttle cargo bay fell in this category, as did the Sortie Can and a RAM. Both could be developed in close coordination with NASA. An American prime contractor would be responsible for managing and integrating the overall system.

Tighter still was the "development of integral systems." Frutkin explained that this "might include the development of some part, element or subsystem of the shuttle itself, such as structural elements of the shuttle, propulsion elements, life support systems, control systems, and so on." This would be organized by United States-European international consortia with an American prime and a European subcontractor.

Frutkin emphasized in June 1970 that nothing was yet fixed in stone. At this early stage, he was laying out general ideas. "No drawings, designs or words should suggest closed doors," he said. "There are no closed doors here. The doors are, in fact, all open wide" [11]. That said, Europe could not hesitate on the threshold indefinitely. Time was of the essence. NASA would proceed with its development program come what may. The sooner the Europeans decided what they wanted to do, the greater would be their impact on the program's overall shape.

The early enthusiasm in Europe for this exciting new opportunity was quickly dulled. The British, who had recently elected a new Conservative government to replace Harold Wilson's Labor, picked up on the difficulty NASA was having in getting Congress to support Paine's original program. The new minister of aviation supply, Freddy Cornfield, noted that there had been considerable changes in the post-Apollo proposal since it was first suggested, and that Congress had still not approved a specific configuration [12]. The new British government was engaged in a comprehensive review of public expenditure. It could not accept "a commitment to share the costs of 10% participation [in post-Apollo], running to as yet unquantifiable but probably very large sums of money, and this in a context of a project too loosely defined to enable any assessment to be made of the benefits in relation to resources involved" [13]. To emphasize the point, the British also remarked that Japan had obtained launch assurances from the United States and the right to manufacture a U.S. launcher under license. "The British therefore failed to see why Europeans should have to participate in the post-Apollo program for the same benefits Japan has obtained for nothing" [14].

Frutkin quickly turned this argument to NASA's advantage. He wrote to Behr to tell him of Britain's reluctance to commit to post-Apollo participation because of the uncertainty of the U.S. commitment to the space shuttle and to continuity in NASA's major programs like skylab. The White House should endorse NASA's FY1972 budgetary and program decisions as soon as possible. They did not, of course.

French representatives expressed a different concern about the wisdom of any kind of European commitment. François-Xavier Ortoli, the minister of industrial development and scientific research, insisted that whatever its final configuration, a 10 percent participation in post-Apollo would probably far exceed the costs of developing a European launcher. In return, the benefits were dubious. U.S. guarantees for providing launchers, especially for communications satellites, were not watertight, and access to technology was restricted. On balance, therefore, it was cheaper and more advantageous technologically and industrially for Europe to go it alone [15].

Technical discussions continued notwithstanding the political turmoil that swirled around post-Apollo in both Europe and the United States. In fall 1971, NASA suggested a number of work packages on the orbiter that would be suitable for European industry. They also presented conceptual designs for the tug they had in mind.

NASA and its contractors identified potential European contributions to the orbiter by breaking down its frame and its vertical tail into discrete components. Fourteen work packages suitable for collaboration and costing some $400 million to develop were defined on this basis (Table 11.1).

TABLE 11.1 WORK PACKAGES ON THE ORBITER SUGGESTED FOR COLLABORATION IN NOVEMBER 1971

Part	Cost
1. Tail assembly (fin, perhaps rudder/speedbrake), *NOT leading edge and thermal protection*	\$20–\$30 million
2. Main wing, *NOT leading edge and thermal protection*	\$65–\$75 million
3. Elevon	\$20–\$25 million
4. Central fuselage, fore and aft	\$100–\$125 million
5. Cargo bay door	\$30–\$40 million
6. Radiator	\$10–\$15 million
7. Landing gear and door	\$20–\$30 million
8. Nose section	\$3–\$5 million
9. Ejection seat	\$10–\$12 million
10. Instrumentation (difficult to integrate)	
Propulsion (without engines)	
11. Orbital maneuvering system (OMS) pods	\$40–\$50 million
12. Reaction control system pods	
13. Air-breathing engine pods	\$20–\$30 million
14. Auxiliary power unit	
	Total Cost: ~\$400 million

The second major element in the post-Apollo package was the space tug, which would transfer payloads from the shuttle's LEO to higher orbits, most notably the geostationary orbit. In April 1970 and again in February 1971, European space officials allotted about \$1 million for conceptual studies of the tug to British and German firms, whose proposals were strongly endorsed at the joint meeting of experts in Huntsville in November 1971. Those present saw the tug as "a logical area for European participation in the post-Apollo activities since it [was] an easily 'separable' item with a relatively clean set of interfaces" that provided significant technology challenges to European firms [16]. On the day Nixon authorized the shuttle in January 1972, RFPs for two Phase A space tug studies costing \$1.4 million were issued in Paris, with more money in the pipeline [17].

Frutkin was firmly persuaded that the tug was "a far a better subject for a technological exercise than an essentially outmoded launch vehicle" [18]. Moreover, the Europeans liked the tug. NASA foresaw as many 30 flights annually for the reusable vessel in its 60-flight-per-year mission model, indicating the need for 3 to 5 new tugs annually—a boon to European industry. It was technologically challenging regarding propulsion, rendezvous, and docking [19]. It also provided the Europeans with direct access to the heart of NASA's space flight planning and operations.

DEALING WITH THE DANGERS OF TECHNOLOGY TRANSFER

Sharing dual-use space technology necessarily raises anxieties about proliferation to the detriment of commercial competitiveness and national security. Propulsion systems for launchers and missiles share much in common. As a spaceplane (see Chapter 1), the shuttle demanded the use of heat-resistant materials (on the leading edge of the wing, for example) that were under development for supersonic aircraft, commercial and military.

National Security Decision Memorandum (NSDM) 72 was released on July 17, 1970 [20]. It had to take into account the findings of two parallel studies that would result in National Security Study Memorandum (NSSM) 71, dealing with Advanced Technology and National Security, and NSSM 72, dealing with International Cooperation in Space. NSDM 72 addressed the "Exchange of Technical Data between the United States and the International Space Community." It stressed Nixon's commitment to international cooperation in ventures like the post-Apollo program and established an interagency group, to be chaired by a NASA representative, to review policy and procedure for technical data exchange between the U.S. and foreign governments and agencies, beginning with Europe. In addition, it specifically asked that those guidelines and procedures "be designed to provide for timely and effective interchange of technical information between the parties, while at the same time insuring the protection of U.S. national interest." Frutkin quickly set up a group that included representatives from the State Department, the DoD, and the National Security Council to formulate suitable technology transfer policies [21]. These policies were defined in consultation with leading aerospace contractors, and they embodied their views on the advisability, or not, of collaborating with the Europeans.

The lynchpin of NASA's argument was that "the thrust of foreign participation in [the] post-Apollo program will be to contribute to the U.S. effort rather than set up a flow from us." This contribution would be secured by selecting projects for development abroad only "where the capacity is already substantial in Europe." Major American aerospace contractors confirmed that this was possible. Grumman Aerospace told NASA that the French firm Marcel Dassault was "one of the most capable manufacturers of high performance aircraft in Europe" and "should be able to contribute any portion of the Shuttle prime structure that France might undertake" except perhaps the main cryogenic tankage. It added that the German firm Dornier was "well capable of handling structural sub-assemblies" for the shuttle and had excellent research and test facilities that could be used during the development of the orbiter [22]. North American Rockwell was contracting with the British Aircraft Corporation (BAC) for shuttle Phase B participation in "structural elements, aerodynamics, flight test instrumentation, and data handling" [23]. McDonnell Douglas was actively pursuing international collaboration with ERNO in

Germany, Hawker Siddeley Dynamics (HSD) in Britain, and Société Nationale Industrielle Aerospatiale (SNIAS) in France [24]. Europe could also make major contributions to the tug and the RAM. In summer 1971, Messerschmitt-Boelkow-Blohm (MBB) and HSD led a group of 10 European companies and presented their design studies [25]. The Convair Division of the General Dynamics Corporation, which had been selected to perform a Phase B RAM study, was subcontracting parts of the work out to MATRA in France (systems design and analysis, and guidance and control), ERNO in Germany (material science and manufacturing in space), SAAB in Sweden (phased array, data processing, and image compensations), and Selenia in Italy (bulk data handling, millimeter wave communication system, and so on) [26].

Charles Donlan, assistant director on the Space Shuttle program, personally confirmed the technological capability of a number of European aerospace companies late in 1971. He made site visits to SNIAS in Paris, ERNO in Bremen, MBB in Munich, BAC in Bristol, and HSD in various locations. At SNIAS, he was struck by their hardware activity relative to the Airbus and Concorde as well as to intermediate-range ballistic missiles. SNIAS was already collaborating with McDonnell Douglas on the shuttle. ERNO had "a strong, if somewhat conservative capability" and was actively involved in Airbus and the third stage of the Europa II rocket. At MBB, Donlan was impressed by the management of large and diverse programs, from aircraft and spacecraft to subway trains. BAC had gained extensive experience on large structures and systems through working on Concorde. But Donlan reserved his most enthusiastic praise for HSD, where he was treated to a "spectacular demonstration of the Harrier capabilities," including short and vertical takeoffs and landings, and low-level, high-g turns at Mach speeds above 0.8. The Trident nuclear missile submarine system, the Blue Streak first stage of the Europa launcher, a British satellite, and R&D work with composites and carbon filters for the Harrier's wings were all observed and discussed. Donlan concluded that the firms he visited represented collectively "an impressive array of talent in the aeronautical and space fields." Thanks to major aerospace programs in Britain and on the continent, Europe obviously had something to contribute to the U.S. effort and could be relied on to meet exacting American standards. Donlan also realized that companies abroad had more experience than their American counterparts in working in different languages across technical interfaces [27]. There was not going to be a one-way flow of technology from the United States to Europe in the post-Apollo program [28]. As Frutkin emphasized over and again, "[...] our principal objective will be to *obtain* a foreign technical contribution, rather than to *provide* technology." Of course, he added, "*Anything* we do by way of serious cooperation with other countries will inevitably enhance their capabilities should they wish to divert them to military purposes" [29]. But that was a risk that had to be taken if NASA was to fulfill the president's directive.

The major American aerospace corporations agreed with him. As Donlan found, European contractors did not need American technology as such. What they wanted was general management and systems engineering know-how that was not easily transferred anyway. Foreign participation would stabilize the program absent an "assurance of adequate and steady funding" from Congress. It would curb the "stimulation of independent and competing programs in Europe." And because the venture was so challenging, the U.S. aerospace industry would benefit far more from its 90 percent share than the Europeans would from their 10 percent effort, emerging "from the post-Apollo enterprise even further ahead of the Europeans than when we started" [30].

The Mounting Opposition to Technology Transfer

NASA and the aerospace industries' persistent efforts to downplay the dangers of technology transfer to Europe ran into a brick wall at the highest levels of the administration. Tom Whitehead, a former RAND systems analyst tasked with evaluating NASA budget and planning proposals, fired the first salvo early in 1970 in a memo to Peter Flanigan. Flanigan was an investment banker and Nixon's deputy campaign manager in the 1968 presidential election campaign. He had oversight responsibilities for space as part of his general duties as assistant to the president for internal economic affairs [31]. Whitehead wrote him that the time had come to rein in NASA's "empire-building fervor." Paine was using international collaboration as an instrument to lock the president into a post-Apollo program that was still not clearly defined, leaving him little alternative but to go along with it. To do this, Paine was reckless enough to give away America's "space launch, space operations, and related know-how at 10 cents on the dollar," a reference to the 10 percent foreign share in the program. What NASA needed now, he wrote, was a new administrator who would present the White House and the Bureau of the Budget (renamed the Office of Management and Budget in July 1970) with "broad but concrete alternatives," "someone who will work with us rather than against us [. . .]" [32]. Paine's days were indeed numbered. His maverick personality and his disregard for the complexities of the budgetary process led Nixon stalwarts to regard him "at best as irrational and at worst as obstinately arrogant" [33]. James Fletcher replaced him in April 1971.

Flanigan tenaciously pursued his campaign to rein in NASA. In February, he met with Ed David, the presidential science adviser; Ehrlichman; and the president himself. They reportedly concluded that post-Apollo cooperation with Europe would not only entail unwanted and undesirable technology transfer but would raise major managerial headaches too [34]. Soon thereafter, Kissinger was also brought into the picture [35]. David repeated these concerns to the national security adviser. He added that by collaborating, the United States would both be unable to reshape the post-Apollo program as needed and would unduly strengthen Europe's capacity to compete commercially in the global satellite market.

The issue was taken up again at a meeting called by Kissinger in April soon after Fletcher took up his post. Flanigan and Whitehead claimed that NASA had inflated the president's wish for international collaboration—he sought little more than to fly a foreign astronaut in space. David joined them in emphasizing again the danger of "high-technology transfer." Cracks began to appear in NASA's position. The new administrator admitted that, in fact, he was opposed to "an 'integrated' program wherein the Europeans would do sub-systems of the shuttle." Fletcher preferred "a separable program where the Europeans would do a complete task such as the space tug or a space station module" [36]. NASA was instructed to prepare a detailed report of the risks of technology transfer and to report back within two months. That report, whose substance was described in the previous section, did nothing to convince the presidential science adviser. As he put it in a long memo to Kissinger in July 1971, "I am not prepared to have the U.S. commit itself to this cooperative program of STS development," nor could he be until the final configuration of the shuttle, and Europe's contribution to it, were settled [37].

The pace of decision-making accelerated once Nixon authorized the shuttle in January 1972. A European delegation that came to Washington in February was stunned to find that its scope for a contribution to the orbiter had been dramatically reduced from 14 to 5 work packages in just a few months (Table 11.2). Europe's potential financial contribution to shuttle development had also dropped sharply, from about $400 million to $100–$115 million for the new items. These were "limit[ed] to subsystems which require least

TABLE 11.2 CHANGE IN WORK PACKAGES OFFERED BY NASA FOR EUROPEAN COLLABORATION FROM LATE 1971 TO EARLY 1972

Orbiter Work Packages Suggested for Collaboration in November 1971	Orbiter Work Packages Suggested for Collaboration in February 1972
1. Tail assembly	1. Tail assembly
2. Main wing	2. ——
3. Elevon	3. Elevon
4. Central fuselage	4. ——
5. Cargo bay door	5. Cargo bay door
6. Radiator	6. ——
7. Landing gear and door	7. Landing gear and door
8. Nose section	8. Nose section
9. Ejection seat	9. ——
10. Instrumentation	10. ——
11.–14. Propulsion (without engines)	11.–14. ——
TOTAL COST: ~$400 million	**TOTAL COST: $100–$115 million**

transfer of technology" [38]. As Frutkin stressed, Europeans were now being offered elements that their firms could carry out "substantially on their own, thus minimizing European need for U.S. technical assistance" [39]. They were also "relatively simple and exclude the most interesting tasks" [40].

By March 1972, Kissinger and David were convinced that Europe should not have any role at all in developing hardware for the Space Shuttle [41]. Their concerns centered on "protecting the technological position of the U.S., maximizing balance of payments and employment benefits for the U.S., and avoiding managerial difficulties that may be encountered in international cooperation in technological activities." NASA was aligned with them. Deputy Administrator Low wrote that "our position is that from a programmatic point of view we would like to develop the Shuttle and all of its ancillary equipment domestically." NASA would seek foreign participation in the *use* of the shuttle (including the tug and the Sortie Can) but not in its development [42]. The only condition under which they would revise their position would be if the State Department came up with overwhelming foreign policy reasons to keep the Europeans deeply engaged. NASA Administrator Fletcher concurred.

The Secretary of State desperately sought a flexible exit strategy that would save America's credibility in the face of its European partners who had, by then, spent $11.5 million on shuttle and tug studies [43]. To no avail. David stepped in and insisted that "the U.S. can accept European participation in the shuttle program, if limited to RAM and Sortie payload modules." It was left to Herman Pollack from the State Department to break the news in June. Europe, led by Germany, went ahead with the Sortie Can, called Spacelab [44]. Led by France, the Europeans also developed their own expendable launcher, Ariane.

The tug was taken off the table along with the orbiter for a number of reasons that were peculiar to it. It was technologically challenging, and American contractors had not done much work on it. Here, there might be considerable flow of sensitive propulsion technology to the Europeans. NASA was beginning to have doubts about the predicted frequency of use of the tug in its hopelessly optimistic mission model for the orbiter (50 percent of as many as 60 flights annually). Upstream of the question of use, NASA was also having doubts about the safety and the technological feasibility of the tug concept itself [45]. The Air Force also weighed in [46]. Although money would be saved, the DoD was reluctant to have foreign powers develop one of their key technologies. They would have to reveal the nature of their missions. Their requirements might be jeopardized by unilateral decisions, technological and industrial deficiencies, and a lack of operational support by the Europeans. Building the tug abroad would also undermine the domestic industrial base in an already-weakened sector that was crucial to national defense. It seemed to NASA, which summarized the situation, that the Air Force would "undoubtedly" manufacture a European tug under license in the United States if it were developed abroad, or it would build its own alternative if it wanted one [47]. Europe's hopes of building a tug production line imploded: No one would buy the tugs.

THE LESSONS

With the wisdom of hindsight, it is evident that Paine's enthusiasm to integrate the Europeans into the core of the STS blinded him to the dangers of technological leakage to European firms and to the opposition that would be generated elsewhere in the administration. The first major lesson to be learned from this is that the agency operates in a complex political context over which it has limited control and—above all—that it is crucial for the administrator to be in tune with that context if NASA is to fulfill its objectives. This was woefully lacking in the formative years of the post-Apollo program. Nixon was genuinely committed to building a new international order. He and Kissinger, Robert Schulzinger writes, "started détente as a recognition of the relative, not absolute decline of U.S. power and the growth of mulipolarity" [48]. The Soviet Union and then China were the focal points of their foreign policy; Europe was less central. In practice, this meant that Nixon was more intent on building international bridges with the communist world than with European allies. He was also not particularly interested in space, whose ratings had plummeted in the eyes of an America far more concerned with matters on Earth. For Paine to push through his agenda, he needed to build strong alliances with the president's closest advisers. He did just the opposite, leading Flanigan to call desperately for a new administrator who would work with the OMB and the White House, not against them [49]. The gap between the NASA administrator and his ambitions, and the president and his, was filled by Flanigan, Whitehead, and David, who were determined to rein in NASA's budget and to help rebuild an American aerospace industry that was facing dire cuts as the Vietnam War wound down. Fletcher, Low, and Frutkin aligned themselves with these priorities.

That said, it is doubtful that a better understanding between Paine and the White House would have led to a different outcome anyway. The second general lesson of this story is that international collaboration involving technological sharing is hostage to major imbalances in technological capability between the partners. In the post-Apollo era, the centrifugal forces against collaboration—technological leakage and increased managerial complexity—easily overwhelmed the counterweight provided by financial and political arguments for an international effort. Successful international collaboration in science and technology demands reciprocity. The advanced technology and industrial contributions that Europe could bring to the table were substantial, but nothing that American firms could not do themselves quickly if they needed to. Europe was not indispensable, nor could it make itself so. Riding the wave of Paine's enthusiasm for international collaboration in the early

Nixon administration, post-Apollo collaboration soon foundered on the asymmetries between the two sides of the Atlantic. It was almost inevitable, then, that the major foreign contributions to the STS—Spacelab and Canadarm—respected Frutkin's mantra that international collaboration should embody clean interfaces and no exchange of funds between the partners. Conversely, the intimacy of technological collaboration (and the degree to which it will protect a program from hostile external forces) is defined by the extent to which there is technological interdependence.

Finally, it is worth noting that *NASA itself* drew lessons from the failed attempt to integrate partners into the technological core of the STS. As an organization (Chapter 3), NASA has an institutional memory and is capable of learning from past experiences. Ken Pederson, who took over from Frutkin as director of the Office of International Affairs in 1979, was determined not to make the same managerial mistakes when he began building an international partnership for the space station in 1982. The most important lessons from the post-Apollo experience, he wrote, were "to avoid making premature commitments or promises and to avoid broad statements that can be misconstrued" [50]. He wanted foreign participation from the outset in the conceptual development of the station but was emphatic that partners should concentrate on mission requirements and not on hardware developments. National industries could be involved in conceptual studies but, to avoid tying NASA's hands when it came to choosing contractors, there would be no industry teaming until late in Phase B, unlike in post-Apollo. Technology transfer would be restricted by limiting cooperative agreements to "discrete hardware pieces with minimal interfaces" [51]. Foreign partners would be officially represented in the United States and would work exclusively through headquarters; foreign visitors would not be allowed to go to NASA centers to avoid being enrolled in inter-center competition. Overall, Pedersen wanted to keep the station program flexible for as long a possible, and above all avoid being locked into a particular configuration prematurely by verbal agreements, financial commitments, and industrial engagements. This would allow NASA to change its priorities if it had to and would give it more time to assess the strengths of European industry. It could better control the contributions that others made, restrict technology transfer, and keep them out of the critical path. In this way, he hoped to avoid the embarrassing situation that had arisen in 1972, when "the government had to walk back from the European perception of the cooperative possibilities" in the post-Apollo program, creating suspicion and distrust that still persisted in some quarters [52]. Pedersen also warned against repeating other management and legal arrangements. It should also not commit itself, as it had in the Spacelab Memorandum of Understanding (MOU), to purchasing a flight unit produced abroad. And it should be prepared for foreign partners who contributed a piece of hardware to want preferential or free access to the whole station.

In sum, a major lesson that NASA itself drew from post-Apollo was, as McCurdy has noted, to reaffirm "the traditional conservative values that had governed international participation within NASA for more than twenty years" [53] and to implement them rigorously. The fact that that lesson was jettisoned later to accommodate Russia does not detract from its initial impact on planning the space station as an international enterprise.

REFERENCES

[1] Transcript of a Conversation between Johnson and Fletcher on January 7, 1972, in *Foreign Relations of the United States, 1969–1976, Vol. E-1, Documents on Global Issues, 1969–1972, Chap. 4, International Cooperation in Space, 1969–1972* [hereafter *FRUS*], Item 274, available at http://www.2001–2009.state.gov/r/pa/ho/frus/Nixon/e1/c15660.htm.

[2] Memorandum for the Record, Low, G. M., "Meeting with the President on January 5, 1972," January 12, 1972, Record No. 12575, Federal Agencies-Presidents, Nixon Administration, Folder Nixon Correspondence (NASA), 1972–1974.

[3] Conversation between Johnson and Fletcher, January 7, 1972.

[4] Prologue, The Paine-Keldysh File, NASA SP-4209 The Partnership: A History of the Apollo-Soyuz Test Project.

[5] Logsdon, J., *Learning from the Leader: U.S. – Japan Collaboration in Space*, unpublished MSS, George Washington University, Washington, DC, undated.

[6] Linbergh, G.,"Candarm and its U of A Connections," *University of Alberta Engineer Magazine*, Summer 2004, Issue 14.

[7] Frutkin, A., Memo to File, "Matters at Stake in Post-Apollo Launch Availability, Phone Call to Colonel Behr, February 25," February 26, 1971, Federal Records Center, Suitland, Accession Number 255–74–734, Record Group NASA 255, NASA Division I Escalation Files [hereafter Record Group NASA 255, NASA Division I Escalation Files], Box 17.VI.D.7.

[8] Memo Behr to Kissinger, "Post-Apollo Space Cooperation," *FRUS*, March 4, 1971, Item 256.

[9] Krige, J., Callahan, A. L., and Maharaj, A., *NASA in the World. Fifty Years of International Collaboration in Space*, Palgrave Macmillan, New York, 2013, Chaps. 4–6.

[10] Frutkin, A., *International Collaboration in Space*, Princeton University Press, Princeton, NJ, 1965.

[11] Frutkin, A. W., "NASA Briefing on Space Station, ESRO, Paris, June 3–4, 1970, Closing Remarks," and "Organizing International Participation in the Post-Apollo Program," June 4, 1970, Record Group NASA 255, NASA Division I Escalation Files, Box 14.I.E.

[12] Heppenheimer, T. A., *The Space Shuttle Decision. NASA's Search for a Reusable Space Vehicle*, NASA SP-4221, Washington, DC, 1999; Hoff, J., "The Presidency, Congress and the Deceleration of the U.S. Space Program in the 1970s," edited by R. D. Launius and H. E. McCurdy, *Spaceflight and the Myth of Presidential Leadership*, University of Illinois Press, Champaign, 1997, pp. 92–132; Logsdon, J. M., "The Space Shuttle Program: A Policy Failure?", *Science*, Vol. 232, No. 4754, May 30, 1986, pp. 1099–1105; Logsdon, J. M., "The Decision to Develop the Space Shuttle," *Space Policy*, Vol. 2, May 1986, pp. 103–119; Williamson, R. A., "Developing the Space Shuttle," *Exploring the Unknown. Selected Documents in the History of the U.S. Civil Space Program, Vol. IV, Accessing Space*, edited by J. M. Logsdon, NASA SP-4407, Washington, DC, 1999, Chap. 2.

[13] European Space Conference. *Annex to (Draft) Minutes of the Meeting held on the Morning of 4th November 1970, CSE/CM* (November 70) PV/1, November 4, 1970, Record Group NASA 255, NASA Division I Escalation Files, Box 15.V.B.

[14] Telegram, Am embassy Brussels to the State Department, "European Space Conference," November 6, 1970, Record Group NASA 255, NASA Division I Escalation Files, Box 16.VI.B.1.

[15] European Space Conference. *Part II of (Draft) Minutes of the Meeting held on the Morning of 4th November 1970, CSE/CM* (November 70) PV/1, November 4, 1970, Record Group NASA 255, NASA Division I Escalation Files, Box 15.V.B.

[16] Report of the Meeting of the Joint Group of Experts on U.S./European Cooperation in Space Programs in the Post-Apollo Period, Held in Washington D.C., 30 November–3 December, 1971, Record Group NASA 255, NASA Division I Escalation Files, Box, 17.IX, p. 12.

[17] Freitag, R. F., "Memorandum of a Telecom," January 6, 1972, Record Group NASA 255, NASA Division I Escalation Files, Box 14.I.C.4.

[18] "Diary Note, Paris Discussion on European post-Apollo Reactions, Meeting with Prof Levy, 9/22/71," Record Group NASA 255, NASA Division I Escalation Files, Box 15. IV.B.3a.

[19] European Space Conference *Report by the Joint ESRO/ELDO Working Group on the Post-Apollo Programme*, WG/COOP/US/23, October 1971, p. 16 in Record Group 255, NASA Division I Escalation Files, Box 16.VI.A.II.

[20] nixon.archives.gov/virtuallibrary/documents/nsdm/nsdm_072.pdf

[21] Letter Packard to Paine, June 3, 1970, Record Group NASA 255, NASA Division I Escalation Files, Box 14.I.G.

[22] Memo attached to letter Lawrence M. Mead (Grumman Aerospace) to Freitag, April 2, 1972, Record Group NASA 255, NASA Division I Escalation Files, Box 16.VI.C.

[23] Letter NAR to Handel Davies, BAC, undated but numbered #320–70, Record Group NASA 255, NASA Division I Escalation Files, Box 16.VI.C, Industrial, Institutional.

[24] A presentation of May 7 is attached to memo from Sh (Sam Hubbard) dated May 19, 1971, and further information is provided in Memo from Sam Hubbard, *McDonnell Douglas Aircraft Company's Activities in International Cooperation,* June 18, 1971; the firm's report, *Space Shuttle Program. International Implementation Development,* Revised June 29, 1971. Both in Record Group NASA 255, NASA Division I Escalation Files, Box 17.VI.D.3; Memo from Hanger, Manager Marketing, Space Shuttle Program to multiple recipients, "SNIAS Space Shuttle Participation," May 18, 1971 and "ERNO Participation in Space Shuttle," May 18, 1971, both in Record Group NASA 255, NASA Division I Escalation Files, Box 17.VI.D.5. Memo Freitag, NASA OMS, "SNIAS Visit to NASA", May 14, 1971, Record Group NASA 255, NASA Division I Escalation Files, Box 16.VI.D.1.

[25] MBB Group Space Tug System Study. *Pre-Phase A/Ext, Final Presentation to ELDO, NASA,* and *European Space Tug. Final Presentation July 71, Pre Phase A study, part 2, Prepared for the European Launcher Development Organisation by Hawker Siddeley Dynamics Limited Leading a group of European companies,* Record Group NASA 255, NASA Division I Escalation Files, Box 16.VI.D.1.

[26] Memo Sam Hubbard to Messrs Frutkin, Morris and Barnes, Phase B RAM Study, April 15, 1971, Record Group NASA 255, NASA Division I Escalation Files, Box17. VI.D.4.

[27] Donlan, C. J.,"Memorandum for the Record" on "Memorandum for M/Mr. Myers," both dealing with his European Visit to XXII International Astronautical Congress and Organizations and Firms Interested in Post-Apollo Participation, both dated November 1, 1971, Record Group NASA 255, NASA Division I Escalation Files, Box 14.II.G.

[28] Memo Sam Hubbard to Messrs Frutkin, Morris and Barnes, Phase B RAM Study, April 15, 1971, Record Group NASA 255, NASA Division I Escalation Files, Box17. VI.D.4.

[29] Frutkin, A., "Notes for Discussion with Dr. Kissinger," June 10, 1970, Record Group NASA 255, NASA Division I Escalation Files, Box 14.I.F.
[30] For their views, see part IV of the study "Technology Transfer in the post-Apollo Program," about July 1970, Record Group NASA 255, NASA Division I Escalation Files, Box 14.I.H.
[31] Hoff, J., "The Presidency, Congress and the Deceleration of the U.S. Space Program in the 1970s," edited by R. D. Launius and H. E. McCurdy, *Spaceflight and the Myth of Presidential Leadership*, University of Illinois Press, Champaign, 1997, pp. 92–132.
[32] Memorandum Whitehead to Flanigan, February 6, 1971, attached to Memo Flanigan for Ehrlichman, in Logsdon, *Exploring the Unknown, Vol. II,* Document I-19.
[33] Memorandum Whitehead to Flanigan, February 6, 1971, attached to Memo Flanigan for Ehrlichman, in Logsdon, *Exploring the Unknown, Vol. II,* Document I-19, p. 95.
[34] Briefing Memorandum from Pollack to Johnson, June 5, 1971, *FRUS*, Item 267.
[35] Memo David to Kissinger, March 17, 1971, *FRUS*, Item 258.
[36] *FRUS*, Item 262.
[37] Memo David to Kissinger and Flanigan, *Post-Apollo Space Cooperation with the Europeans*, July 23, 1971, *FRUS*, Item 268.
[38] Unsigned memo, "Possible Spectra of Alternatives," January 24, 1972, attached to Arnold Frutkin, "Diary Note. Post-Apollo Coordination, January 20, 1972, Record Group NASA 255, NASA Division I Escalation Files, Box 14.II.H.
[39] Frutkin, A. "Approaches to Foreign Participation in the Shuttle Program," Memorandum, January 18, 1972, Record Group NASA 255, NASA Division I Escalation Files, Box 14.II.H.
[40] Frutkin, A. "Post Apollo Action Alternatives," Memorandum, February 29, 1972, Record Group NASA 255, NASA Division I Escalation Files, Box 14.II.H.
[41] Information memorandum from Pollack to Rogers, *Post-Apollo Cooperation in Jeopardy*, March 17, 1972, *FRUS*, Item 277.
[42] Memo Low to Fletcher, "Position Paper on Post Apollo International Cooperation," 27 March, 1972, NASA Historical Archives, Record No. 14462, International Cooperation and Foreign Countries, International Cooperation, Folder International Space Documents from Dwayne Day, 1959–1975.
[43] Rogers, W. P. Secretary of State, to the President, Memorandum, April 29, 1972, *FRUS*, Item 279. See also Doc I-26, Logsdon, gen. ed., *Exploring the Unknown, Vol. II.*
[44] Lord, D. R., *Spacelab. An International Success Story*, NASA SP-487, Washington, DC, 1987, tells the history of its development.
[45] Sakss, U.J., Memorandum for File, "Causse/Dinkespiler Visit April 19 to Discuss Post-Apollo Questions," May 2, 1972, Record Group NASA 255, NASA Division I Escalation Files, Box 14.I.1.
[46] The briefing charts spelling out the Air Force's arguments for and against using a tug developed in Europe was attached to Memo, "Air Force Use of European Tug," from Sam Hubbard to Barnes, Morris and Frutkin, May 17, 1971, Record Group NASA 255, NASA Division I Escalation Files, Box 16.VI.D.1.
[47] The briefing charts spelling out the Air Force's arguments for and against using a tug developed in Europe was attached to Memo, "Air Force Use of European Tug," from Sam Hubbard to Barnes, Morris and Frutkin, May 17, 1971, Record Group NASA 255, NASA Division I Escalation Files, Box 16.VI.D.1.
[48] Sulzberger, R. D., "Détente in the Nixon and Ford Years, 1969–1975," edited by M. P. Leffler and O. A. Westad, *The Cambridge History of the Cold War. Vol. II Crises and Détente,* Cambridge: Cambridge University Press, 2010, pp. 373–394, at p. 374.
[49] Cited by Hoff, op. cit., p. 100.
[50] Pedersen, K. S., "The Changing Face of International Space Cooperation. One View of NASA," *Space Policy*, Vol. 2, May 1986, pp. 120–135, at p. 131.

[51] Memo Kenneth Pedersen to John Hodge, *Strategy for International Cooperation in Space Station Planning*, undated, but around August 1982, reproduced in John M. Logsdon, ed., *Exploring the Unknown, Vol. II,* Doc I-31, pp. 90–100, at 99.
[52] Memo Kenneth Pedersen to John Hodge, *Strategy for International Cooperation in Space Station Planning*, undated, but around August 1982, reproduced in John M. Logsdon, ed., *Exploring the Unknown, Vol. II,* Doc I-31, p. 92.
[53] McCurdy, H. E., *The Space Station Decision. Incremental Politics and Technical Choice*, Johns Hopkins University Press, Baltimore, MD, 1990, p. 103.

Chapter 12

Living and Working on the Shuttle: The Challenge of Routine

Amy E. Foster

Introduction

If there was one word to describe what NASA intended the Space Shuttle to be (or at least how NASA sold the vehicle), it is "routine" [1]. Throughout this book, we have seen that the shuttle was meant to provide a cost-effective, regular means of getting into LEO. Whether the shuttle lived up to the promises—or the hype—does not change one important fact: that this vehicle created a living and working environment that was very different from the environment in any other space vehicle designed by NASA.

Throughout the first decade of the U.S. space program, the astronauts flew in small, cramped capsules that offered little working space and no privacy. Until the shuttle and Skylab, engineers paid only minimal attention to the concept of habitability in the design of NASA's space vehicles, not because of any unwillingness to address an astronaut's comfort but because the mission and the constraints placed on that mission by the space race took precedence. However, with the end of the Apollo program, and subsequently the space race, NASA's new emphasis on science meant that design elements more conducive to living and working in space, instead of just surviving it, became more important as well. Not only that, but the emphasis on "routine" flight opened up the idea that being an astronaut no longer meant being a "fighter jock" or a single combat warrior [2]. This opened the door for astronaut candidates who had yet to break into such an elite profession, particularly women,

---

but it also expanded opportunities for scientists, physicians, and eventually teachers. This chapter explores how the Space Shuttle program redefined the historical concepts of habitability for NASA, what it meant to live and work in space, and who flew as astronauts. NASA's attempt to sell and design the shuttle in a way that made spaceflight "routine" actually resulted in technology that redefined what it meant to fly in space.

Habitability as a Measure of "Routine"

When President Richard Nixon announced his support for the development of the Space Shuttle on January 5, 1972, NASA was simultaneously working on the development of Skylab (1973–74), America's first experimental space station [3]. The conceptualization of the Skylab project took place between 1965 and 1969, earlier than the shuttle. However, the real design and construction of Skylab occurred simultaneously with the conceptualization of the shuttle in the late 1960s and early 1970s. Issues of habitability and similar component design resulted in overlap between the two programs. These two programs pushed NASA engineers to think more about the physiological and psychological needs of its astronauts. For the three-man crews on Skylab, their missions would last no less than four weeks (Skylab 2) and upwards of three months (Skylab 4), which made private space for sleeping and downtime significantly more important than it was even on the longest Apollo mission (Apollo 15, which lasted 12 days, 17 hours, and 12 minutes) [4]. The Space Shuttle was different from Skylab in that its missions were expected to last only between 7 and 30 days [5]. But the shuttle complement was larger than that on Skylab, numbering between four and seven crewmembers once the shuttle became operational after its fourth flight in 1982 [6]. (Although the standard largest complement was seven astronauts, eight astronauts flew on board STS-61-A in 1985.) For both Skylab and the Space Shuttle, the design engineers focused more on making the astronauts content with their living conditions instead of just able to survive them.

"Habitability" became the new buzzword for both Skylab and the shuttle. The discussion was not new, however; the issue had worked its way into the discussion of design as early as 1964, when Frank B. Voris, the chief of human research in the Biotechnology and Human Research division of NASA's Office of Advanced Research and Technology, wrote for public distribution a pamphlet about the hazards and safety concerns for astronauts. His greatest concerns focused on the long-term effects of weightlessness on the body's cardiovascular and skeletal systems, radiation exposure, metabolism changes, how to handle bodily wastes, and to a lesser extent psychological problems. Admittedly, though, Voris argued that both lessons learned from isolation studies on the ground and the astronauts' extensive training would prepare them so well that their emotional reactions to apprehension, fear, and stress were likely to be negligible [7]. Although most of these concerns were inconsequential for the short-duration missions being flown in the 1960s (at least the engineers of the Apollo era were forced by technical and time constraints to make them a lower priority), as NASA turned its focus toward longer-duration flight, these issues of astronaut health and well-being—both physical and psychological—became more significant.

Nevertheless, the Skylab engineers faced fewer challenges to making the laboratory a comfortable living and working environment for the astronauts than the shuttle engineers. Skylab's orbital workshop, where the astronauts spent almost all of their time, measured just over 295 cubic meters (10,426 cubic feet) [8]. In comparison, the flight and mid-deck of the shuttle measured only 71.5 cubic meters (2525 cubic feet), which was one-quarter the size and which had to accommodate an extra two to four crewmembers on most of the shuttle flights between 1982 and 2011 [9]. Making the very best use of limited space while still providing for the needs of the astronauts became an even greater priority when engineers designed the shuttle.

So what factors defined "habitability" aboard the shuttle? As with Skylab, the elements that the physicians focused on most were the astronauts' diets, exercise, and hygiene (not surprisingly, because physicians emphasize these issues with their Earth-bound patients). For the engineers, the most pressing factors were privacy and sleeping accommodations. With a crew of seven astronauts in a living space equivalent to an average recreational vehicle for a couple of weeks, privacy became a commodity. What made privacy and "habitability" a more pressing issue for the shuttle engineers was that for the first time in NASA history, they had to design a vehicle for a mixed-sex crew. The concern was not that these male and female co-workers would be working together; that was far from new. The concern was that they would be *living* together.

In 1976, NASA announced a new call for astronauts, the eighth group to be selected in the agency's 20-year history. The previous seven astronaut classes were not all cut from the same cloth (two groups—Group IV and Group VI—were selected as scientist-astronauts); however, the make-up of these groups was very similar: Caucasian and male [10]. With the new call for astronauts, NASA specifically targeted women and ethnic minorities to apply. But why then? Why was there little to no emphasis on recruiting women or racial minorities before the mid-1970s? The answers are complex, but the major reason does have a lot to do with the construction of the shuttle and how it redefined spaceflight.

DIVERSITY AND THE SPACE SHUTTLE

The discussion of racial inequality in NASA's astronaut corps has gotten only modest attention from historians, although there are a few exceptions [11].

The noticeable lack of women in the astronaut corps before 1978 has gotten significantly more [12]. This has much to do with the historical evolution of the astronaut selection criteria, the public outcry in the 1960s by women who felt discriminated against, and political fallout from the launch of Soviet cosmonaut Valentina Tereshkova on June 16, 1963.

For the first three astronaut classes, NASA hardly changed the selection criteria for its candidates: in addition to passing a Class 1 military-flight status physical exam, all the astronauts had to be test pilots, most of which came from the military [13]. Group II astronauts Neil Armstrong and Elliot See were the first astronauts to qualify as "civilian" astronauts, meaning neither was serving as active duty military when they were selected. Armstrong was a test pilot for NASA's Flight Research Center, and See was working as a flight test engineer for General Electric. Both had previous military experience, though. With the selection of the first scientist-astronauts, NASA relaxed the medical requirements (because the Mercury program had effectively shown that spaceflight was not as rigorous or physically taxing as had been imagined before 1961), and it focused more on candidates with doctoral-level training or advanced experience in science, engineering, or medicine [14]. These selection criteria, by default, restricted most women and many minorities from qualifying for the astronaut corps because of their minority representation in test pilot school and scientific fields, which served as pipeline careers into the astronaut corps.

Two African American men came closest to becoming astronauts in the 1960s. Captain Edward Dwight applied for astronaut training in March 1963 after finishing eighth in his class at the experimental test pilot school at Edwards Air Force Base; however, the selection board subsequently did not choose him. Robert Lawrence had been selected to participate in the Air Force's Manned Orbiting Laboratory (MOL). This program was essentially the military version of NASA's spaceflight program. However, Lawrence died in a routine training flight on December 8, 1967, 18 months before the DoD cut the MOL budget, cancelling the program. Although he hadn't flown as part of NASA's program, the U.S. House acknowledged him as "the ninth U.S. astronaut to die on the ground" [15]. Although still few in number, some racial-minority men did meet the minimum qualifications for the astronaut corps. That could not be said about women—Caucasian or not. Until the 1970s, when the U.S. military opened pilot training and test pilot school to women, no woman had that military test pilot training or experience. The Army and Navy opened their doors to women in pilot training in 1973. The Air Force followed suit in 1976. Capt. Jacqueline "Jackie" Parker became the first woman to enter the Air Force's test pilot school as a pilot in July 1988. In the 1960s, when most of the astronauts being selected were chosen as pilots, this seriously hampered women's abilities even to qualify for astronaut candidacy [16].

The opportunities for women specifically, but also for racial minorities, were better in the sciences and engineering than in the upper echelons of the military (with test pilot status representing the top of the "ziggurat," as described by author Tom Wolfe) [17]. When NASA announced its call for astronauts in 1976, it solicited applications for pilots and, for the first time, mission specialists. Unlike the pilot-candidates and the scientist-astronauts of the 1960s, the mission specialists were not required to have any pilot training, which created new opportunities for those women who had a background in science, engineering, and medicine but who had never learned to fly or had not served in the military. In part because NASA saw the Shuttle program as a platform for new and groundbreaking research, it wanted active corps members with expertise in those scientific, engineering, and medical fields, not just ad hoc training like that the Apollo astronauts had received in geology before their flights [18]. (Dr. Leon Silver, a geologist from Caltech, trained the Apollo crews, beginning with Apollo 13.)

In January 1978, NASA announced the selection of the first class of Space Shuttle astronauts, which included six Caucasian women, three African American men, and one Asian American man. Donna Miskin, a journalist for the *Washington Star*, identified the look of the new astronauts as "a NASA affirmative action poster" [19]. The women of this class were Anna Fisher, Shannon Lucid, Judith Resnik, Sally Ride, Margaret Rhea Seddon, and Kathryn Sullivan. Known as Group VIII, this was the largest class of astronauts that NASA had selected to date, with 14 pilot-astronaut candidates and 21 mission-specialist candidates, for a total of 35. The size of the class addressed not only the need to replace the Gemini and Apollo astronauts who were either moving into management or retiring, but also to expand the astronaut corps because the shuttle was being designed for missions with a larger crew. The diversity of the class—in terms of sex, ethnicity, and professional background—reflected NASA's changing agenda, one that ideally emphasized America's technological achievements that made it possible to live and work in space (Fig. 12.1).

Although NASA's upper administration—to include NASA Administrator James Fletcher and JSC Director Christopher Kraft—had been openly discussing and planning to integrate women into the astronaut corps at least as early as 1972, transforming NASA and the shuttle in light of that realization came slowly. In part, NASA's new emphasis on "routine" meant that living and working in space should be close to living and working on Earth; no astronaut should be inconvenienced or limited because of her anatomy or his size. Further, NASA envisioned a shirt-sleeve working environment. Engineers had to address issues related to the design of the orbiter's hardware (the most famous being the toilet) as well as personal gear (such as spacesuits for EVA, and urine collection devices for launch and landing) for mixed-sex crews. However, given

Fig. 12.1 NASA's first six women astronauts pose with a mockup of a personal rescue enclosure (PRE) or "rescue ball" in the crew systems laboratory at the JSC. The group includes mission specialists, from left to right, Margaret R. (Rhea) Seddon, Kathryn D. Sullivan, Judith A. Resnik, Sally K. Ride, Anna L. Fisher, and Shannon W. Lucid. (NASA photo no. GPN-2002-000207)

the relatively small size of the orbiter's crew compartment, one of the engineers' most challenging design criteria was how to provide privacy.

Allen Louviere, chief of the Engineering Technology Branch at JSC, identified in a January 1972 memo a list of preferences for privacy expressed by the astronauts. They included an "enclosed hygiene/toilet component," "provisions for privacy without isolation," and "movable partitions, modularized appointments, and variable lighting ... to satisfy personal arrangements and accommodate different functions, e.g. dining, recreation, sleeping, etc" [20]. When engineers designed Skylab, providing for this level of privacy (separate sleeping berths for each astronaut, designated dining facilities, and a private shower) came fairly easily because of the available space within the crew module. Skylab's habitable space, on the other hand, measured nearly four times the size of the orbiter's crew compartment and had to accommodate a crew only half the size.

Engineers took their solution from the practice used on naval vessels—"hot bunking." Early in the Shuttle program, the orbiters were designed to carry three horizontal bunks and one vertical bunk installed against one wall of the mid-deck. The procedure of "hot bunking" meant that two astronauts working opposite schedules shared the bunk. This guaranteed each astronaut a quiet and private space to call his or her own, which lessened the larger public concerns about privacy among a mixed-sex crew. However, not every flight used these bunks because not every flight had a split work schedule.

Further, when the crew was actually large enough to have a split work schedule, that was when the crew compartment felt the most cramped. This ultimately led to the decision by the astronauts themselves to abandon the bunks altogether. What the astronauts found was that this attempt at creating a "routine" living environment actually conflicted more with life on the shuttle than it complemented it.

Where technology failed to improve living conditions on the shuttle, crew dynamics came through. Beginning with STS-1 until March 1988, when George S. W. Abbey took over as NASA's deputy associate administrator, he was responsible for selecting all the shuttle crews. Aside from engineering efforts to create aboard the orbiter a living and working environment that provided privacy and an improved sense of habitability that was overlooked in the space race days of the Mercury, Gemini, and Apollo programs, positive crew dynamics could do as much to make living and working on the shuttle "routine" and comfortable. This was particularly true when it came to mixed-sex crews, an observation made by both Rhea Seddon and Kathy Sullivan.

Shuttle crews trained together for a solid year, assuming no major launch delays [21]. Seddon said about the bond that formed between the crewmembers and even their families over that period, "A lot of social activity revolved around the crew that you were assigned to. People socialize together, did a lot of things together, and with only three flights I would have only been involved with that in three crews. But since my husband [Robert "Hoot" Gibson] flew five times, I really got involved in eight crews. There is a lot of camaraderie being a part of a crew" [22]. That dynamic translated to issues of privacy as well. Sullivan recalled [23]

> On one crew of mine—because we didn't have bunks—everybody mainly used the potty or the airlock when they were going to peel off their shirt. On another one of my crews, the guys or I would just say, "I'm going to change my shirt." The dynamic of that crew was that statement was a gentle announcement of a little preference for some privacy [but also] that there was not going to be some big hassle or embarrassment if somebody did turn around or come by me with my top off or a guy in his skivvies.

The situation was far from routine, but crews made the situation work for them.

The New Scientific Mission

When Abbey and his successors chose the crews for the shuttle missions, two driving factors contributed to the decision. Crew coherence, yes, but the mission agenda narrowed the selection process for the director of flight operations. How was the shuttle mission defined? What made it stand out when compared to the programs that preceded it? The answer was science.

NASA had proposed the shuttle to Congress and President Nixon as a "space truck" to the space station that it had hoped to build concurrently [24]. However, when the OMB slashed funding, leaving NASA to choose either the Shuttle program or the space station, the lack of funding meant redefining the Shuttle program's mission. Since December 1969, Deputy Administrator George Low had been working to reinvigorate NASA's public image, which needed a shot in the arm after the excitement over the success of Apollo 11 had died down. The growing concern in the United States focused on Earth-based problems. NASA's new mission in the 1970s mirrored that self-reflective attitude, turning its attention to science and ways in which the space program could help solve health and environmental problems on Earth.

NASA had been working on those environmental problems as part of its robotic and satellite programs, particularly with the development of the Earth Resource Technology Satellite (ERTS), which had been monitoring changing environmental conditions since 1972 [25]. (This program is better known to the general public as Landsat.) Low pushed his center directors, particularly his friend at NASA Langley, Ed Cortright, to identify what scientific and technological challenges NASA could address [26]. By emphasizing the contributions that NASA could make to society through research and development—particularly those contributions that could be developed using the shuttle—Low saw a way to not only reinvigorate public excitement for the space program, but also to justify the development of this new launch vehicle.

In 1975, the Ad Hoc Subcommittee on the Scientist-Astronaut, which served as part of NASA's advisory council, submitted to Low a report about the value of scientists in the astronaut corps and the efficacy of incorporating them into the Shuttle program. The report asserted that by "bridging the sometimes wide gap between scientific and flight operations points in view," the scientist-astronaut "can contribute to a productive Space Shuttle science program" [27]. Mostly as a way to keep the scientist-astronauts and the incoming mission specialists actively engaged in their core disciplines, NASA's administration created the "Life Sciences" and the "Space and Applications" Astronaut Offices at JSC [28]. NASA's growing focus on science—and less on the geopolitics that had been the foundation of the space race and the Apollo program—meant expanding the astronaut corps with the introduction of new mission specialists. It also meant keeping those astronauts actively involved in the scientific and engineering disciplines that would prove useful to the scientific missions NASA was planning for the shuttle. These two new offices helped to do that.

Historically, the selection of the Group VIII astronauts gets remembered for being the first class to include women. However, the make-up of the class in terms of the skill sets that it brought to the new program and its size reflected NASA's desire for more astronauts who were capable of serving as primary investigators and experiment operators aboard the shuttle. NASA initially used the term “experiment operator” to refer to the astronauts in flight who would be primarily responsible for running the experiments and managing the experimental packages. By the mid-1970s, though, NASA switched the term from “experiment operator” to “mission specialist.” A total of 21 (three-fifths of the class) of the astronauts chosen in 1978 represented science, technology, engineering, and mathematics (STEM) disciplines. This was certainly true of the six women, all of whom possessed either a Ph.D. in a scientific or engineering specialty, or were physicians. Fisher was trained as an emergency room physician. Lucid held a Ph.D. in biochemistry. Resnik’s Ph.D. was in electrical engineering. Ride earned her doctorate in astrophysics, Seddon was a surgeon, and Sullivan held her doctorate in marine geology [29].

Since 1978, NASA has selected 11 additional classes of astronauts, totaling 78 more pilots, 118 mission specialists, 32 international mission specialists, and 3 educator mission specialists [30]. The educator mission specialist class was introduced in 2004 with the selection of Group XIX. Barbara Morgan, best known for her role as the back-up to Christa McAuliffe as part of the Teacher-in-Space program, was selected in 1998 as part of the Group XVII class of astronauts. Although her training and experience was as a teacher, Morgan was selected as a full-fledged mission specialist, not as an educator mission specialist. At the time of publication, NASA was preparing to announce the names of the Group XXI astronauts, scheduled for June 2013. In every class of astronauts selected beginning in 1978, the mission specialists outnumbered the pilots, an indication of NASA’s desire to put science aboard the shuttle first (Fig. 12.2).

Fig. 12.2 Astronaut Robert L. Curbeam, Jr., mission specialist, works with the Bioreactor Demonstration System (BDS) on the Space Shuttle *Discovery*'s mid-deck in 1997. (NASA photo no. STS085-E-5011)

THE CHANGING IMAGE OF WORKING IN SPACE

Over the course of 30 years and 135 missions, the crews of the Space Shuttle launched dozens of commercial and military satellites; deployed, repaired, and serviced the HST; conducted hundreds of scientific and medical experiments; made significant contributions to constructing and servicing the ISS; and collaborated with the former Soviet Union to extend the life of and support the Russian space station, Mir. It was the shuttle's design that made all this possible. With a cargo bay 60 feet long and 17 feet wide and a larger crew capacity than any other vehicle built by NASA, the shuttle expanded the limits of what astronauts could accomplish [31].

Shortly after President Nixon announced his support for the Shuttle program, University of Michigan astronomer James A. Loudon wrote to the *New York Times*, "For the first time, scientists will be able to perform experiments in space without spending years in irrelevant pilot training first," which had been a condition required of the first two classes of scientist-astronauts but had been abandoned for the incoming mission specialists. He continued, "Very simply, the shuttle's effect will be to make space very much more available, with results for a dozen different branches of science that are now incalculable" [32]. Spacelab made much of this possible.

Spacelab was the combined effort of NASA and the ESRO, later called the ESA. Initially designed as small "sortie cans" that could be carried in the shuttle's cargo bay, Spacelab developed into a larger program with its own program office by October 1973 [33]. As part of the agreement between NASA and the ESRO, the latter got to name the program. That conglomerate of 11 European nations responsible for building the hardware changed the name from "sortie lab" to "Spacelab" in October 1973. Generally, Spacelab referred to a pressurized module that provided a "shirt sleeve" environment in which astronauts could work. However, the Spacelab system also included small modules and pallets that could house experimental packages. This combination of hardware gave NASA engineers considerable flexibility when it came to designing scientific missions for the shuttle. Although only the missions that flew the laboratory module were officially considered "Spacelab" missions, between 1981 and 1998, over 30 missions made use of some Spacelab hardware [34].

NASA tested the shuttle's capabilities as a scientific platform even before the vehicle had completed its test flights. (After four flights, the STS was declared operational.) The first Spacelab package was launched on March 22, 1982, aboard Space Shuttle *Columbia* (STS-3). Called OSS-1 after the Office of Space Science, the platform contained nine different experiments that spanned the scientific gamut from life sciences to space plasma physics. In addition, the experiments on OSS-1 represented the work of six different universities (five American institutions and one British institution) as well as the Naval Research Laboratory and NASA's Goddard Space Flight Center [35]. As NASA's new "routine" vehicle, its mission included serving not only as a scientific platform but also the launch vehicle of choice for the United States, if not the world (Fig. 12.3). Whether it was a military or commercial satellite or a scientific experiment that needed to launch, NASA promoted the shuttle as the best choice, mostly because of its crew of scientists and engineers who were on board to oversee and troubleshoot any potential problems. Of course, this proved particularly significant when it came to the HST.

The history of the HST chronicles both the highs and lows of the shuttle era. Nevertheless, in terms of the shuttle's contribution to the story, it epitomized what designers and astronauts had hoped for when they imagined working aboard the shuttle. In April 1990, the crew of *Discovery* deployed the space telescope from the shuttle's cargo bay, and then watched as the satellite's internal thrusters pushed it into a higher orbit to begin its observational mission. Two months to the day after Hubble's deployment, engineers discovered that a flaw in Hubble's primary mirror was causing the pictures it was taking and returning to Earth to be out of focus. Although the images were still of much greater detail than was possible with any land-based telescope in the world, after 13 years of planning and development—Congress approved funding for the HST in 1977 and development began in earnest in

Fig. 12.3 Almost the entire crew performs operations in the shirt-sleeve environment of the Spacelab Science Module in the cargo bay of the Earth-orbiting Space Shuttle *Columbia* on STS-94 in 1997. Payload specialist Gregory T. Linteris is looking over flight plan data in the right foreground. Astronaut Donald A. Thomas (upper right), mission specialist, works nearby. Astronaut Michael L. Gernhardt (nearer camera in left corner), mission specialist, and Roger K. Crouch, payload specialist, share an Inflight Maintenance (IFM) chore. Astronaut Janice Voss, payload commander, oversees the IFM. (NASA photo no. STS094-372-008)

1981—the media turned the flaw into a major black mark on the agency's record of achievement [36].

Understandably, the engineers who had worked on Hubble and the NASA family as a whole found the situation embarrassing. Having the shuttle, however, gave them the means and opportunity to fix the problem, a capability that no other launch vehicle in the U.S. fleet or in the world could manage. On December 2, 1993, the crew of STS-61 launched aboard *Endeavour* to attempt the first-ever satellite repair in space. After capturing Hubble, two teams of mission specialists (Story Musgrave and Jeffrey Hoffman worked together, and Kathryn Thornton and Tom Akers made up the second team) performed a series of five spacewalks over several days to install the Corrective Optics Space Telescope Axial Replacement (COSTAR) to serve essentially as a set of corrective lenses for the telescope. They also installed new solar panels to keep Hubble's batteries charged, two new pairs of gyroscopes to keep the telescope aligned on its target, and a new wide-angle camera [37].

Over the next 15 years, the astronauts performed four more servicing missions to Hubble, with the last mission performed in May 2009. Each time, the mission specialists upgraded computers, hardware, and software to extend the telescope's life and capability. The extent of the astronomical knowledge that the Hubble telescope will ultimately provide to scientists before it is finally decommissioned is vast. Most recently, the images sent back from Hubble include photographs of 7 primitive galaxies that formed around 13 billion years ago, somewhere between 300 and 650 million years after the Big Bang [38]. The shuttle has been a key factor in the ongoing success of this major scientific program.

In 1998, the astronauts along with their international partners began constructing the ISS, fulfilling the purpose envisioned for the shuttle back in the late 1960s and early 1970s. Before that happened, though, shuttle crews participated in a similar, although perhaps a more important, initiative: flying to the Russian space station Mir. In 1994, NASA and the Russian Space Agency began a partnership that involved sharing knowledge and resources. The Soviet Union had launched the first section of Mir in 1986. As the United States developed plans to build its own space station, a lot could be learned from the Russian experience. Further, after the collapse of the Soviet Union, the Russian Space Agency welcomed an experienced partner to keep Mir maintained, crewed, and supplied. Unlike the Apollo-Soyuz Test Project in 1975, this partnership signified a real working relationship with the Russians.

Over the course of 5 years, 7 American astronauts—Norman Thagard, Shannon Lucid, John Blaha, Jerry Linenger, Michael Foale, David Wolf, and Andy Thomas—spent nearly 1000 days living on Mir and working alongside Russian cosmonauts. In exchange, the shuttle performed nine separate docking missions to Mir, delivering new team members, bringing home departing

Mir team members, and resupplying the station and delivering new hardware. The NASA History Office has created an in-depth website for the Shuttle-Mir program, including a graphic timeline [39]. Often, Russian cosmonauts flew aboard the shuttle, serving as mission specialists and the technical experts on Mir for that mission. For NASA, the experience served as a dry run for the upcoming construction of the ISS. Equally important, though, was the development of that close working relationship with the Russians. The ISS was an *international* project, with contributions from the United States, Russia, ESA, and 14 individual countries. Although the two-way partnership of the Shuttle-Mir program could not mimic the more complex relationships of the ISS project, that learning experience—including the technical aspects of making the shuttle work with technology not designed by Americans—was invaluable (Fig. 12.4).

Fig. 12.4 Astronaut Jerry L. Ross, STS-88 mission specialist, is pictured during one of three space walks that were conducted on the 12-day mission between December 4 and 15, 1998. Perched on the end of *Endeavour*'s RMS arm, astronaut James H. Newman, mission specialist, recorded this image. Newman can be seen reflected in Ross' helmet visor. The solar array panel for the Russian-built Zarya module can be seen along the right edge. This was just the first of about 160 spacewalks totaling 1920 work-hours required to complete the ISS. (NASA photo no. STS088-355-015)

LESSONS LEARNED

After 30 years of flight, the shuttle remains the most complicated vehicle that NASA has ever flown. NASA administrators sold it as "routine" flight that would make space available and affordable. The latter was a failure, and those men and women who processed and maintained the fleet knew that flying the shuttle was never routine (something emphasized and reemphasized by the Rogers Commission after the *Challenger* accident and by the CAIB) [40]. However, even if the shuttle never reached the potential promoted by NASA's leaders, it did redefine what it meant to live and work in space.

First, the shuttle redefined what it meant to be an astronaut. The era of the Mercury 7—the manly, white, Protestant, "fighter jock" Cold Warriors—was over. The new astronaut corps included men and women, people of all ethnic backgrounds, as well as professional backgrounds, which gave the corps a more egalitarian feel.

Second, it redefined what spaceflight meant. The new focus on science and utility (the job of launching and repairing satellites) transformed not only the mission profiles of the 1960s into something more concrete, but it redefined the balance of power in the astronaut corps. The "scientist" dominated. By the time the shuttle was declared operational in 1982, the number of mission specialists was quickly outpacing the number of pilot-astronauts. Although it was the pilots who still commanded the shuttle, the scientists defined the raison d'être.

Last, if we think once again about the idea of the shuttle as "routine," that idea captured the changing desire at NASA for its mission to be something more than a political tool in the fight against the Soviet Union. As the geopolitical dynamic changed, even before the collapse of the Soviet Union, the shuttle opened up new opportunities for partnership, in large part because of its capacity to carry larger crews (seats were available for more than just American astronauts) and the changing mission profile. A lasting legacy of the shuttle era is the new era of international cooperation.

The Shuttle program could not live up to the promises made about it in the early 1970s. However, what rises to the top when we look at how NASA, its engineers, and its astronauts faced the challenges of living and working in space is a success story, one that redefined what it meant to fly in space and one that was in many ways a social experiment that reflected changing attitudes about gender, diversity, and international relations.

REFERENCES

[1] NASA. *Space Shuttle*, NF-79/6-77, Government Printing Office, Washington, DC, 1977, p. 1.
[2] Wolfe, T., *The Right Stuff*, Bantam Books, New York, 1980, pp. 101–105.
[3] Compton, W. D., and Benson, C. D., *Living and Working in Space: A History of Skylab*, NASA, Washington, DC, 1983.

[4] "NASA – Skylab," http://www.nasa.gov/mission_pages/skylab/index.html; also "NASA – Apollo," http://www.nasa.gov/mission_pages/apollo/index.html.

[5] NASA. *Space Shuttle*, p. 4; *The Shuttle Era*, NF-127/3-81, Government Printing Office, Washington, DC, 1981, p. 7.

[6] NASA, "Space Shuttle Basics," http://spaceflight.nasa.gov/shuttle/reference/basics/orbit.html; also http://www.nasa.gov/mission_pages/shuttle/shuttlemissions/archives/sts-61A.html (accessed 9 August 2012).

[7] Voris, F. B., *Medical Aspects of Space Flight*, National Aeronautics and Space Administration, Washington, DC, 1964, pp. 11–12.

[8] Belew, E. F. (ed.), *Skylab, Our First Space Station.* National Aeronautics and Space Administration, Washington, DC, 1977, p. 20, http://history.nasa.gov/SP-400/contents.htm. Accessed last 12 August 2012.

[9] NASA. *Life Aboard the Space Shuttle*, KSC 337–82, Government Printing Office, Washington, DC, 1983, p. 2; For a listing of the Shuttle crew members by mission, see http://www.nasa.gov/mission_pages/shuttlemissions/index.html (accessed 12 August 2012).

[10] Atkinson, J. D., and Shafritz, J. M., *The Real Stuff: A History of NASA's Astronaut Selection Program.* Praeger, New York, 1985, p. 83.

[11] Dunar, A. J., and Waring, S., *Power to Explore: A History of Marshall Space Flight Center, 1960–1990*, NASA SP-4313, Washington, DC, 1999, pp. 115–125; McQuaid, K., "'Racism, Sexism, and Space Ventures': Civil Rights at NASA in the Nixon Era and Beyond," edited by Dick, S. J., and Launius, R. D., *Societal Impact of Spaceflight*, NASA SP-2007-4801, Washington, DC, 2007, pp. 421–449; Moss, S., "NASA and Racial Equality in the South,1961–1968," M.S. thesis, Texas Tech University, 1997.

[12] Weitekamp, M., *Right Stuff, Wrong Sex: America's First Women in Space Program*, Johns Hopkins University Press, Baltimore, 2004; Foster, A. E., *Integrating Women into the Astronaut Corps: Politics and Logistics at NASA, 1972–2004,* Johns Hopkins University Press, Baltimore, 2011.

[13] Armstrong, N., and See, E., Astronaut Biographies, NASA Johnson Spaceflight Center, http://www.jsc.nasa.gov/Bios/htmlbios/astronbio_former.html (accessed 28 August 2012).

[14] Atkinson and Shafritz, *The Real Stuff*, p. 54.

[15] U.S. Congress, House, *Congressional Record*, December 12, 1967, H16747, referenced in Atkinson and Shafritz, *The Real Stuff*, p. 109. Seven of the remaining MOL astronauts transferred to NASA as Group VII in August 1969. See "NASA Astronauts Fact Sheet," http://ntrs.nasa.gov/archive/nasa/casi.ntrs.nasa.gov/19750007522_1975007522.pdf. Accessed 28 August 2012.

[16] Franke, L. B., *Ground Zero: The Gender Wars in the Military*, Simon & Schuster, New York, 1997; also "Women in Aerospace," http://www.womeninaerospace.org/about/history_achievements.html (accessed 4 September 2012).

[17] Rossiter, M., *Women Scientists in America: Before Affirmative Action*, and *Women Scientists in America: Forging a New World Since 1972*, Johns Hopkins University Press, Baltimore, MD, 1995 and 2012.

[18] Chaiken, A., *A Man on the Moon*, Penguin Books, New York, 1998; pp. 392, 393, 398, 399.

[19] Miskin, D., "In the New Astronaut Class," *Washington Star*, 19 November 1978, p. C-5.

[20] A. J. Louviere to W. D. Ray, 21 January 1972, EW Reading file, Box 007-25, Johnson Space Center Archives, University of Houston-Clear Lake, Houston, TX.

[21] Cooper, H. S. F., Jr., *Before Lift-Off: The Making of a Space Shuttle Crew*, Johns Hopkins University Press, Baltimore, MD, 1987.

[22] Seddon, M. R., interview with author, Nashville, TN, 12 February 2002.
[23] Sullivan, K., interview with author, Columbus, OH, 19 November 2002.
[24] Logsdon, J., "The Decision to Develop the Space Shuttle," *Space Policy*, Vol. 2, No. 2, 1986, pp. 103–119.
[25] "Landsat Science," http://landsat.gsfc.nasa.gov/about/landsat1.html.
[26] Low, G. to Cortright, E. M., 9 June 1971, File 87-12, Box 57, George M. Low Papers, Rensselaer Polytechnic Institute archives, Troy, NY.
[27] Naugle, J. E. to Newell, H., 27 June 1975, Files 87-12, Box 14, Low Papers.
[28] Low, G. M., "Memorandum: Astronaut Policy," 25 March, 1974, File 87-12, Box 15, Low Papers.
[29] "NASA – Astronaut Biographies," http://www.jsc.nasa.gov/Bios/index.html [accessed 3 November 2012].
[30] "Astronaut Bio: Barbara R. Morgan (7/2010)," NASA Johnson Spaceflight Center, http://www.jsc.nasa.gov/Bios/htmlbios/morgan.html; also see "Astronaut Selection," http://astronauts.nasa.gov/content/timeline.htm (accessed 14 November 2012).
[31] "NASA-The Orbiter," http://www.nasa.gov/returntoflight/system/system_Orbiter.html.
[32] Louden, J. A., "Why We Really Want a Space Shuttle," *New York Times (1923-Current File)*, 28 March 1972. p. 42.
[33] Shayler, D. J., and Burgess, C., *NASA's Scientist-Astronauts*, Praxis Publishing, Chichester, UK, 2007, p. 285; Rumerman, J. A., *NASA Historical Data Book, Vol. 5: NASA Launch Systems, Space Transportation, Human Spaceflight, and Space Science 1979–1988*, SP-4012, NASA, Washington, DC, 1999, p. 464.
[34] Shayler and Burgess, *NASA's Scientist-Astronauts*, p. 291.
[35] Rumerman, *NASA Historical Data Book*, p. 399.
[36] Johnson, G., "The Trouble with Hubble," *New York Times*, 1 July 1990, p. E7; "NASA Blamed in Hubble Flaw," *New York Times*, 26 November 1990, p. A17; NASA—Historical Milestones of the Hubble Project website, http://www.nasa.gov/mission_pages/hubble/story/timeline.html. Accessed 16 December 2012.
[37] Wilford, J. N., "Shuttle Nearing Space Telescope for Repair Job," *New York Times*, 4 December 1993, p. 8; Wilford, J. N., "Shuttle Releases Hubble Telescope: Repaired and Re-equipped, Eye on the Stars is Sent into Orbit with Hope," *New York Times*, 11 December 1993, p. 11.
[38] NASA – NASA's Hubble Provides First Census of Galaxies Near Cosmic Dawn, http://www.nasa.gov/mission_pages/hubble/science/galaxy-census.html. Accessed 17 December 2012.
[39] Shuttle-Mir Program, NASA History Office, http://spaceflight.nasa.gov/history/shuttle-mir/history/h-timeline.htm. Accessed 17 December 2012.
[40] Lipartito, K., and Butlers, O. *A History of the Kennedy Space Center*, University Press of Florida, Gainesville, FL, 2009; "Report of the Presidential Commission on the Space Shuttle Challenger Accident," http://history.nasa.gov/rogersrep/genindex.htm; *Columbia Accident Investigation Board Report*, http://caib.nasa.gov/.

Chapter 13

WOWING THE PUBLIC: THE SHUTTLE AS A CULTURAL ICON

LINDA BILLINGS

INTRODUCTION

From first flight to the present, the Space Shuttle has been an important symbol of U.S. technological capability, universally recognized as such by both the American people and the larger international community. NASA's Space Shuttle remained until retirement one of the most highly visible symbols of American excellence worldwide. This chapter will consider the evolving place of the Space Shuttle as a cultural icon from its first conceptualization until the present.

From 1982 to 2012, the Space Shuttle was the public face of NASA. In a way, for people outside the space community, the Space Shuttle *was* NASA. If people didn't know what N-A-S-A stood for, or why the United States had a space program, they all knew about the Shuttle. They saw it on television and on the front page of their daily newspapers. They heard its astronauts talk about their trips into space. They read books about the shuttle to their children. They bought shuttle posters, shuttle toys, shuttle T-shirts, you name it. They shared in the national pride over this impressive technoscientific achievement. And they participated in the national mourning over the loss of two shuttle missions, crews, and spaceships destroyed.

The Space Shuttle *means* something to people, but exactly *what* does it mean? How did we experience it? What does it stand for? What is the legacy of the space shuttle as a cultural icon? How will we remember it? Will it endure in public memory?

For me, the Space Shuttle means *power*: the brute power of a super-rocket launch, seen and heard and felt in the bones at NASA's KSC press site, a few miles down the road on the sand of Cocoa Beach, and in the mediated environment of an IMAX theater screening of "The Dream Is Alive." For me, it stands for the brainpower invested in conceiving and building such a thing, and for the political and industrial power it took to do it. For me, it means freedom, too: the freedom of flight, freedom from Earth's gravity, freedom from the routines of everyday life, freedom from limits. For me, it is a truly awesome thing.

For most people, the shuttle has been an entertaining, exciting, and inspiring public spectacle. In official culture—at NASA headquarters, in Congress, and at the White House—the shuttle represents U.S. political and technological prowess, power, leadership, and accomplishment. In popular culture, the shuttle carries these meanings, too, but along with other meanings as well, like adventure and escape.

It may seem odd to consider the Space Shuttle apart from its human crews. After all, the shuttle is iconic primarily because it is a *human* space transportation system. However, this chapter will focus primarily on the object, and in large part on imagery of the object, because most people have known the shuttle through its imagery. The visual rhetoric of the Space Shuttle—that is, the messages and meanings that the Space Shuttle has conveyed, both intended and construed—is powerful, and worth considering.

What Is a Cultural Icon?

The Space Shuttle indisputably is a cultural icon, familiar to people all over the world as a bright-shining-white representation of achievement, progress, prowess, power, and freedom. For Americans and others, it stands for U.S. political and technological leadership, specifically in space and more broadly in the world arena. It conveys escape from, even transcendence of, "this mortal coil"—the trials, tribulations, and limitations of everyday life.

So what exactly *is* an icon? What becomes an icon? How? And why? An icon is, simply, a picture, an image, a representation. One specific definition of "icon" refers to a certain type of religious object: an artistic representation of some sacred person, such as a saint, and venerated itself as sacred. Given the intense and sometimes acrimonious competition to obtain one of the four retired space shuttles (including the test object *Enterprise*) for public display, it is reasonable to argue that these shuttle-objects are, in effect, revered as sacred in themselves. In scholarly research, however, icons and symbols are studied as different phenomena; in popular discourse, "icon" and "symbol" are generally treated as synonyms, and in this chapter, I will follow suit.

The terms "icon" and "iconic" are overused in popular discourse. On any given day, news headlines will be peppered with the terms, describing people, places, buildings, and other objects—from the "iconic Empire State Building" and "Coke's iconic bottle" to the "iconic Land Rover Defender," "London: one of the world's iconic capitals," and cultural "icons" such as Oprah and Elvis. In July 2012, I conducted a Google News word search for "iconic" and found the word in 102,000 headlines. A similar search for "icon" at the same time produced 13 million hits—for example, "Mariah Carey to be named icon," "Alexander Cockburn, left journalist icon," and "music icon Molly Meldrum." In scholarly research, a wide range of people, characters, and objects have been examined as cultural icons [1]. In space studies, scholars have analyzed astronauts [2] and space suits [3] as such. The Space Shuttle itself has its turn here.

PUBLIC IMAGE AND POLITICAL REALITY

To understand what the shuttle has meant, and will continue to mean, it is important to consider the cultural context in which the Space Shuttle was conceived, born, flew, and retired. The shuttle was a "modern" object that came into being in a postmodern world and retired in a post-postmodern milieu. The cultural movement known as modernism was a response to the economic, social, and political conditions of industrial development. Wikipedia characterizes it as "a socially progressive trend of thought that affirms the power of human beings to create, improve, and reshape their environment with the aid of practical experimentation, scientific knowledge, or technology". Modernism's object of worship was science and technology. In the heydays of modern U.S. scientific and technological enthusiasm, the nation was dedicated to conquering and exploiting the natural environment to construct a machine-made, managed world. Although the idea of the Space Shuttle was modern, by the time that idea was executed into reality, this "enthusiasm" was on the wane. The shuttle was a massive technocratic accomplishment, conceived before but born after the postmodern critique of Big Technology that unfolded in the 1960s and 1970s [4]. Thus, the Space Shuttle was out of its time from the beginning. As Walter McDougall has explained, the U.S. space program was "part of a package that Americans bought after Sputnik in the belief that the United States must adopt the technocratic model to get back on top...." By the time NASA was up and running, the United States had already "exceeded its technical and financial grasp. In time the original model for civilian technocracy, the space program, became dispensable" [5].

Vested interests shaped the shuttle program in ways that, first, benefited those interests, and, second, advanced the civilian space program. Although the sometimes-ugly private politics of the shuttle program are well known, at the same time the public image of the shuttle has remained relatively pristine. The *Challenger* and *Columbia* disasters seemed to do as much to rally public interest as they did to fuel questions about the shuttle's purpose. By 2012, NASA and its constituent interests had recovered from disasters and failures and restored the shuttle program to "greatness," with the losses of *Challenger* and *Columbia* and all the rest of the Space Shuttle's lesser troubles now integrated into the cultural narrative of the shuttle.

Consider, for example, these remarks to the House of Representatives by Democratic Congresswoman Eddie Bernice Johnson, representing the 30th district of Texas, on the occasion of NASA's last shuttle mission [6]:

> The Space Shuttle has been a source of pride and inspiration for the American people. It sparked interest in many fields of engineering and science which benefitted the United States economy, inspired successive generations, and contributed to our leadership in science and technology. We must continue to provide our children and grandchildren with a similar source of inspiration. As the chapter on the Space Shuttle closes later this year, a new chapter in the book of human exploration begins.

And consider the words of then–NASA Administrator Michael Griffin at the launch of shuttle mission STS-122 in 2008 [7]:

> Societies as well as individuals must be willing to risk. We risk our treasure, our pride and, above all else, the ineffable quality of our attention to and involvement with those things that we set as priorities, the things that make us who we like to think we are. It is always easy, all too easy, to step away from those things that we do, in President John Kennedy's perfect words, "not because they are easy, but because they are hard." But if we as a society ever step away from this endeavor, then we will have become a society unworthy of those who are out there on the pad today, poised to risk their all to bring it about.

For Griffin, the shuttle was the means at hand for executing "the exploration and development of the space frontier. . .truly the most technically challenging endeavor of our generation and many to follow" [8]. Others begged to differ. To *National Review* columnist John Derbyshire, for example, the shuttle was "the folly of our age. . .. The imaginative appeal of space travel is irresistible," he admitted, but the shuttle system itself was ill-conceived: "Its flights accomplish nothing and cost half a billion per" [9]. Ex-shuttle astronaut Tom Jones countered in AIAA's *Aerospace America*, declaring that the shuttle "remains vibrantly productive more than 15 years after its debut," that the resumption of shuttle flights following the loss of *Columbia* symbolized "the renewal of our determination to explore," and that space exploration itself is "a societal imperative" [10]. The editor of *Popular Science* has described the shuttle as "a flying contradiction. On the one hand, it is an elegant and ambitious feat of engineering," but on the other hand, it was also "an ungainly and impossibly delicate piece of hardware." The shuttle was conceived as "a daring and innovative project . . . a bold experiment. But it kept flying far too long" [11].

On April 12, 2011, at a celebration of the 30th anniversary of the launch of STS-1, NASA Administrator Charlie Bolden told his shuttle workforce they had "inspired a generation, helped make the world a better place and...given us a roadmap for future exploration." Remarking on the "many proud moments" the shuttle had delivered to the American people, Bolden said [12]:

> This flagship program has become part of the fabric of our nation's history. It's helped us improve communications on Earth and to understand our home planet better. It's set scientific satellites like Magellan and Ulysses speeding on their missions into the solar system and launched Hubble and Chandra to explore the universe. [It's] given us tremendous knowledge about a reusable spacecraft and launch system from which future government and commercial systems will benefit. It's enabled construction of the International Space Station, our anchor for future human exploration.... And it's provided "first ever" astronaut flight and command opportunities for women and minorities.

Perhaps the most widely known American woman in space after astronaut Sally Ride is educator-astronaut Christa McAuliffe, who will undoubtedly remain a part of the shuttle's legacy. Although Ride, with a Ph.D. in physics from Stanford University, embodied excellence, McAuliffe embodied, apparently quite intentionally, the "typical American"—or, in the words of one commentator, "representative mediocrity" [13]. "In NASA's view," writes cultural studies scholar Constance Penley [14]:

> McAuliffe was perfect, the all-American girl next door: pretty but not too pretty, competent but no intellectual, a traditional mother and teacher whose...husband was her high school sweetheart. She led a Girl Scout troop, volunteered...and taught catechism.

Penley notes that her point "is not to fault McAuliffe for her lack of skills or knowledge or to fault NASA for attempting to be popular. Rather, the issue is why the Teacher in Space scheme was so wrongly conceived and disastrously executed."

Her "wonderful smile...was one of the qualities that won over the selection board," writes historian Bettyann Kevles [15]. NASA "milked the selection process for all of the media attention it could win—which was a lot" [16]. NASA announced McAuliffe's selection as its teacher-in-space at a White House event. "When that shuttle lifts off," NASA Administrator James M. Beggs remarked, "our winning candidate will soar with it—and right into the hearts and minds of young people and their teachers throughout the country. And we will not only have won one for the Gipper, but for American teachers and Americans everywhere" [17].

McAuliffe "became an instant celebrity...living proof that a woman who taught children in school could fly in orbit. Her very public flight—and the media circus leading up to it—would put NASA in a spotlight that spilled over into every American classroom" [18]. Thus, McAuliffe's flight on *Challenger* was a highly anticipated and widely viewed public event, at the Cape and via the mass media. The launch was televised live, and thus millions saw her die, in effect, when *Challenger* blew up barely a minute after liftoff.

McAuliffe became a focus of collective mourning, which has evolved into a perpetual memorializing machine in the form of the Christa McAuliffe Center at Framingham State University in New Hampshire (McAuliffe's alma mater), which provides K-12 teacher support and training [19]. In addition, numerous K-12 schools across the country are named after her; there is a "Christa McAuliffe Sabbatical for gifted teachers" [20]; the "Christa McAuliffe Reach for the Stars Award" for innovative K-12 social studies teaching is offered by the National Council for the Social Studies; and "Christa McAuliffe Scholarships" are offered in several states (such as the Baltimore [MD] Community Foundation and the Lincoln [NE] Education Association).

According to polling data, public support for the Space Shuttle program was highest in the immediate aftermath of the *Challenger* and *Columbia* disasters, in 1986 and 2003, respectively. A CBS News poll conducted in 2005 found 59 percent of respondents favored continuing the Space Shuttle program [21]. A *USA Today*/Gallup poll conducted in 2006 found that 48 percent of respondents deemed the Shuttle program "worth the money," whereas 48 percent said the money "would have been better spent elsewhere" [22]. Roger Launius has studied public opinion about the Space Shuttle and found that [23]:

> Despite the loss of [*Challenger* and *Columbia*], most agree that the Space Shuttle is a magnificent machine. . . . A massively complex system—with more than 200,000 separate components that must work in synchronization with each other and to specifications more exacting than any other technological system in human history—the Space Shuttle must be viewed as a triumph of engineering and excellence in technological management. Any assessment of the Space Shuttle that does not recognize this basic attribute of the system is both incomplete and inaccurate. Because of its technological magnificence, the Space Shuttle has become an overwhelmingly commanding symbol of American excellence in the world community. Ask almost anyone outside the United States what ingredients they believe demonstrate America's super-power status in the world, and in addition to military and economic might they will quickly mention the Space Shuttle—as well as NASA's larger space exploration program—as a constant reminder of what Americans can accomplish when they put their minds to it.

The ugly politics behind the *Challenger* and *Columbia* disasters are now a matter of public record. Internal and external problems contributing to the loss of *Challenger* and *Columbia* and their crews are well documented in the reports of their respective investigation boards [24]. These tragedies are a part of the American cultural landscape, for sure. Yet they are also somewhat insulated from our collective memory of the shuttle. Public response to the carefully staged and prolonged spectacle of the retirement of NASA's remaining orbiters during 2011 and 2012 indicated that most people remember, or choose to remember, the Space Shuttle in its glory.

Space Exploration as Spectacle: The Space Shuttle as a "Really Big Show"

The Space Shuttle has, indeed, been a spectacular adventure—a long-lived public spectacle. Space Shuttle launches are inherently spectacular. Getting an orbiter off the launch pad and into Earth orbit requires some seven and a half million pounds of thrust. The orbiter itself dwarfs all familiar objects around it, though its Vertical Assembly Building at KSC dwarfs the orbiters themselves. Shuttle numbers themselves are "wow" factors: rate of liquid fuel consumption during takeoff—62 million gallons per minute; rate of solid fuel consumption during takeoff—660,000 pounds per minute; speed in orbit—17,000 miles per hour. And so on [25]. Before the novelty wore off, television networks would interrupt their scheduled programs to broadcast shuttle launches. The Space Shuttle could not have become the cultural icon it is without the mass media. Only small subsets of Earthlings have been able to view shuttle launches in person. The vast majority of observers have seen the shuttle only by means of mediated images. Thanks to increasing numbers and varieties of media outlets and mass communication technologies and techniques, and the general ubiquity of mass media in our cultural environment, the spectacle of the Space Shuttle was widely accessible.

During the era in which the shuttle came into being, cultural theorists were promulgating the view that in heavily mediated and technology-dominated contemporary society, images of reality were supplanting reality itself. For most observers, the reality of the Space Shuttle was the image of it. Space Shuttle launches were inarguably spectacles, ritualized public performances conveying official national values—global dominance, technological excellence, and pride of country. These ritualized performances were demonstrations of a technological might that epitomized U.S. political and economic might. Though they were anything but, NASA's presentation of orbital operations as humdrum day-to-day living and working in space was a way of rationalizing the drive to extend human presence into space, making the unnatural seem natural. The shuttle made astronauts heroes in the mythical sense, carrying them on symbolic journeys of departure from the ordinary, initiation into the extraordinary, and return to Earth with new knowledge and insight [26]. The shuttle served as a bridge between humanity's real life of Earth-bound ignorance and an ideal life of enlightenment, transcendence, and understanding.

Even before the shuttle's first launch, it was cause for spectacle. Robert Kahl, director of operations for Rockwell International/Space Systems in Palmdale, California, at the time when the orbiters were being built, described the public spectacle created simply by the overland and air transport of orbiters, before any of them had launched into space [27]:

> We used to transport [Enterprise] overland. We had a huge transporter that we would put the vehicle on and literally transport it down city streets. I can remember prepping the city streets prior to Enterprise's first trip down. ... We removed stop signs, streetlights, telephone poles, trees to make it wide enough to accommodate the width of the wings for a 35-mile overland trip out to Edwards. I can remember that first day that we transported Enterprise. The people lined up on the streets. The nucleus of the public didn't really know what the Space Shuttle was or what it was going to look like. The oohs and ahs and the applause was—goose bumps, it was pretty amazing.

Kahl's account foreshadowed the flood of breathless public accounts of the retired orbiters' final journeys to their new museum homes (see below).

NASA's KSC, home of the shuttle, and the commercially operated Kennedy Space Center Visitors Complex promoted the shuttle to the public as a cultural icon of Big Technology, the epitome of the "built environment" we now inhabit. Go to KSC's website, click on "visit Kennedy," and you'll be directed to the Visitors Complex site, where you'll learn how to "embark on mankind's ultimate journey": "A stay in Florida isn't complete without a trip to Kennedy Space Center Visitors Complex. Just east of many popular Orlando, Florida, vacation destinations and theme parks, NASA's launch headquarters is the only place on Earth where you can tour launch areas, meet a veteran astronaut, see giant rockets, train in spaceflight simulators, and even view a launch" [28].

Given that the Space Shuttle was an icon of the "built" environment and the modern conquest of nature, it's ironic that droves of tourists came to gawk at NASA's orbiters—and its launch pads, crawlers, and processing facilities—amidst the natural splendor of a wildlife refuge, established in 1963 as an overlay of KSC. The only reference to the shuttle's natural environment on KSC's website is under "visitor information," in an answer to the question, "What is the Kennedy Space Center's fishing policy?" which explains that the Merritt Island National Wildlife Refuge is in charge of fishing there. According to its keeper, the U.S. Fish and Wildlife Service, the Refuge "provides a habitat for more than 1,500 species of plants and animals" [29]. Photographers, both amateur and pro, in still and moving images, love to frame the Space Shuttle in wild nature, contrasting it with the technological feat—the conquest of nature—that the shuttle represents.

In addition to White House officials, members of Congress, and international dignitaries, NASA liked to invite celebrities to shuttle events. When NASA rolled out its *Enterprise* flight-test orbiter in 1976, it arranged for the cast of the TV series "Star Trek" to be there, beginning an enduring tradition of bringing film stars, pop stars, and other celebrities to shuttle launches and other notable shuttle-related events—for instance, prominent feminists for the launch of astronaut Sally Ride's first mission. NASA often arranged other celebrity tie-ins for shuttle missions—for example, meals for shuttle crews prepared by Martha Stewart and Emeril Lagasse, and wake-up songs by popular music groups. NASA regularly flew objects of symbolic value on shuttle missions—flags, pins, and other trinkets to be passed out to VIPs as "flown in space"; the Copley Medal of England's Royal Society, flown on STS-121 in 2006 and later presented to physicist Stephen Hawking by then-NASA Administrator Michael Griffin; and a Talmud that survived the Holocaust and that was flown on *Columbia*'s last mission.

The Shuttle program has also yielded a mountain of commercial trinkets, collectibles, and memorabilia that are sold in gift shops at NASA facilities and science and technology museums. The Space Store, a commercial online gift shop, offers a myriad of tchotchkes shaped like or sporting the image of a Space Shuttle orbiter: models, toys, DVDs, and CDs; clocks; books; mugs; patches and pins; cookie cutters; T-shirts and flight jackets; party favors; piñatas; paper plates, cups, and napkins; Christmas tree ornaments; games and playing cards; and other memorabilia and "collectibles" such as a shuttle mission patch set, a Space Shuttle sticker set of all 135 missions, and an "Acrylic Presentation with flown artifacts from all Space Shuttles." An "Official Space Shuttle Program Limited Edition Commemorative Edition" set of all 135 official Space Shuttle mission pins "was developed for NASA and commissioned by NASA's Space Shuttle Program Office," says The Space Store [30]. Also available is a "Pride of America" poster, signed and numbered, featuring glorious photos of the shuttle orbiters *Discovery*, *Atlantis*, and *Endeavour*, each with the American flag waving high in the foreground. Among The Space Store's "top 12 best sellers" for June 2012 were a shuttle "final mission" patch and "end of program" lapel pin, a thermal-tile box set, a "junior space explorer Space Shuttle," and a "space mission Space Shuttle set."

Technical and popular accounts of the Space Shuttle program abound. A keyword search of the Library of Congress catalog for "Space Shuttle" books yields some 300 entries, including dozens of congressional hearing records, numerous conference proceedings, a considerable number of books and reports about the *Challenger* and *Columbia* disasters, and many popular books with titles ranging from the straightforward *Space Shuttle* to the more embellished *Space Shuttle: A Triumph of Manufacturing*, *Space Shuttle: The Quest Continues*, *Space Shuttle: America's Wings to the Future*, and *Space Shuttle: A*

Quantum Leap. However, only a few cast a critical eye on it [31]. Most offer what has been established as a standard narrative of the Space Shuttle program—a story of realizing dreams of extending human presence into space, building and sustaining an immensely complex technological system, conquering obstacles, transcending failures and losses, making human spaceflight routine, proving U.S. leadership and excellence, and making America proud [32]. Typical of the lot is *The Space Shuttle Operators Manual*, a book written by curators of the National Air and Space Museum (NASM), which promises readers: "You are about to embark on a spectacular adventure, blazing a trail for future space travel in the world's greatest flying machine" [33].

The Shuttle as Visual Rhetoric

Like conventional (verbal) rhetoric, visual rhetoric—images, artifacts, and performances—is a means of public persuasion and plays a powerful role in shaping culture [34]. As an analytical perspective, visual rhetoric "focuses on the symbolic processes by which visual artifacts perform communication." Three primary characteristics of artifacts that perform visual rhetoric are that they "must be symbolic, involve human intervention, and be presented to an audience for the purpose of communicating" [35]. Visual rhetoric encompasses "presented elements," such as space and color, used so vividly in the case of the shuttle to convey the red-white-and-blue ideology of American exceptionalism, and "suggested elements" such as "concepts, ideas, themes, and allusions that a viewer is likely to infer" [36]. What the Space Shuttle meant to its audiences, both at home and abroad, certainly encompasses the political and ideological meanings that its advocates intended it to convey. However, as rhetorical scholar Sonja Foss observes, "Once an artifact is created... it stands independent of its creator's intention" [37]. For many, the shuttle likely conveyed more intellectual, emotional, and spiritual values and meanings as well. The ways in which people respond to the visual rhetoric of the shuttle is both individual and collective. Such responses are "a process of accrual in which past experiences merge" with the visual evidence itself to construct meaning [38]. The purpose, or intended meaning, of the shuttle as visual rhetoric was not the same as its function, or the actual meaning it communicated.

In an increasingly visual and mediated culture that is saturated by television, film, and other sorts of still and moving imagery, all thrown into hyperdrive by Web 2.0, the Space Shuttle proved an opportune icon for human spaceflight, American might, and the U.S. aerospace-industrial complex. As visual rhetoric, the shuttle communicated a set of beliefs: that humans are "destined" to be in space (thus justifying the existence of a U.S. human space flight program), that technological progress is good and necessary (thus justifying the maintenance of a space industrial base), and that the United States is and must remain the world's leader, in space and on Earth (thus justifying U.S. economic and foreign policies). The object on the launch pad was a visual take on the American flag: white orbiter, red external tank, and blue-sky backdrop. The fact that the space shuttles were winged is a significant element of their visual rhetoric. Throughout human history, wings have symbolized spirituality, imagination, and thought. Wings may represent "the light of the sun of justice, which always illuminates the mind of the righteous." And, of course, wings signify mobility. The shuttle's dual representation of mobility and enlightenment can be interpreted as an expression of "the possibility of 'progress in enlightenment' or spiritual evolution" [39].

By and large, the visual rhetoric of the Space Shuttle is panegyrical. The only iconic photograph of a space shuttle mission that does not fit this panegyrical mold is the one of the *Challenger* disaster—and this photograph does not depict the orbiter itself, only the aftermath of its destruction. This iconic photo of the end of *Challenger* has been compared to other enduring images of dramatic events, such as the Kent State shootings and the flag raising over Iwo Jima [40]. The political economy that produced the Space Shuttle intended it to communicate national might, prowess, power, and leadership [41]. In 1983, after shuttle launches had commenced, NASA flew its flight-test orbiter *Enterprise* across the Atlantic to show off at the Paris Air Show, then took it to Germany, Italy, England, and Canada before returning it to the United States, where it ended up a focal point of the NASM's Stephen F. Udvar-Hazy Center in Virginia [42]. This campaign was not so much to benefit "all mankind" as to show off U.S. power and expertise.

Shuttle photography is ostensibly documentary, but fundamentally it is staged because NASA carefully orchestrated access to the best photo/viewing opportunities. Shuttle launches always took place against a gorgeous backdrop of a clear bright-blue day or midnight-blue night sky, with nothing competing for attention in the frame except the icon itself. This circumstance arose because it was necessary to locate the launch site near the open ocean and away from people and their structures, to tightly restrict human access for safety reasons, and to avoid launches in bad weather. Nonetheless, it has resulted in a visual public record of the shuttle that includes no images of orbiters' shuttles looking gray and bedraggled on the launch pad in the rain. Thus coffee table books about the Space Shuttle are full of technology "cheesecake"—spectacular color photos of the shuttle just off the launch pad, in processing, and on orbit. In these books and elsewhere, almost every shuttle photograph, whether taken by NASA photographers, photojournalists, or amateur photographers, looks as if it was staged on a Hollywood set: color saturated, perfectly lighted, and framed [43]. Photos of the shuttle on orbit are especially dramatic. With the sun shining on it without atmospheric interference, the shuttle appears perfectly lighted, as if it were on a Hollywood set.

It is worth noting that the visual rhetoric of the Space Shuttle in the *Challenger* accident investigation report [44] is radically different from the rhetoric in the *Columbia* accident investigation report [45]. The *Challenger* report has no image on its cover. It opens [46] with a lovely aerial view of KSC, featuring *Challenger* on the launch pad ready for its last flight. The next time readers see the shuttle in this report is in a gruesome chronological sequence of 35 images [47] showing first the earliest evidence of smoke leaking from the right SRB before the vehicle left the launch pad and last the breakup of the SRBs, ET, and the orbiter itself. The rest of the imagery in this report is grisly, too: shots of scorched parts of *Challenger* recovered after the accident. In contrast, the *Columbia* report is full of crisp, bright, and dramatic color images of the shuttle in its glory, rolling out to, standing on, or leaving the launch pad. Nine of these photos are close-ups in full-page format [48]. Images of the breakup of *Columbia* are small, blurry, and from a distance. Images of parts of the orbiter recovered after the accident are thumbnail size and poorly lit [49]. Although the text of the *Columbia* report is indisputably damning, the visual rhetoric of the shuttle within it belies the text.

Over the Space Shuttle's lifetime, and especially in its early years, it was a popular subject of artists, and NASA made sure to provide them with access to it. In the 1960s, NASA Administrator James E. Webb was the first to invite artists "to record the strange new world of space," thus initiating what became an organized NASA Art Program. NASA notes, "Although an intensive use of photography had long characterized NASA's work, the Agency recognized the special ability of the artist's eye to select and interpret what might go unseen by the literal camera lens. No civilian government agency had ever sponsored as comprehensive and unrestricted an art program before" [50]. Most of the shuttle imagery produced for the NASA Art Program is moodily impressionistic/expressionistic—see, for example, "On the Pad and Launch" (Ron Salmon), "Shuttle Rollout from VAB" (Nicholas Solovioff), and "The Landing—Columbia 3" (Jack Perlmutter). Others are dramatically photorealistic, like "STS-4 at 00.52 Sec" (Ron Wicks). Another evokes the grand nineteenth-century luminist landscapes of the American West: In "Florida Homecoming," Kent Sullivan paints a tiny orbiter gliding over pristine Florida wilderness; it is the only manufactured object in the frame and is dwarfed by the expansive landscape, with a rainbow arching over it [51].

The late Robert McCall, perhaps the most famous U.S. space artist, adored the Space Shuttle as a subject, treating it with the reverence accorded a religious object. He painted the shuttle as the apotheosis of technological progress and American achievement. The shuttle was by no means McCall's sole subject, however. He is perhaps best known for his paintings of numerous wild visions of human life in space, such as his poster art for the films "2001: A Space Odyssey" and "Star Trek: The Motion Picture." McCall Studios describes the artist as "NASA's visual historian," claiming he "has done more than any other artist to enable Americans to visualize their nation's presence in space." McCall himself characterized his vision thus: "The future is bright and filled with promise for us all. And the human spirit driven as it is, with an insatiable desire to know, to explore, and to understand will continue forever to reach upward and outward" [52]. McCall's space art is monumental, in scale and in subject matter. His masterful blending of saturated colors is unique and shows itself especially well in his red-white-and-blue depictions of the Space Shuttle. His paintings of the shuttle glorify the object as a stand-in for the American flag, symbolizing nationalism, patriotism, conquest, and greatness. "Opening the Space Frontier—The Next Giant Step" (1979), a mural McCall painted for NASA's JSC, depicts, from left to right, the progression of NASA's spacecraft, from the Mercury capsule to the Space Shuttle, with space-suited astronauts standing in the foreground, atop clouds, as if they were gods. One is posed heroically, holding the American flag, recalling the famous image of U.S. soldier John Bradley raising the flag on Iwo Jima. In the Iwo Jima photo and the McCall mural, the flag flies backward toward the left of the frame. In the mural, the flag provides a dramatic backdrop for the astronauts and their spacecraft, appearing to reach over the scene like a protective shield [53].

"Hail Columbia—April 12, 1981" (1990), a McCall painting in a private collection, grandly depicts the launch of NASA's first Space Shuttle mission. The stars and stripes of the flag are laid over the blue-sky backdrop of *Columbia*'s takeoff. Atop a blinding cone of rocket exhaust, *Columbia* is gloriously erect, pointed straight up toward the heavens, and clearly the center of attention [54]. "Columbia's Glorious Return" (1981, artist's collection) depicts the orbiter landing in the Mohave Desert on April 14, 1981, the broad red and white stripes of the flag rolling out on the runway toward the viewer, like a celebrity red carpet. The flag's bright-white stars are spread across the sunlit blue sky behind the orbiter. *Columbia* is about to touch down, nose up as if saluting the flag [55]. In the painting "Challenger's last flight" (1987, private collection), *Challenger* points heavenward against a red-white-and-blue backdrop. It's atop a bright-white plume of rocket exhaust that splays out against a reddening deep-blue sky. The stars in the sky are not the stars of the flag but seven celestial stars, representing the seven people who died on *Challenger*. In an inset bisecting this backdrop is a nebula expressing "the aspirations of all those courageous explorers who risk their lives to discover new worlds" [56]. In "The spirit of NASA" (1982, Disney Corporation collection), astronauts Bob Crippen and John Young suit up for the first flight of the Space Shuttle in the foreground. In the background, the flight unfolds from left to right: *Columbia* stands ready on its launch pad, takes off, and returns to Earth. Between launching and landing, the American flag flies high and pointed toward space [57].

In addition to posing for a myriad of artists and photographers, the shuttle has been a favorite subject of editorial cartoonists over the years. I searched for and reviewed space shuttle cartoons at Google Images, the website of the American Association of Editorial Cartoonists, www.thedailydose.com, and www.cartoonstock.com. In a blog post around the time of the last shuttle mission, NBC News political cartoonist Daryl Cagle took "a look back at the Space Shuttle" as a subject for cartoonists, noting that he and his peers "have covered the Space Shuttle program... [as] a symbol of progress and achievement, for the last 30 years" [58]. In the wake of the *Challenger* and *Columbia* accidents, cartoonists made the shuttle an object of reverence. Dana Summers, editorial cartoonist for the *Orlando Sentinel*, explained how he decided to draw his cartoon about the loss of *Challenger*: "The day the shuttle exploded and the shock wore off, my first thought was that I had 26/27 hours to come up with an idea for a cartoon that should convey the feelings of our readers" — he drew seven doves leaving *Challenger* in space [59]. From 2009 through 2012, numerous editorial cartoons blamed President Obama for the decision made by President George W. Bush to retire the space shuttles. They showed retired orbiters for sale in used-car lots, about to make a deal with the "Cash for Clunkers" program, serving as a shelter for unemployed aerospace workers, and so on. Evoking the idea of space flight as escape from the mundane, one poignant cartoon depicts a "last bird's eye view of Atlantis from ISS," with the astronauts aboard the orbiter wondering, "Do we *have* to go back to Earth?" as they look down at a planet plagued with problems such as "Euro debt," the News Corp. hacking scandal, and African famine and war [60].

Along with the visual rhetoric, NASA packaged its Space Shuttle fleet with plenty of verbal rhetoric. The names of orbiters were chosen explicitly to convey discovery and global reach [61]. According to KSC, the orbiters were

> named after pioneering sea vessels which established new frontiers in research and exploration. ... Another important criterion in the selection process was consideration for the international nature of the Space Shuttle program. The name of NASA's newest orbiter, Endeavour, was selected from names submitted by schoolchildren around the world.

Enterprise "was originally to be named Constitution (in honor of the U.S. Constitution's Bicentennial). However, viewers of [TV's] "Star Trek" started a write-in campaign urging the White House to select the name Enterprise," recalling Captain Kirk's mission statement: "Space, the final frontier.... These are the voyages of the Starship Enterprise, its five-year mission to explore strange new worlds, to seek out new life and new civilizations, to boldly go where no man has gone before."

Challenger was named after the nineteenth century British naval research vessel *HMS Challenger*, which sailed the Atlantic and Pacific. *Columbia* was named after a late-eighteenth-century Boston-based sloop captained by Robert Gray, who led his ship and crew on the first U.S. circumnavigation of the globe. "Other sailing ships have further enhanced the luster of the name Columbia," according to KSC officials. "The first U.S. Navy ship to circle the globe bore that title, as did the command module for Apollo 11. … On a more directly patriotic note, 'Columbia' is considered to be the feminine personification of the United States."

Discovery was named after one of two ships used by British explorer James Cook in the 1770s on voyages in the South Pacific. "Other famous ships have carried the name Discovery," KSC notes, "including one used by Henry Hudson to explore Hudson Bay [and] search for what was hoped to be the Northwest Passage from the Atlantic to the Pacific in 1610 and 1611. Another [*Discovery*] was used by the British Royal Geographical Society for an expedition to the North Pole in 1875."

Endeavour was named after another of Cook's ships. *Atlantis* was named after the primary research vessel of the Woods Hole Oceanographic Institute in Massachusetts from 1930 to 1966, and the first U.S. vessel used for oceanographic research.

THE BEGINNING OF THE END

In 2011, following an intense competition for the artifacts, NASA announced that its orbiters would go on public display as follows: Atlantis at Kennedy Space Center, Endeavour at the California Science Center in Los Angeles, Discovery at NASM's Stephen F. Udvar-Hazy Center in Virginia, and Enterprise at the Intrepid Sea, Air and Space Museum in New York. Through 2012, as NASA wrapped up its shuttle program, it orchestrated a prolonged public celebration of the shuttle.

Valerie Neal, a curator of human spaceflight and keeper of the orbiters *Enterprise* and *Discovery* for NASM, offered these thoughts on the place of the shuttle in culture as it retired from flight [62]:

> As the icon for 30 years of human spaceflight conducted by the United States, the Space Shuttle symbolises [*sic*] impressive technical and social achievements. … The towering complete launch vehicle and the winged orbiter merely by their appearance suggest a rich technical legacy. … Nothing like the Space Shuttle has ever flown, and it may be that nothing like it ever will. That may be its ultimate legacy: the shuttle symbolizes an ambitious effort to make spaceflight as routine as airline travel. In many ways, it did.

With the placement of retired shuttle orbiters in public museums around the United States, a new chapter in the history of the Space Shuttle program officially began: the construction of a collective memory of the shuttle, supervised not only by NASA but also by curators, though conducted, as always, by "The People." As NASM curator David H. DeVorkin has explained, the value of collecting and preserving space artifacts is that they provide validation that something happened. They enable people to celebrate events, engendering a "sense of transcendence promoted by physically encountering an object that made history." They provide "illumination... stimulation...inspiration—evidence of challenges met or exceeded, handicaps overcome, struggles vindicated" [63].

Mass media around the world widely and excitedly covered this new phase in shuttle history. NASA planned showy "flyovers" of each city receiving an orbiter. For example, *Discovery*, piggybacked on a Boeing 747 aircraft, made three so-called "victory laps" around Washington's National Mall, drawing people into the streets for a look and generating lots of media coverage of this spectacular photo opportunity. The *Washington Post* reported on *Discovery*'s flyover with a color photo at the top of page 1, virtually dripping with symbolism—flying above the U.S. Capitol, with the Washington Monument looming in the background, against a backdrop of a cloudless blue sky. A photographer couldn't have asked for a better setup [64]. The front-page photo was accompanied by a splashy front-page story in the Style section, including five more color photos of *Discovery* [65]. Calling it "the world's most famous spaceship," the *Post* compared the flight-worn *Discovery* to the flight-test orbiter *Enterprise*:

> Pristine, shiny white, never launched, Enterprise is virginal. Discovery, by contrast, is very well loved. Her siding is singed, seared, burned and battered, badly in need of a wash.... With 39 flights in 27 years, Discovery was NASA's hardest-working Space Shuttle. It played every conceivable orbital role: science platform, satellite launcher, telescope repair station, space station delivery truck.

In a related story, the *Post* ran a large color photograph of *Discovery* and *Enterprise* posed "cute," nose to nose on a Dulles runway. As part of the carefully orchestrated spectacle of *Discovery*'s arrival at NASM's Udvar-Hazy Center, opera star Denyce Graves "serenaded [*Discovery*] into its new job" with "The Star-Spangled Banner...the U.S. Marine Drum and Bugle Corps played, fans and astronauts greeted the shuttle and a full day of events commenced to greet" the orbiter [66].

Endeavour was greeted by similar fanfare in Los Angeles. The *Los Angeles Times* reported that at the opening of the California Science Center's exhibit on October 30, one six-year-old, "who wants to be an astronaut, said he couldn't really describe his reaction. 'I just feel like I'm gonna burst on the floor'. ..." [67]:

> Public officials were also on hand to mark the pavilion's opening at a morning ceremony, including Gov. Jerry Brown, L.A. Mayor Antonio Villaraigosa and Inglewood Mayor James T. Butts. The family of late astronaut Sally Ride was also in attendance, along with actresses June Lockhart ("Lost in Space") and Nichelle Nichols ("Star Trek")....

> Villaraigosa called Endeavour's return to California—which included a three-day, cross-country trip and 12-mile crawl through city streets—a "spectacular" story that would inspire children to be scientists, mathematicians and engineers. "This isn't just a ribbon-cutting for Endeavour's home," he said. "This is a ribbon-cutting for the future of L.A."

On November 2, 2012, the orbiter *Atlantis* was welcomed to the Kennedy Space Center Visitors Complex with speeches, a ceremony, and a parade including the Merritt Island High School Color Guard, the Titusville High School Marching Band, and current and former astronauts and shuttle workers. "The public seemed to love all the pomp and circumstance of parades taking Space Shuttles Endeavour and Atlantis to their new homes to begin their educational missions," a NASA Office of Communications official observed. "The crowds were huge and enthusiastic for the bi-coastal events. While clearly the orbiters were the 'money shot,' it was great to have so many NASA exhibits at both events describing current and future programs. It was also great to have so much hardware present representing our new cargo and crew spaceships" [68].

LESSONS LEARNED: POLITICS, IMAGE, AND MEMORY

The politics and economics of Space Shuttle development and operations were just as complex—and arguably more interesting—than their science and technology. From the beginning, NASA, in concert with its industrial partners and other advocates, worked hard to entrench the civilian space program in the U.S. political economy. The Space Shuttle system was planned and developed like any other element of the program, by NASA and the aerospace industry in concert with the White House and Congress. Although the idea of a winged ship that would ferry people from Earth to space and back predates NASA by decades, the idea of building a U.S. space shuttle finally began to jell in September 1969, when a White House Space Task Group reported to President Richard Nixon on long-range planning options for NASA's human space flight program. Options ranged in cost from an $8 billion to $10 billion per year program encompassing a human mission to Mars, space stations, and a reusable space ferry or shuttle; a scaled-down version of option 1 cost less than $8 billion per year, and a $4 billion to $7 billion per year program included a space shuttle and an Earth-orbiting space station. President Nixon approved the Space Shuttle program but not the space station that was supposed to go along with it. Thus, the Space Shuttle, conceived as a shuttle *to* and *from* somewhere, came into being as a shuttle with no specific place to go [69].

Nonetheless, NASA and the aerospace community had constructed and sold a need for a reusable space transportation system to maintain continued U.S. access to, leadership in, and dominance of space, and they worked continuously to sustain political support. Proponents argued that a shuttle system would reduce the cost of space transportation and enable more routine launch operations. Virtually every NASA field center played a role in Space Shuttle development and operations, ensuring broad political and industrial support. Even before shuttle missions began to fly, NASA and its contractors could show interested officials how Shuttle program funding would be distributed across almost all 50 states of the union [70].

Historically, the Shuttle program was perpetually behind schedule and over budget. As early as 1971, the White House OMB was directing NASA to lower shuttle development costs. Nonetheless, shuttle development studies projected that NASA would be launching Space Shuttle missions at a rate of 48 per year by 1980. The reasoning was that a high launch rate would justify the cost of shuttle fleet operations and maintenance and would keep down costs per launch. NASA launched its first Space Shuttle mission on April 12, 1981, a two-day flight with a crew of two. In 1984, the agency initiated plans to build a space station. In 1984, NASA Administrator James M. Beggs told the U.S. Space Foundation [71]:

> The shuttle has more than proved its worth as America's all-purpose space vehicle.... Over the next 9 months we are planning to fly a mission a month. Next October we plan to fly two missions during the month, leading to an accelerated flight schedule that will take us to the 100-mission mark by 1989. By the end of the century if the market continues to expand, which we fully expect to happen, we will be flying about 40 shuttle missions a year.

In 1985—the year before the January 1986 *Challenger* accident—NASA reached its highest-ever annual shuttle launch rate of nine missions. Throughout its lifetime, the shuttle was characterized, by NASA officials and others, as an unproven research vehicle *and* as a "space truck" capable of routine operations. NASA declared the shuttle system "operational" after its first four flights, but even in its last years of operation, this "operational" system continued to be subject to problems that required launch delays, upgrades, and other modifications and repairs. It was not so much the workers who actually designed and built the Space Shuttle system who imbued it with meaning. The bureaucrats, politicians, industry lobbyists, and other policymakers did so, toward ensuring that it would actually be built and that it would keep going, year after year, no matter what the cost [72].

From beginning to end, NASA justified the shuttle system as a necessary element of a long-term human exploration program. Roger Handberg examined NASA's "fixation" on sustaining a human space flight program "despite numerous obstacles, political, financial, and technological. This fixation has endured," he concluded, "even though the political environment for its...vision has grown progressively less supportive over the years.... Organizational reform and subsequent policy changes have not negated that fundamental direction laid down during the agency's infancy..." [73]. And from beginning to end, the purpose of the shuttle has been debated, from one administration to the next and one Congress to the next, and within the aerospace community. In fact, the shuttle's mission changed over time. NASA had advanced its shuttle plans by claiming the shuttle would serve as the nation's space transportation system, launching military and commercial as well as scientific satellites, ultimately supplanting the U.S. Air Force's fleet of ELVs. The Air Force and commercial ELV builders resisted giving up their turf to NASA, and after the *Challenger* accident, NASA got out of the business of launching military and commercial payloads. NASA also marketed the shuttle as a laboratory for scientific and commercial experiments in microgravity. Infrequent launches, expensive launch delays, orbiter losses, and research advances on the ground effectively squelched that market.

As the Shuttle program came to a close, Joe Pelton commented in the journal *Space Policy* that the shuttle, "for better or worse, has dominated the U.S. space program for some 30 years and is now an American icon. The Space Shuttle orbiters have flown over 120 missions and certainly accomplished some amazing feats, including the deployment of the ISS, the launch and double repair of the Hubble Telescope, a number of classified missions for the U.S. defense establishment and the cementing of international cooperation in space." However, the Shuttle program was also marked by "major lost opportunities and flawed policy decisions that have had multi-billion dollar consequences," and "Congress, the White House, and NASA leadership have all played a role. If there have been failings, they have not been by NASA alone, but the entire U.S. space policy leadership" [74].

Because of, or in spite of, such analyses, will people remember the Space Shuttle program as a failure? Most people have known NASA by its representations—its space-walking heroes and their spaceships, the HST, and anthropomorphized rovers on Mars. Out of this rich pastiche of spectacles, people make their own meanings. With the end of shuttle flights, the process of remembering and memorializing the Space Shuttle has begun. NASA and the curators of its retired orbiters are retooling them as tools for education. Is this how we will remember the Space Shuttle? How will we construct our collective memory of it? Will we remember the iconic shuttle as the object we now can see: a beat-up, retired workhorse, laid out horizontally, permanently locked to Earth? Or will it endure as the upward- and outward-bound white knight of a spaceship we all know? Ironically, the most famous visual image of the Space Shuttle program does not include the icon itself, only smoke trails left by the breakup of the orbiter *Challenger* shortly after launch. Over time, people may or may not remember all the images of pieces of the exploded *Challenger* and *Columbia*, fallen back to Earth, collected, and laid out like a gruesome collection of body parts.

It is possible, but not likely, that people will remember the shuttle that way. It is more likely that most people will remember, celebrate, and commemorate the shuttle as the cultural icon it was made to be: a symbol of U.S. leadership, power, and accomplishment, a stand-in for the American flag. Although historians will continue to ponder the complex story of NASA's Shuttle program, including the public hype and the political maneuvering involved in initiating and sustaining it, people will continue to make their own meanings out of the shuttle. Scholars now have the opportunity to examine how the retired Space Shuttle orbiters, and the overall Space Shuttle program they were a part of, are presented to public audiences—how they are displayed, described, characterized, framed, and contextualized. I hope they will also explore how "the people" remember the shuttle.

REFERENCES

[1] Foss, S., K., "Ambiguity as Persuasion: The Vietnam Veterans Memorial," *Communication Quarterly*, Vol. 34, No. 3, 1986, pp. 326–340; Nelkin, D., and Susan Lindee, M., *The DNA Mystique: The Gene as a Cultural Icon*, University of Michigan Press, Ann Arbor, 1995.

[2] Launius, R. D., "Heroes in a Vacuum: The Apollo Astronaut as Cultural Icon," *Florida Historical Quarterly*, Vol. 87, No.2, Fall 2008, pp. 174–209.

[3] Shaw, D. B., "Bodies out of This World: The Space Suit as Cultural Icon," *Science as Culture*, Vol. 13, No, 1, 2004, pp. 123–144.

[4] Ellul, J., *The Technological Society* (Translated from the French by John Wilkinson. With an introd. by Robert K. Merton) Alfred A. Knopf, New York, 1964; Schumacher, E. F., *Small Is Beautiful: A Study Of Economics As If People Mattered*, Blond and Briggs, London, UK, 1973. http://science.ksc.nasa.gov/shuttle/resources/orbiters/enterprise.html.

[5] McDougall, W. A., *...the Heavens and the Earth: A Political History of the Space Race*, Basic Books, New York, 1985, p. 422.

[6] Congressional Record, Friday April 15, 2011, p. E751, Extension of remarks, speech of Hon. Eddie Bernice Johnson of Texas in the House of Representatives.

[7] Administrator's Remarks At the Launch of STS-122 Atlantis, Griffin, M. D., Administrator, National Aeronautics and Space Administration, February 7, 2008, http://www.nasa.gov/pdf/210582main_Admin_Briefing_7_Feb_08.pdf.

[8] Griffin, M., "Continuing the voyage: the spirit of Endeavour," Remarks to The Royal Society of the United Kingdom, December 1, 2006.

[9] Derbyshire, J., "The folly of our age: the space shuttle," *National Review*, June 16, 2006, http://old.nationalreview.com/derbyshire/derbyshire200506160749.asp.

[10] Jones, T., "The Critics Return to Flight," *Aerospace America*, August 2005, pp. 14–16.

[11] Meigs, J., "Was the Space Shuttle Really Worth It?" *Popular Mechanics*, July 8, 2011, http://www.popularmechanics.com/science/space/nasa/was-the-space-shuttle-really-worth-it?click_main_sr.

[12] Bolden, C. F., Jr., *Remarks by Administrator Bolden at the 30th Anniversary of STS-1*, April 12, 2011, http://www.nasa.gov/news/speeches/admin/index.html.

[13] Penley, C., *NASA/Trek: Popular science and Sex in America*, Verso, New York, 1997, p. 24.

[14] Penley, C., *NASA/Trek*, p. 24.

[15] Kevles, B. H., *Almost Heaven: The Story of Women In Space*, Basic Books, New York, 2003, p. 100.

[16] Kevles, *Almost Heaven,* p. 102.
[17] Beggs, J. M., remarks prepared for delivery: teacher-in-space announcement, White House; June 19, 1985, p. 93, https://mira.hq.nasa.gov/history/ws/hdmshrc/all/main/Blob/41042.pdf?upp=0&m=44&w=NATIVE%28%27SERIES+ph+any+%27%27Beggs%27%27%27%29.
[18] Kevles, *Almost Heaven,* p. 93.
[19] http://www.christa.org/about_us.html.
[20] http://www.nhcf.org/page.aspx?pid=629.
[21] Newport, F., "Space Shuttle Program a 'Go' for Americans," *Gallup News Service*, August 10, 2005, http://www.gallup.com/poll/17761/Space-Shuttle-Program-Go-Americans.aspx.
[22] Carroll, J., "Public Divided Over Money Spent on Space Shuttle Program," *Gallup News Service*, June 30, 2006, http://www.gallup.com/poll/23545/Public-Divided-Over-Money-Spent-Space-Shuttle-Program.aspx#1; Billings, L., "50 years of NASA and the Public: What NASA? What Publics?" *NASA's First 50 Years: Historical Perspectives*, edited by S. J. Dick, NASA SP-2010-4704, Washington, DC, 2010, pp. 151–182.
[23] Launius, R. D., "Public Opinion Polls and Perceptions of US Human Spaceflight," *Space Policy*, Vol. 19,2003, pp. 163–175.
[24] *Report of the Presidential Commission on the Space Shuttle Challenger Accident*, June 6, 1986, Washington, DC, and *Columbia Accident Investigation Report*, Volumes I–VI, August-October 2003, Washington, DC.
[25] Sheppard, A., "Looking Back on the U.S. Space Shuttle Program by the Numbers," *Popular Mechanics*, March 9, 2011, http://www.popularmechanics.com/science/space/nasa/us-space-shuttle-program-by-the-numbers?click_main_sr.
[26] Campbell, J., *The Hero with a Thousand Faces*, World Publishing Co., Cleveland, OH, 1956, 1970.
[27] NASA STS Oral History Project, Edited Oral History Transcript: Robert H. Kahl, Interviewed By Rebecca Wright, Downey, CA, August 25, 2010, http://www.jsc.nasa.gov/history/oral_histories/STS-R/KahlRH/KahlRH_8-25-10.htm.
[28] http://www.kennedyspacecenter.com/index.asp.
[29] http://www.fws.gov/merrittisland/.
[30] www.thespacestore.com.
[31] Launius, R., "A Shelf of Indispensable Books on the Space Shuttle," http://launiusr.wordpress.com/2009/08/14/a-shelf-of-indispensable-books-on-the-space-shuttle/.
[32] Bizony, P., *The Space Shuttle: Celebrating Thirty Years of NASA's First Space Plane*, Zenith Press, Minneapolis, MN, 2011.
[33] Joels, K. M., Kennedy, G. P., and Larkin, D., *The Space Shuttle Operators Manual*. Ballantine Books, New York, 1982, 1988.
[34] Olson, L. C., Finnegan, C. A., Hope, D. S. (eds.), *Visual Rhetoric: A Reader in Communication and American Culture*, Sage, Los Angeles, CA, 2008.
[35] Foss, S., "Framing the Study of Visual Rhetoric: Toward A Transformation of Rhetorical Theory,"*Defining Visual Rhetorics*, edited by C. A. Hill and M. Helmers, Lawrence Erlbaum Associates Inc., Mahwah, NJ, 2004, pp. 304–305.
[36] Foss, S., "Framing the Study of Visual Rhetoric: Toward A Transformation of Rhetorical Theory,"*Defining Visual Rhetorics*, p. 307.
[37] Foss, S., "Framing the Study of Visual Rhetoric: Toward A Transformation of Rhetorical Theory,"*Defining Visual Rhetorics*, p. 308.
[38] Foss, S., "Framing the Study of Visual Rhetoric: Toward A Transformation of Rhetorical Theory,"*Defining Visual Rhetorics*, p. 306.
[39] Cirlot, J. E., *A Dictionary of Symbols*, Philosophical Library, New York, 1962, pp. 354–355.
[40] Hariman, R. and Lucaites, J. L., *No Caption Needed: Iconic Photographs, Public Culture, and Liberal Democracy*, University of Chicago Press, Chicago, 2007.

[41] Handberg, R., *Reinventing NASA: Human Space Flight, Bureaucracy, and Politics*, Praeger, Westport, CT, 2003.
[42] http://science.ksc.nasa.gov/shuttle/resources/orbiters/enterprise.html.
[43] Reichhardt, T. (ed.), *Space Shuttle: The First 20 Years*, DK Publishing, New York, 2002), photos on pp. 18–19, and 26–27.
[44] *Report of the Presidential Commission on the Space Shuttle Challenger Accident.*
[45] *Columbia Accident Investigation Report.*
[46] *Report of the Presidential Commission on the Space Shuttle Challenger Accident*, p. 11.
[47] *Report of the Presidential Commission on the Space Shuttle Challenger Accident*, pp. 22–35.
[48] *Columbia Accident Investigation Report*, pp. 8, 10, 18, 20, 98, 175, 206, 228, and 242 in Vol. I.
[49] *Columbia Accident Investigation Report*, pp. 40–41, 47, 72.
[50] http://er.jsc.nasa.gov/seh/spaceart.html.
[51] http://er.jsc.nasa.gov/seh/shuart.html.
[52] www.mccallstudios.com.
[53] *The Art of Robert McCall* (introduction, Ray Bradbury; captions, Tappan King), Bantam, New York, 1992, p. 3.
[54] *The Art of Robert McCall*, p. 19.
[55] *The Art of Robert McCall*, p. 20.
[56] *The Art of Robert McCall*, p. 23.
[57] *The Art of Robert McCall*, p. 29.
[58] http://cartoonblog.nbcnews.com/_news/2011/07/08/7041985-a-look-back-at-the-space-shuttle?lite.
[59] http://www.orlandosentinel.com/news/space/os-dana-summers-challenger-cartoon-01262011,0,3823078.photogallery.
[60] http://archives.thedailydose.com/2011/the-daily-dose-Thursday,_2011-07-21,_10:00.html.
[61] http://science.ksc.nasa.gov/shuttle/resources/orbiters/orbiters.html.
[62] Neal, V., "Viewpoint: The Space Shuttle–'Magnificent Flying Machine'," *BBC News*, July 21, 2011, http://www.bbc.co.uk/news/science-environment-14173630?print=true.
[63] DeVorkin, D. H., "Space Artifacts: Are They Historical Evidence?" *Critical Issues in the History of Spaceflight*, edited by S.J. Dick and R. D. Launius, NASA SP-2006-4702, Washington, DC, 2006, pp. 598–599.
[64] "A Connecting Flight to the Voyage Home," *Washington Post*, April 18, 2012, p. A1.
[65] Vastag, B., "One Giant Lap: As Discovery Soars Toward Museum Retirement, Throngs in the D.C. Area Salute the Aerial Icon's Final Mission Accomplished," *Washington Post*, April 18, 2012, pp. C1, C10–11; Vastag, B., "A Battered Space Hero – Discovery, 'Champion of the Shuttle Fleet' – Comes to Rest at the Udvar-Hazy Center," *Washington Post*, April 15, 2012, p. E1.
[66] Trescott, J., and Vastag, B., "In Final Voyage, Discovery Rolls to a New Home: Fans Turn Out to Udvar-Hazy Center for Shuttle Ceremony," *Washington Post*, April 20, 2012, p. C9.
[67] "Space Shuttle Endeavour Exhibit Officially Opens to Public," *Los Angeles Times*, October 30, 2012, http://latimesblogs.latimes.com/lanow/2012/10/space-shuttle-endeavour-exhibit-opens.html.
[68] Ladwig, A., NASA Communications Update, Public Outreach Division, Weekly Update – 11-02-12.
[69] Review of U.S. Human Spaceflight Plans Committee, Seeking a Human Spaceflight Program Worthy of a Great Nation, October 2009, p. 28.

[70] Billings, L., "Space Shuttle," *Encyclopedia of Science and Technology Communication,* edited by S. J. Dick and Susanna Hornig Priest, Sage, Los Angeles, CA, 2010, pp. 826–831.
[71] Beggs, J. M., NASA Administrator, suggested remarks, United States Space Foundation Symposium, Colorado Springs, Colorado, November 27, 1984, https://mira.hq.nasa.gov/history/ws/hdmshrc/all/main/Blob/42253.pdf?upp=0&m=93&w=NATIVE%28%27SERIES+ph+any+%27%27Beggs%27%27%27%29.
[72] Heppenheimer, T. A., *History of the Space Shuttle, Volume 1: The Space Shuttle Decision, 1965–1972*, Smithsonian Institution Press, Washington, DC, 1998.
[73] Handberg, R., *Reinventing NASA,* p. 2.
[74] Pelton, J. N., "The Space Shuttle – Evaluating an American Icon," *Space Policy*, Vol. 26, No. 4, 2010, pp. 246–248.

Chapter 14

Retiring the Space Shuttle: What Next?

John M. Logsdon

Introduction

In the early morning of July 21, 2011, Space Shuttle orbiter *Atlantis* rolled to a "wheels stop" position at the Shuttle Landing Facility at Florida's Kennedy Space Center (KSC) (Fig. 14.1). With the completion of its 135th mission, the Space Shuttle program came to a close, over 39 years after President Richard Nixon had announced on January 5, 1972 [1]

> I have decided today that the United States should proceed at once with the development of an entirely new type of space transportation system designed to help transform the space frontier of the 1970s into familiar territory, easily accessible for human endeavor in the 1980s and 1990s. This system will center on a space vehicle that can shuttle repeatedly from earth to orbit and back. It will revolutionize transportation into near space by routinizing it.

The earlier chapters of this book have provided authoritative accounts of all aspects of the Space Shuttle program. They have chronicled its challenges, successes, and failures. This chapter asks a different question: What will follow the four-decade Space Shuttle program, both in terms of replacing the shuttle's capability for carrying humans to low-Earth orbit (LEO) and, more fundamentally, as the centerpiece of U.S. efforts with respect to human spaceflight?

For the better part of the past three decades, the need for a shuttle replacement has been discussed, and occasionally acted upon, but to date without success [2]. As part of its analysis of the February 1, 2003, breakup

Fig. 14.1 The final landing of the Space Shuttle program.

on reentry of orbiter *Columbia*, the Columbia Accident Investigation Board's (CAIB) August 2003 report noted that there had been no "sustained national commitment over the past decade to improving access to space by developing a second-generation space transportation system." The CAIB concluded that "*the United States needs improved access for humans to low-Earth orbit as a foundation for whatever directions the nation's space program takes in the future* [emphasis in original]." The CAIB report suggested that it was "*in the nation's interest to replace the Shuttle as soon as possible as the primary means for transporting humans to and from Earth orbit* [emphasis in original]." Finally, it contained the following indictment: "*Previous attempts to develop a replacement vehicle for the aging Shuttle represent a failure of national leadership* [emphasis in original]" (Fig. 14.2) [3]. Roger D. Launius, in his comprehensive and insightful 2004 analysis of U.S. policy towards space access, used even stronger language than the CAIB. He suggested that "the lack of a firm decision to develop a Shuttle replacement represents the single most egregious failure of space policy in history" [4].

These are harsh judgments. Are they still justified a decade after they were articulated? Is regaining the ability to carry humans to LEO indeed still the "foundation" for the future of the U.S. space program, as the CAIB suggested? Has there been a continued failure of national leadership with respect to maintaining the U.S. capability to launch its own astronauts to LEO, at a time when Russia has maintained that capability and China has developed it?

Or, with the renewed emphasis on deep space exploration that has become the central element of U.S. space policy, might human transportation to LEO now be on the path to becoming almost a commodity, like space transportation for satellites and other cargo, to be obtained from non-government U.S. providers or from other governments or non-U.S. providers? Has the shift in

Fig. 14.2 The 13 member CAIB in the CAIB boardroom with the official STS-107 portrait on the wall behind them. Seated from left to right are: G. Scott Hubbard, Dr. James Hallock, Dr. Sally Ride, Board Chairman Admiral (retired) Hal Gehman, Steve Wallace, Dr. John Logsdon, and Dr. Sheila Widnall. Standing from left to right are: Dr. Doug Osheroff, Maj. General John Barry, Rear Admiral Stephen Turcotte, Brig. General Duane Deal, Maj. General Ken Hess, and Roger Tetrault. CAIB Photo by Rick Stiles 2003 (http://history.nasa.gov/columbia/people_pics.html)

priority from space operations in Earth orbit to a focus on human exploration of distant destinations reduced the centrality to U.S. interests of replacing the Space Shuttle and regaining a U.S. capability to transport crews to LEO? Is the post-Shuttle delay in regaining U.S. crew transportation capability, though embarrassing at a time when Russia and China have that capability, acceptable in national interest terms, as long as there is progress towards creating new human space transportation systems?

This chapter will suggest that the latter of these two alternatives is a more realistic depiction of the current situation. It will argue that the *Columbia* accident and the consequent decision, announced by President George W. Bush in January 2004, to retire the shuttle from active service marked a dividing line in U.S. thinking, not only with respect to the character of future U.S. space efforts but also with respect to a replacement human transportation system. Before the accident, most proposals for replacing the Space Shuttle focused on an advanced technology substitute, retaining most of the shuttle's capabilities but employing new technologies to overcome some of the shuttle's negative features, such as high operating costs, irreducible risks, and only partial reusability. Since the 2004 announcement that the shuttle would be retired, the focus has been on a less challenging undertaking—developing a transportation system optimized only for transporting crews to and from Earth orbit safely, reliably, and at less cost than the shuttle. Even with this lessening of ambitions, however, replacing the shuttle's crew transportation capability has proven to be less than straightforward.

The record of U.S. human spaceflight to date can thus be plausibly divided into two distinct periods, with considerable uncertainty regarding the character of the next period:

- **The Age of Apollo, 1958–1972**, when the United States learned the basics with respect to launching humans into space through Projects Mercury and Gemini and quickly applied that learning in Project Apollo to the bold venture of sending astronauts to the moon as a symbolic demonstration of U.S. leadership in the bilateral Cold War competition with the Soviet Union. Once that demonstration was successful, there was no compelling reason to continue deep space missions using Apollo systems, and those systems and their capabilities were rapidly abandoned.
- **The Space Shuttle Era, 1972–2011**, when the United States developed and operated a partially reusable, technologically advanced system that could launch humans and heavy cargo into LEO; could carry out a variety of space operations, most centrally the assembly and outfitting of an orbiting laboratory, the International Space Station; and could return both crew and cargo to Earth. Thirty years of experience operating the shuttle did not provide convincing evidence that the benefits provided by the totality of the system's capabilities outweighed the shuttle's costs and risks, and as the Space Shuttle approached retirement, there was no support for replacing it with an equally or more capable vehicle. Unlike the situation with respect to Apollo systems, there is ongoing an effort to use Shuttle-derived hardware for the next stage in the U.S. human spaceflight program.

The nature of that next stage in government-sponsored human spaceflight remains uncertain, although it seems clear that it will focus on activities beyond LEO. *Thus, a brief answer to the question "What next, after the Space Shuttle?" is new systems for new destinations,* not *a second-generation shuttle.*

Why No Shuttle Replacement since Columbia?

Before the *Columbia* accident in February 2003, there had been a number of failed attempts to replace the Space Shuttle with an alternative crew-carrying system. Space policy analyst Scott Pace recently estimated that NASA between 1990 and 2010 spent over $20 billion on space transportation systems that were later cancelled, many of which were intended as a shuttle replacement. None of these systems ever flew*.

In the wake of the accident, there was a fundamental rethinking of the goals of U.S. civilian space policy. One element of this rethinking was a decision to retire the Space Shuttle from service after it had completed a particularly critical task, one only it could perform: launching the elements

*Personal communication from Scott Pace.

of the ISS for in-orbit assembly and outfitting. In 2004, completing the ISS was planned for no later than 2010; this meant that the shuttle would be retired much earlier than had, until then, been anticipated.

When President George W. Bush announced new goals for the United States in space in January 2004, he also announced the decision to retire the shuttle. The president declared that NASA would "develop and test a new spacecraft, the crew exploration vehicle" and would "conduct the first manned mission no later than 2014." He assigned a dual role to this new spacecraft: "The crew exploration vehicle will be capable of ferrying astronauts and scientists to the space station after the shuttle is retired. But the *main purpose* [emphasis added] of this spacecraft will be to carry astronauts beyond our orbit to other worlds." It is worth noting that even as shuttle retirement was announced, replacing it with another system capable of carrying crew to LEO was given secondary priority to the system's deep space capabilities, and that a gap between the end of the shuttle program and the availability of a replacement crew-carrying system was accepted as part of the new Bush space policy.

Retiring the shuttle and developing a new crew exploration vehicle were only a part of President Bush's setting out of a new direction for the U.S. space program, the "Vision for Space Exploration." That vision focused on a "return to the moon by 2020, as the launching point for missions beyond." A primary goal was a "sustainable and affordable human and robotic program to explore the solar system and beyond" [5]. Setting this goal represented a fundamental shift in U.S. space policy, which since the start of the Space Shuttle era had been focused on activities in LEO.

Not made explicit in the president's statement, but implied by the budget projections provided by NASA at the time of the Bush announcement, was the possibility that the United States would withdraw from active utilization of the ISS soon after 2015, with no plans for a successor outpost; this also made the need for rapidly developing a capability for carrying a crew to LEO less than pressing. For the few years between shuttle retirement and withdrawal from ISS utilization, the United States could purchase transportation services to the station from Russia, while concentrating NASA's resources on developing systems for travel to destinations farther from Earth. Space leadership would be achieved through resuming deep space exploration, not by replacing the Space Shuttle with an equally or more capable vehicle.

Although President Bush had outlined ambitious goals for the U.S. civilian space program, he did not match them with a realistic financial commitment, saying that "NASA's current 5-year budget is $86 billion. Most of the funding we need for the new endeavors will come from reallocating $11 billion within that budget. We need some new resources, however. I will call upon Congress to increase NASA's budget by roughly a billion dollars, spread out over the next 5 years." He added "Future funding decisions will be guided by the progress we make in achieving these goals" [6]. In fact, the Bush administration after 2004 did not request even the promised $1 billion in increased NASA funding.

There was thus an important difference between earlier attempts to replace the Space Shuttle and that which took place in the aftermath of President Bush's 2004 announcement. At the time of the *Columbia* accident, NASA was focusing its future efforts on activities in LEO and planning to continue to use the shuttle as its primary crew transportation system through at least 2020, perhaps longer; according to NASA's thinking, any short-term alternative system for crew transportation would be a complement to the shuttle, not its replacement. The situation was different with the firm decision to retire the shuttle by 2010. This decision, as part of reshaping U.S. space policy to emphasize deep space exploration, was perceived as making a shuttle replacement to transport crews to the ISS a matter of limited urgency, with a lower priority than developing the new launchers and spacecraft to carry out the "return to the moon" mission. (In fact, because the pace of shuttle flights after return to flight in 2005 was slower than anticipated and because shuttle advocates were successful in convincing NASA to add an extra flight to the shuttle manifest, the final three shuttle flights actually slipped to 2011).

A new NASA Administrator, Michael Griffin, took office in April 2005. His immediate goal was to identify the hardware systems needed to implement the Vision for Space Exploration; these systems became embodied in a new program called Constellation (Fig. 14.3). For the crew exploration vehicle, NASA selected a spacecraft called Orion, to be launched by a shuttle-derived expendable launch vehicle designated Ares I; these were to be the first two elements of the Constellation program. As the president had suggested, transport to ISS was a secondary purpose for these systems; their primary role was a part of the "system architecture" needed for travel beyond Earth orbit.

Development of these two vehicles did not proceed as rapidly as NASA had hoped. After announcing the Vision for Space Exploration in 2004, over the next several years the White House requested significantly less funding for NASA, and thus for Orion and Ares I, than NASA had anticipated. The lack of adequate funding combined with the technical and managerial challenges that arose in both programs led by the end of 2008 to slips in the development schedule.

The status of the Constellation program and indeed the overall approach to the future of human spaceflight were matters of concern to the Obama administration as it took office in January 2009. One of its early space-related actions was to appoint a blue-ribbon "Review of U.S. Human Spaceflight Plans Committee," quickly known as the Augustine Committee after its chair, retired aerospace executive Norman Augustine. The committee carried out an intense three-month examination of the U.S. human spaceflight program. It found that although the original Constellation schedule had called for Orion

Fig. 14.3 NASA's Constellation program consisted of several components: especially the Ares I crew launch vehicle topped by the Orion spacecraft (right) and the Ares V heavy cargo launch vehicle. (NASA photo no. JSC2007-E-00152)

and Ares I to be available by 2012 or shortly thereafter, initial flights had slipped to at least 2017. With U.S. participation in the ISS then anticipated to end in 2015, this meant that both systems would be available too late to serve the mission of replacing the shuttle as the means for transporting crew to the space station. The committee was particularly critical of the Ares I vehicle; as its report was released in October 2009, Augustine described the booster as "the wrong launcher at the wrong time."

The Augustine Committee concluded, however, that the benefits of extending U.S. participation in the ISS to at least 2020 justified the costs of such an extension, even if that meant delaying a return to the moon until after 2020. Such an extension would create a continuing need for transporting crew to the ISS. The committee observed that such transportation service did not have to be provided by the government-operated Ares I/Orion system. Rather, "as we move from the complex, reusable Shuttle back to a simpler, smaller capsule, it is appropriate to consider turning this transport service over to the commercial sector." The committee added "there is little doubt that the U.S. aerospace industry, from historical builders of human spacecraft to the new entrants, has the technical capability to build and operate a crew taxi to low-Earth orbit" [7]. (The term *crew taxi* suggested the lower expectations for the shuttle replacement system.)

On February 1, 2010, NASA announced a new approach to human spaceflight, and it closely reflected the results of the Augustine Committee's analysis. The Constellation program was to be cancelled, and with it the government-developed Ares I/Orion combination that was to serve as the crew-carrying replacement for the Space Shuttle. As an alternative, NASA proposed to enter into several NASA-industry partnerships to develop new human transportation systems on a "commercial" basis. The use of the term *commercial* for what in essence a new approach to a public-private partnership in developing space hardware has led to unneeded confusion about the character of the change being proposed. The arrangement is "commercial" in the sense that NASA proposed to buy a privately developed transportation service rather than specify the characteristics of the system to provide that service and then closely oversee its development. NASA also hoped that several firms would be competing as service providers. However, at least in the early years, NASA would be the only customer for the transportation service, and so there would not be a market-based commercial situation. NASA hoped to provide $6 billion over the next five years as its financial investment in these partnerships; in addition, NASA would set the requirements that a private-sector provider of transportation services would have to meet to qualify for a subsequent contract for carrying NASA and its station partners' astronauts to the ISS. The private sector would have primary responsibility for designing and developing systems to meet those requirements.

This proposal for what was designated as the Commercial Crew program met at best a skeptical, and at worst a hostile, reception from NASA's Congressional authorization and appropriation committees, from the aerospace contractors that had been part of Constellation, and from now-former Administrator Mike Griffin. In particular, highly respected Apollo-era astronauts Neil Armstrong, James Lovell, and Eugene Cernan expressed doubts about the ability of newer entrepreneurial space firms to meet the demanding requirements of launching humans into orbit. In April 2010, the three signed an open letter suggesting that the decision to abandon a government-developed crew transportation system would severely cripple American capabilities. The letter said that commercial development of such a system "cannot be predicted with any certainty, but is likely to take substantially longer and be more expensive than we would hope" and "without the skill and experience that actual spacecraft operation provides, the United States is far too likely to be on a long downhill slide to mediocrity." The letter was widely distributed and reported [8]. The Apollo astronauts' opposition to the path proposed by the Obama administration persisted. On September 22, 2011, Armstrong told the House Committee on Science and Technology that "America cannot maintain a leadership position without human access to space, After a half century in which Americans were being launched into Earth orbit and beyond, Americans find themselves uncertain of when they can reasonably expect our astronauts to travel to the International Space Station or other off the earth destinations in other than a foreign built and commanded spacecraft."

Armstrong insisted that America "must find ways of restoring hope and confidence to a confused and disconsolate work force." He said "the reality that there is no flight requirement for a NASA pilot-astronaut for the foreseeable future is obvious and painful to all who have, justifiably, taken great pride in NASA's wondrous space flight achievements during the past half century." At the same hearing, former NASA Administrator Griffin suggested, "A real space program may, and indeed should, offer a stable market to be addressed by commercial providers, but it cannot be dependent upon such providers for strategic capabilities." Griffin added, "A real space program recognizes that this nation has interests that rise above the fortunes of individual private contractors, and it protects those interests. The proper role of government is to reward winners, not to pick them, nor to step in as an investor in enterprises which cannot pass the tests that the capital markets impose" [9].

Reflecting this criticism and Congress's own reservations regarding the path that NASA took in human transportation, Congress appropriated only $300 million for the Commercial Crew program in FY2011 and $406 million in FY2012, a reduction from the $1.35 billion that the White House had requested over those two years. But the criticism did not alter administration policy, which has gradually gained broader support. Even the space policy paper released by the presidential campaign of Mitt Romney in September 2012 seemed to endorse a commercial approach to crew transportation, saying, "NASA will look whenever possible to the private sector to provide repeatable space-based services like human and cargo transport to and from low Earth orbit" [10].

Future funding for the program is also likely to be less than what NASA hopes for; Congress approved a funding level of $525 million for FY2013. As a result, the program is proceeding at a slower than technically possible pace; NASA now estimates that initial astronaut flights to the ISS on a commercially provided spacecraft will not occur until sometime in 2017, or even 2018. Those first NASA-sponsored flights are likely to be preceded by test flights that are funded by the competing companies and that carry non-government crewmembers.

In reality, if the United States is to regain its own capability to transport humans to orbit, it will be through the results of the NASA Commercial Crew program; no other alternatives are being pursued. In August 2012, NASA announced that three firms had been selected to receive a third round of program funding. NASA hopes that at least two of these firms will develop and demonstrate crew-carrying spacecraft that meet NASA's stringent technical and safety requirements for transporting government astronauts to the ISS. Once such a demonstration is achieved, NASA intends to contract for crew transportation services, thereby becoming an "anchor tenant" for what it hopes will be a new business sector: commercial human space transportation.

The three firms selected by NASA for Commercial Crew program funding were as follows:

- Boeing, with its CST-100 capsule. Boeing is, of course, a very established company and is the repository of most U.S. industrial experience with astronaut transportation. Boeing plans to launch the CST-100 spacecraft on a human-rated Atlas 5 booster (Fig. 14.4).
- Space Exploration Technologies (SpaceX), with its Dragon capsule. SpaceX is an entrepreneurial space company that quelled significant skepticism regarding its capabilities with several successful flights of the Dragon capsule, carrying cargo, but not yet a crew, to the ISS. SpaceX plans to launch the Dragon on its own Falcon 9 booster.
- Sierra Nevada, with its Dream Chaser spaceplane. Unlike the CST-100 and Dragon space capsules, which will land via parachutes, Dreamliner is a lifting body that will be able to glide to a landing on a conventional runway. Sierra Nevada plans to launch the Dream Chaser on an Atlas 5 booster. Sierra Nevada received approximately half the NASA funding provided to Boeing

Fig. 14.4 In the Orbiter Processing Facility-3 (OPF-3) at NASA's KSC in Florida, a pressure vessel for Boeing's CST-100 spacecraft is unveiled at a 2011 ceremony. (NASA photo)

and SpaceX, reflecting the limited funding available for the program and the higher technical risks associated with the Dream Chaser concept.

Barring unforeseen developments, then, the United States will, by 2017–2018, finally have available from one to three U.S.–manufactured systems for carrying humans to LEO. These systems will be "shuttle replacements" only in a limited, although important, sense, in that they will provide a way to launch humans into orbit and return them to Earth. They will not have the other capabilities of the Space Shuttle, such as the ability to launch and return heavy payloads, or to carry out in-orbit operations such as satellite servicing and retrieval. For the foreseeable future, these capabilities must either be provided by robotic systems or will be missing.

Lessons from Attempts to Replace the Space Shuttle

We can draw a variety of lessons that are relevant to understanding recent U.S. space policy from the quarter century of discussions and false starts with respect to developing a replacement for the Space Shuttle, and particularly from the experience of recent years as shuttle retirement became imminent.

The Need to Avoid Technological Over-Optimism

Before the *Columbia* accident in 2003, the national leadership had given the aerospace community two major opportunities to develop a shuttle replacement; these opportunities were accompanied by significant (although not adequate) funding commitments. The first of these opportunities, the National Aerospace Plane program (NASP), was begun in 1986 and initially justified on national security grounds; NASA, a junior partner in the undertaking, was not able to continue it as a development effort leading to a flight-test vehicle, once DoD funding was withdrawn [11]. The second opportunity was the Single-Stage-to-Orbit (SSTO) X-33 effort initiated by NASA in 1996 in response to NASA's internal studies and then to the 1994 National Space Transportation Policy. After spending $1.5 billion on the effort, NASA in 2002 abandoned the program after it ran into major technological difficulties.

With the benefit of hindsight, it is possible to see that these two efforts were very likely doomed to failure from their outset. In both cases, the approach selected depended on being able simultaneously to bring to an adequate level of maturity a variety of challenging technologies in areas such as aerodynamics, guidance and control, materials, and propulsion. Those responsible for both efforts within the DoD, NASA, and the aerospace industry assured their leaders that they could overcome these technological challenges and move forward rapidly and with affordable costs. These assurances were at variance with what actually transpired.

The costs of a lack of historical perspective and unchecked technological optimism, bordering on hubris, have been high. Roger Launius has suggested that the X-33 program and the NASP before it "have been enormous detours for those seeking to move forward with a replacement for the Space Shuttle. Expending billions of dollars and dozens of years in pursuit of reusable SSTO technology, the emphasis on this approach ensured the tardiness of development because of the strikingly difficult technological challenges" [12]. The CAIB agreed, suggesting that one reason for the "failure of national leadership" related to the absence of a replacement for the Space Shuttle was "continuing to expect major technological advances" in a replacement vehicle [13].

How are nontechnical decision-makers to be protected against the enthusiasm of technological optimists? That is a topic well beyond the scope of this chapter, but clearly, in the case of X-33 and NASP, the necessary checks and balances were missing or not influential.

The post-*Columbia* attempts to replace the crew-carrying capabilities of the Space Shuttle seem to have taught us to avoid unwarranted technological optimism. The Ares I/Orion elements of the Constellation program did not utilize cutting-edge technology; in fact, they were designed to build on technologies derived from the Space Shuttle and even the Apollo programs. The major obstacles to keeping the effort on schedule arose from inadequate funding and the lack of recent workforce experience in designing and developing human-rated hardware. With respect to the contenders participating in the Commercial Crew program, low technological risk has been a positive feature in evaluating their prospects.

It is fair to ask whether this reaction to the technological failures of X-33 and NASP has led to an overcorrection in avoiding technological risk in future crew-carrying systems. Clearly, the benefits of reusability, advanced propulsion systems, and a variety of in-orbit capabilities would be worth having at some future time. Investments in developing the technological foundation for such attributes should be part of the government space investment portfolio, so that systems embodying advanced technologies can be incorporated in the space transportation systems of the future, whether publicly or privately developed.

CHANGING THE STATUS QUO IS HARD

During its 30-year flight history, the Space Shuttle program was supported by a multibillion-dollar annual budget and employed tens of thousands of people in various locations across United States. The Shuttle program was the focus of much of the activity at Johnson Space Center (JSC), with the large astronaut corps located there; Marshall Space Flight Center (MSFC); and KSC. Each center was staffed by a large complement of civil servants and support contractors. Major and smaller aerospace firms across the United

States worked on the Shuttle program; the program thus employed a large contractor workforce.

It was not surprising, then, that throughout the Shuttle program's history a politically active coalition of NASA center, contractor, local and regional, and congressional supporters developed. These supporters argued that because the shuttle was a vehicle that continued to be superior in capabilities to any technologically feasible replacement, the preferred course of action was to invest scarce funds in upgrading and modernizing the shuttle rather than seeking to replace it. Even after President Bush in 2004 announced the decision to retire the shuttle, and up almost to the time of the final shuttle flight, some argued that the Space Shuttle should be kept flying until a replacement system was ready for use.

The existence of an organized coalition of public and private interests with a stake in the Space Shuttle program was an entirely legitimate phenomenon. The whole system design of the American political process is intended to allow organized interests to contend for a favorable policy outcome. In this case, however, there was no organized alternative interest group pushing for an early shuttle replacement, and thus before *Columbia*, the default outcome of annual policy debates was likely to favor the pro-shuttle position, or, at a minimum, not result in outcomes opposing it. Although, for example, there was opposition from the scientific community and some members of Congress in the 1980s and 1990s to the space station program, there was no similar consistent opposition to the Space Shuttle. Only after the *Columbia* accident did the power of the pro-shuttle coalition diminish to the point that it was unable to achieve its desired outcome of continued flights until a replacement was ready.

Especially in the decade before the *Columbia* accident, the combination of limited NASA budgets and uncertainty about when the shuttle might be replaced created an ambivalent policy attitude towards the Space Shuttle program. This outcome was perhaps the worst possible situation—not enough funding for optimum operation of the shuttle, but also inadequate political commitment behind an effort to replace it. It was most fundamentally a reflection of the place that human spaceflight then held in the list of national priorities—something that most Americans wanted to see continue but were unwilling to invest enough resources in to do well.

Recognizing the strength of the pro-shuttle coalition, the Constellation program was designed to draw as much as possible upon the shuttle's technological heritage. That way, many in the NASA workforce and major aerospace contractors would be able to transition from Space Shuttle to Constellation work, particularly as that work would be managed primarily by the MSC, KSC and JSC and their traditional contractor partners. In addition, as the assembly of the ISS neared its completion, shuttle supporters argued for keeping the system in operation until its replacement was ready, rather than accepting a multiyear gap during which the United States would have to depend on Russia for transporting U.S. astronauts to the ISS.

Support for extending shuttle operations continued even after the Obama administration cancelled Constellation but was not powerful enough to counter the momentum towards retirement. Once it became clear that the Space Shuttle was indeed at the end of its lifetime, the pro-shuttle coalition shifted its attention to imposing on NASA a mandate to continue the Orion program as a "multi-purpose crew vehicle" and to developing a shuttle-derived heavy lift Space Launch System vehicle that would provide continuing work for many in NASA and its contractors. Compared to the powerful political support for these long-term programs, which has resulted in Congress appropriating multibillion dollar budgets for them, little influential backing has been given to the effort to develop new crew-carrying capabilities through the Commercial Crew program; the program's budget after Congressional reductions has thus been barely enough to support meaningful progress.

Several of the entrants in the Commercial Crew program were new participants in the political system, and so they couldn't organize one or more support coalitions powerful enough to counter the lasting influence of the long-standing alliance between NASA, Congress, and the traditional aerospace industry. This alliance has provided the support for the Space Shuttle and the ISS over the past three decades, and it now gives priority to developing the capabilities for exploration beyond LEO. Attempts by the Obama White House and the NASA political leadership to change the *status quo* have met with only limited success. The president, rather than fight for his new approach to space activities, accepted the Congressional directives with respect to Space Launch System and Multi-Purpose Crew Vehicle development contained in the NASA FY2010 Authorization Bill. The current effort to replace the Space Shuttle's crew-carrying capability, rather than being a high-priority NASA program, has struggled to gain political support and forward momentum. Congress has seemed more willing to see funds channeled to Russia for crew transportation services rather than to support U.S. firms that are not part of the traditional pro-space coalition. All of this is evidence of the difficulties inherent in changing the space *status quo*.

PRIORITIES HAVE CHANGED

When President Richard Nixon approved the development of the Space Shuttle in January 1972, he was in effect giving overriding priority for the foreseeable future to U.S. space activity in LEO. Nixon gave less attention to the specific purposes that the shuttle would serve than to keeping alive the U.S. program of human spaceflight with a new development program that would, in the short term, also produce jobs in states critical to his 1972 reelection. Between 1972 and 2004, the priority assigned to LEO activities remained in force, even though space advocates were consistent in their proposals for journeys to more distant destinations. The Space Shuttle and its joined-at-the-hip handmaiden, the ISS, dominated U.S. human spaceflight efforts from the time of the first shuttle flight in 1981 to the 2011 completion of the ISS assembly. As earlier chapters in this volume have attested, the shuttle's many capabilities were essential to carrying out a program of human spaceflight focused on LEO activities.

For the first two decades of Space Shuttle operations, the concept of a shuttle replacement was thus for the most part interpreted as developing a new LEO-limited system that was superior in some respect to the shuttle's impressive capabilities. However, with priority in the NASA program shifting away from LEO activities to deep space exploration, there is less need for a highly capable vehicle for LEO operations and crew transportation. Post-*Columbia* plans for replacing the Space Shuttle have recognized this change. As noted above, the Constellation program focused on developing capabilities for crew transport to LEO, and these abilities were based on existing technologies and were secondary in priority to using the same systems as part of deep space travel. Participants in the Commercial Crew program have also offered crew-carrying solutions with low technological risk and capabilities limited to crew transportation.

Thus, the concept of "shuttle replacement" has changed in meaning over recent years, as the overall goals of the civilian space program have shifted. Rather than giving priority to technological advances vis-à-vis the many shuttle capabilities, current plans emphasize safe, reliable, and affordable crew transport. This makes the challenge of replacing the shuttle less daunting and, in principle, less expensive to develop. But it also means that, although U.S. politicians rhetorically bemoan dependence on Russia for providing crew transportation services for U.S. astronauts, there is comparatively limited political support behind the effort to replace the technological marvel that was the Space Shuttle with a much more modest "crew taxi."

SUBSTITUTING MEANS FOR ENDS

There are some significant similarities between the situation as the Space Shuttle is retired and that in the 1969–1972 period, when the decision to develop the shuttle was made. In both instances, decisions were made with respect to developing major new hardware systems in advance of policy and public consensus regarding future space goals and why those systems were needed to achieve them. Policy analyst James Vedda wonders whether the U.S. space program is "poised for a replay of the 1970s." He points out "many similarities," such as [14]

> across-the-board retrenchment in program ambitions and resources; termination of a flagship human spaceflight capability; an expected gap

> of a few years before a new spaceflight capability could be ready; development of a new launch vehicle using a disproportionate amount of agency resources; and inadequate strategic planning for what will be done with the new vehicle, or indeed, in exploration and development generally.

This lack of strategic planning for future space activities is the most striking parallel between the situation when it was decided to develop the Space Shuttle and the situation as the shuttle has been retired. In both instances, policy makers have approved the development of a *means* for carrying out a future program, rather agreeing on the broader national goals that such a future program would serve.

As the shuttle was approved, there was a generalized expectation that lower-cost and routine access to LEO would allow the development of a wider variety of space missions, including both already established undertakings and new, innovative applications of the capability to access LEO and to operate there. There was little critical analysis of this expectation, even though the economic case for the Space Shuttle depended on a much higher flight rate than had historically been the case. The economic analysis think tank Mathematica analyzed the economic dimensions of the shuttle program and clearly made this point in its report to NASA. Mathematica noted, "A new, reusable Space Transportation System should only be introduced if it can be shown, conclusively, what it is to be used for and that the intended uses are meaningful to those who have to appropriate the funds, and those from whom the funds are raised . . ." [15]. This advice was ignored; the Shuttle program went forward without any agreement on the long-term purpose of U.S. human spaceflight other than keeping U.S. astronauts flying in space.

As noted several times already, the focus of future U.S. human spaceflight activities has shifted to travel beyond Earth orbit since the 2003 *Columbia* accident and the consequent articulation of a new "Vision for Space Exploration" in 2004. The Augustine Committee in its 2009 report recognized "a strong consensus in the United States that the next step in human spaceflight is to travel beyond low-Earth orbit." In 2010, Congress approved, and the president signed, a NASA authorization bill that specified, "The long term goal of the human space flight and exploration efforts of NASA shall be to expand permanent human presence beyond low-Earth orbit...and to do so, where practical, in a manner involving international partners." These bold but vague statements offer little guidance in planning an exploration program for the coming decades. The Augustine Committee also commented, "Planning for a human spaceflight program should begin with a choice about its goals—rather than a choice of possible destinations" [16].

As was the case in 1971–1972, this advice has been largely ignored. In 2010, President Obama set a mission to a particular destination—a near-Earth asteroid—in the 2025 time period as the first step in a new human spaceflight program, with missions to the vicinity of Mars in the 2030s as the longer-term destination. NASA identifies its current planning as "capability-based," suggesting that the systems it is developing will be able to carry astronauts to a variety of destinations, without a clear rationale for selecting among those destinations in terms of a broader U.S. space strategy. Enthusiasm for a near-Earth asteroid mission as the first step in a long-term exploration effort has waned since 2010, and so NASA has begun to emphasize that its new systems will also provide the capability to operate in cislunar space during the 2020s, prior to undertaking missions to an asteroid or to Mars or its moons.

One difference in the situation in the early 1970s and at the current time is the greater recognition now of the need for a strategic framework within which human spaceflight missions would take place. The momentum of Project Apollo carried the NASA human spaceflight effort forward through the development of the Space Shuttle and the approval of a space station without significant discussion of the longer-term goals served by those efforts. Not until 1984, as Congress created a National Commission on Space, was there serious questioning of the purposes served by sending humans into space. The 1986 report of that commission set out a bold strategic vision for the U.S. space program [17]:

> The Solar System is our extended home. Five centuries after Columbus opened access to "The New World" we can initiate the settlement of worlds beyond our planet of birth. The promise of virgin lands and the opportunity to live in freedom brought our ancestors to the shores of North America. Now space technology has freed humankind to move outward from Earth as a species destined to expand to other worlds.

This call for space settlement as the strategic goal for U.S. space efforts had little immediate impact, coming at the time of the 1986 *Challenger* accident and the subsequent focus, extending for the next quarter century, of operating the Space Shuttle safely and completing an LEO laboratory, the ISS. However, as the Augustine Committee reviewed the U.S. human spaceflight program in 2009, it concluded, in resonance with the 1986 report, that "the ultimate goal of human exploration is to chart a path for human expansion into the solar system" [18]. Uncertain whether pursuing such an ambitious goal was indeed the proper strategy to guide U.S. human spaceflight, both Houses of Congress in 2010–2011 asked the National Academies of Sciences and Engineering/National Research Council to carry out blue-ribbon studies of NASA's strategic direction. While the Senate-requested study, being carried out by a Committee on Human Spaceflight, is not scheduled to be completed until 2014, the House-requested study, carried out by the Committee on NASA's Strategic Direction in its December 2012 report observed that "There is no national consensus on strategic goals and objectives for NASA. Absent such a consensus, NASA cannot reasonably be expected to develop enduring strategic priorities for the purpose of resource allocation and planning" [19].

It is uncertain whether the results of these or any other studies can form the foundation for a long-term national commitment to space exploration. The question remains, as it did when Mathematica raised it in 1972 with respect to shuttle development, whether developing a new capability for human exploration, leading to eventual space settlement, "should only be introduced if it can be shown, conclusively, what it is to be used for and that the intended uses are meaningful to those who have to appropriate the funds, and those from whom the funds are raised...."

It is a tough challenge to forge through the play of American politics a sustainable national consensus that supports an exploration effort leading in the long term to humanity becoming a multiplanet species. The past history of similar attempts is not promising. Veteran space policy participant Mark Albrecht notes that "three presidencies . . . passed since President George H. W. Bush [in 1989] called for a post–Cold War exploration program to keep America advancing into the new frontier. . . America could muster neither the will nor the wallet to answer the call." Albrecht notes that a similar fate befell President George W. Bush's 2004 call for a similar effort and is skeptical that the political response to current space exploration will be different. He suggests that both China and perhaps India "have taken up the gauntlet of space exploration" and are "prepared to accept the costs and risks of human space exploration along with the costs of modernization, education, and infrastructure developments on the ground. . . .The question for this generation of Americans is, 'will we'?" He doubts that the answer will be positive [20]. While China is indeed pursuing a slow but steady development of its human spaceflight capabilities, there is little indication that India is willing to commit the resources required to achieve similar capabilities.

Perhaps a formal national commitment to space exploration and eventual settlement is not in fact essential. The Space Shuttle program was able to sustain itself for 40 years despite the lack of a strategic framework within which it functioned, primarily because it served specific powerful political interests while at the same time engaging the American public with the sporadic excitement of humans going into space. Might the current space exploration program follow a similar path? Can the United States in the coming decades lead the way in human journeys to Mars based on year-by-year decisions to sustain the human spaceflight program rather than a fundamental national commitment to space exploration? That possibility may be the most basic parallel between the time when the Space Shuttle was begun and the time when it ended its service to the country.

Crew Transportation: Commodity or Strategic Capability?

This chapter has suggested that the capability to carry humans to LEO is approaching the status of a commodity that in the near future may be able to be purchased from several U.S. suppliers, and that a public-private partnership to bring that capability, operating on a commercial basis, into being is an appropriate path for the country to pursue. The alternative view, articulated by former NASA administrator Mike Griffin among others, is that crew transportation is a strategic capability that serves important national interests such as global leadership in space and national security broadly defined. He and others also feel that long-term dependence on U.S. industry operating on a commercial basis to supply that capability is unacceptable, given the uncertain overall market for human transportation and thus the possibility that one or more U.S. suppliers could leave the market.

Assured access to space for crucial nonhuman government payloads, particularly those related to national security, has been deemed a strategic necessity for many years. To ensure that the United States possesses that capability, the DoD has for many years provided funding to maintain two expensive launch vehicles, the Atlas 5 and Delta 4, even as the once-anticipated commercial market for those vehicles did not emerge. There is no similar national policy consensus on the importance of launching humans into space; the lack of such a consensus is an underlying factor in the failure of past efforts to replace the Space Shuttle. It is beyond the scope of this chapter to assess whether the various rationales offered for a continuing U.S. government program of human spaceflight add up to making such an effort important enough to U.S. interests to be deemed strategic. The fact is, however, that in the 40 plus years since the end of the Apollo program, policy and political support for human spaceflight has been consistent—but barely adequate—to maintain a viable program. The leaders of the U.S. government have not provided financial and policy support for human spaceflight in a manner befitting an activity essential to U.S. strategic interests.

This could change. The U.S. government *is* developing, as a government-managed program, the capability to carry humans into space through its current Orion spacecraft and Space Launch System projects. That capability is aimed at carrying those humans beyond LEO and reflects the 2009 Augustine Committee finding that "There is now a strong consensus in the United States that the next step in [government-sponsored] human spaceflight should be to travel beyond low-Earth orbit" [21]. Whether this consensus is strong enough to support a robust program of space exploration and subsequent development remains to be seen.

Perhaps the issue of replacing the crew-carrying capability of the Space Shuttle comes down to a question of the future of U.S. government activities in LEO. Once utilization of the ISS is ended, will there be another government-financed orbital outpost? If not, will other orbital activities involving human presence be carried out under government sponsorship? If the answer to these questions is in the negative, then there is no requirement for a shuttle replacement, and the 40-year Shuttle era will be seen in retrospect as just one finite step in the progressive development of humanity's space activities.

REFERENCES

[1] Nixon, R. "Statement Announcing Decision To Proceed With Development of the Space Shuttle," January 5, 1972. Online by Peters, G. and Woolley, J. T., The American Presidency Project. http://www.presidency.ucsb.edu/ws/?pid=3574. (retrieved 21 July 2012).

[2] Logsdon, J. M., "A Failure of National Leadership': Why No Replacement for the Space Shuttle?" edited by S. J. Dick and R. D. Launius, *Critical Issues in the History of Spaceflight*, NASA SP-2006-4702, Government Printing Office, Washington, 2006.

[3] *Columbia Accident Investigation Board Report*, Vol. 1, NASA and GPO, Washington, DC, August 2003, pp. 209–211.

[4] Launius, R. D., "After *Columbia*: The Space Shuttle Program and the Crisis in Space Access," *Astropolitics*, Vol. 2, Autumn 2004, pp. 278–279.

[5] Bush, G. W., "Remarks at the National Aeronautics and Space Administration," January 14, 2004. Online by Peters, G., and Woolley, J. T., The American Presidency Project. http://www.presidency.ucsb.edu/ws/?pid=72531. (retrieved 21 July 2012).

[6] Bush, G. W., "Remarks at the National Aeronautics and Space Administration.

[7] Review of U.S. Human Spaceflight Plans Committee, Seeking a Human Spaceflight Program Worthy of a Great Nation, October 2009, pp. 89–90, 13, 70. Report is available online at www.nasa.gov/pdf/396093main_HSF_Cmte_FinalReport.pdf. (retrieved 21 July 2012).

[8] http://www.guardian.co.uk/commentisfree/cifamerica/2010/apr/15/obama-nasa-space-neil-armstrong (retrieved 18 October 2012).

[9] http://science.house.gov/press-release/armstrong-cernan-stress-importance-ambitious-human-spaceflight-program (retrieved 18 October 2012).

[10] http://www.spacepolicyonline.com/news/romney-space-policy-still-short-on-specifics (retrieved 18 October 2012).

[11] Butrica, A. J., *Single Stage to Orbit: Politics, Space Technology, and the Quest for Reusable Rocketry*, Johns Hopkins, Baltimore, MD, 2003, Chap. 4.

[12] Launius, R. D., "After *Columbia*, p. 291.

[13] *Columbia Accident Investigation Board Report*, p. 211.

[14] Vedda, J. A., *Becoming Spacefarers: Rescuing America's Space Program*. Xlibris Corporation, Bloomington, IN, 2012, pp. 83–84.

[15] Logsdon, J. M., Williamson, R. A., Launius, R. D., Acker, R. J., Garber, S. J. and Friedman, J. L., *Mathematica*, "Economic Analysis of the Space Shuttle System," January 31, 1972, reprinted in Logsdon, J. M., et al., *Exploring the Unknown: Selected Documents in the History of the U.S. Civil Space Program, Vol. IV: Accessing Space*, NASA SP-4407, Government Printing Office, Washington, 1999, p. 241.

[16] Review of U.S. Human Spaceflight Plans Committee, p. 9.

[17] National Commission on Space, *Pioneering the Space Frontier*, Bantam Books, New York, 1986, p. 5. The report is available online at www.history.nasa.gov/painerep/begin.html. (retrieved 21 July 2012).

[18] Review of U.S. Human Spaceflight Plans Committee, p. 22.

[19] Committee on NASA's Strategic Direction, National Research Council, *NASA's Strategic Direction and the Need for a National Consensus*, National Academies Press, Washington, December 2012, p. 5. Report is available online at http://www.nap.edu/catalog.php?record_id=18248. (retrieved 28 March 2013).
[20] Albrecht, M. J., *Falling Back to Earth: A First Hand Account of the Great Space Race and the End of the Cold War*, New Media Books, Washington, DC, 2011, p. 191.
[21] Review of U.S. Human Spaceflight Plans Committee, p. 21.

Chapter 15

EPILOGUE

JOHN KRIGE

This book looks back over 30 years of the Space Shuttle's history. It is a tribute to an outstanding technological and social achievement. However, it is not merely an exercise in nostalgia. Its added value lies in our attempt to identify key aspects of the shuttle's history and to draw lessons for the future from them. This epilogue pinpoints some of the Space Shuttle program's achievements and what we can learn from them.

There is no one best way of putting humans into space. Each solution adopted, capsule or spaceplane, involves a compromise between technological and social considerations (Launius). The concept of the shuttle—a reusable plane that carried humans into space and that brought them back again to a landing strip on Earth—was not novel. It had fired the imaginations of engineers, science fiction writers, and the public ever since the 1920s, if not before. The imperatives of the Cold War obliged the engineers at NASA, previously NACA, to deviate from their preferred paradigm. To beat the Soviets to the moon, they replaced the nosecone that shielded a nuclear warhead on an ICBM with a blunt capsule that carried three men on a multistage rocket (Launius). That mission accomplished, they reverted to the spaceplane concept, even though it adds drag at lift-off and adds nothing to in-orbit performance. The attraction of at least a fully reusable spaceplane was that it would return all but the payload and fuel to Earth after each mission. By contrast, Apollo returned less than 0.25 percent of its original mass in a single small crew capsule, leaving the rest on the moon or to burn up on descent. Another major advantage of a plane over a capsule was on reentry and landing, both aerodynamically and in terms of elegance: stepping from a cockpit onto a red carpet is more appealing than climbing unceremoniously out of a capsule (Launius)!

The shuttle was originally conceived as the primary means of ferrying parts, people, and supplies to an orbiting space station (McCurdy). Its final configuration decoupled it from this mission, and it was reconceptualized as a multipurpose vehicle that could also launch scientific, commercial, and military satellites into LEO (Hersch). Its architecture was the result of compromises between multiple stakeholders, and it was made only partially reusable to cut development costs. To satisfy the demands of the DoD, which wanted to use it to orbit large reconnaissance satellites, it had a bigger cargo bay than NASA originally planned. It had a cross range maneuvering capability of 1265 statute miles in the interests of national security. This required a huge delta wing that increased the area of the most vulnerable parts of the shuttle's body: the leading edges of the wings and the underbelly (Launius). Once it was operational, the defense and national security agencies imposed a culture of classification, need to know, and limited access to meetings when their payloads were flown. To NASA's relief, these agencies no longer used the shuttle after the *Challenger* accident. Secrecy was alien to NASA's civilian culture of openness (Hale, Hersch).

NASA experimented with many shuttle designs as it did with several management schemes. Key responsibility for development and then operation was located at headquarters, located at lead centers, or distributed pluralistically (Lambright). This had some important advantages. For example, NASA insisted on splitting the shuttles' hardware contracts from the software contracts, locating control over the latter in the Johnson Space Flight Center, where a dedicated team of programmers could draw on and extend their experience gained from previous human spaceflight programs (Leveson). This also helped maintain secrecy for national security payloads. The lead-center concept also had disadvantages, notably inter-center rivalry. None of the macro-management schemes persisted over the life of the shuttle, and none was deemed appropriate for all eventualities. The default model, redeployed first after the *Challenger* accident and then after *Columbia*, was a reversion to centralizing decision-making power at NASA headquarters. This was the model that had worked so well for the Apollo program. There is no one best way to manage the vast technological system that is putting humans safely into space.

Contrary to popular expectations, the Space Shuttle was an ongoing R&D project. Engineers sought to constantly improve its performance by using new technological possibilities as they arose. Consider the SSME. It was the only reusable large liquid propellant rocket engine in the world and operated at very high chamber pressures (Biggs). It was controlled by a digital computer and software programs that were also the first of their kind in the world (Leveson). In clusters of three engines, each of the SSMEs provided a half million pounds of thrust to place space shuttles and crews into LEO. During the SSME's development, the team at the MSFC and Rocketdyne had to face countless minor problems and two major setbacks with turbopumps (violent gyration of the rotor, and high-pressure explosions of LOX), and these problems increased the pressure to stay on schedule and to deliver a safe and reliable engine. In operation, the SSME was removed from the orbiter after each flight for invasive inspections and maintenance: a pretext for making other improvements at a cost to the flight rate (Hale). The *Challenger* accident, tragic as it was, provided a breathing space during which 71 changes were made to reduce risk and improve margin. New turbopumps built by Pratt and Whitney reduced rotating parts from 50 to 28 and the number of welds from 300 to 4 (Biggs). The stand-down on the shuttle also allowed for wide-ranging software improvements, including the elimination of life-critical errors (Leveson). The SSME's configuration, like that of the shuttle hardware and software, was never completely frozen even after the shuttle was declared operational in 1982.

The search for the most suitable TPS stimulated innovation in universities, government laboratories, and industries alike (Jenkins). When the project got under way, it was generally assumed that the ablative heat shields used on capsules in the 1960s would be replaced by a metallic system. This was prone to oxidation, however, and instead ceramic tiles were chosen for the body and reinforced carbon-carbon for the leading edge of the wings and for the nose. NASA engineers pursued many different materials and bonding systems, including one developed at NASA Ames Research Center using a new material named polybenzidimazole (PBI). PBI was never used on the shuttle but became a product of choice in high-performance protective apparel like firefighter suits. The development of the shuttle's TPS saw new ideas and products migrating within the aerospace industry (many lessons were learned from Dyna-Soar; see Launius) and between the space program and civil society. In the end, the TPS, despite its long development and complex nature, performed well, although at a much higher operational cost than anticipated.

Software proved to be one of the most important if underestimated components of the shuttle system. The Space Shuttle program benefitted from lessons learned in the Apollo program, notably engineering for real-time systems (Leveson). Even though it was recognized that software is more difficult to develop than hardware, NASA's original budget for shuttle software was just 10 percent of the actual development cost ($200 million). Once operation began, mandatory changes to software were required for each flight to accommodate different payloads, to perform hardware upgrades, and to remove flaws in the software itself (Hale). By 1992, NASA was spending $100 million a year on maintaining software (Leveson). It was under constant pressure to reduce the size of the software, such that the astronauts eventually used small portable computers to augment the onboard graphics displays, which were cluttered and difficult to read. Maintaining the software's integrity required extensive planning, continuous improvement, exhaustive testing, and careful documentation at every stage. The software culture at NASA was not can-do—on the contrary. It called for dedication, self-discipline, and rigor, and it provided a stability and need for professionalism that seemingly made software development on the shuttle particularly appealing to women (Leveson).

All three of these major technological developments (the SSME, the TPS, and the software) illustrate the importance of long-term (government) investments in major space projects. The continuity gave programmatic stability, provided invaluable opportunities for learning, and led to the accumulation of experience in a body of skilled scientists, engineers, and managers who were able to push the knowledge envelope with confidence as their understanding of complex processes in a particular domain increased. The fact that the shuttle was human rated *and* that it often demanded state-of-the-art knowledge and technology forced shuttle teams to learn how to balance reliability with novelty, and to combine self-discipline with creativity.

To sell the shuttle to an indifferent public, a skeptical Congress, and determined cost-cutters in the OMB, NASA promoted the shuttle as an economical, routine, and safe means of access to space: the name itself expressed those qualities. The system proved to be none of these. The cost of a flight climbed steadily from about $330 million in the early 1980s to over $1 billion in the 2000s (McCurdy). NASA was never able to sustain promised annual flight rates. The two tragic high-profile accidents of the *Challenger* and *Columbia* orbiters scuttled any idea that this was a risk-free environment. The challenge that NASA and its contractors faced was not to eliminate risk altogether—that was impossible—but to reduce it to an acceptable level. That level evolved with experience and over time, and it was the negotiated outcome of the need to strike a balance between multiple considerations including cost and schedule, and the demands of patrons and of clients (Waring).

Accidents were not unexpected. Indeed, the burn-through on the Viton O-rings that caused the destruction of the *Challenger* had been of concern for years, and the TPS was a constant source of anxiety (Hersch). NASA managers normalized these dangers, as they did hundreds of others, defining them as in-family maintenance problems, not life-critical hotspots (Waring). They erred, not in making such choices, but in the arrogance born of their success with Apollo, and their tendency to sideline dissenting voices from worried engineers. The complex communication networks, made more difficult by the policy of outsourcing safety matters to contractors after 1992, added to the messiness of the decision-making process under the stress of launch and reentry. It is unfortunate that measures were not taken within the first four days after launch to send a shuttle up to remove the crew from *Columbia* (Waring). It is unrealistic to expect the shuttle never to have failed. It was a double tragedy that it failed on highly visible missions that were being followed in real time by millions. It is ironic that public support for the Space Shuttle program was highest immediately after the two orbiter accidents (Billings).

The shuttle democratized access to space. In the 1960s, astronauts were manly, white test pilots called upon to beat the Soviets in a Cold War struggle for space supremacy. The Space Shuttle program was open to women and ethnic minorities, it included many scientist-astronauts alongside shuttle pilots, and it often carried international crews (Foster). Engineers had to design an onboard environment that was habitable and that allowed for mixed-sex crews living and working in a spaceplane, rather than just surviving in a capsule. Life on board the shuttle was uncomfortable but certainly more luxurious than in the earlier capsules. Tensions arose between highly trained career astronaut pilots (Hale) and payload specialists who were far less experienced (Hersch). Technological solutions to living in a cramped space for two weeks were complemented by building crew member camaraderie that helped solve problems of privacy (Foster). By routinizing access to space, in the sense of making it accessible to nonprofessional astronauts, the shuttle paved the way for the commercialization of human spaceflight and an era of space tourism.

Expendable launch vehicles could have performed some of the shuttle's missions, doubtless at lower cost, but not all. The shuttle was a multipurpose vehicle. Thousands of experiments proposed by very different scientific communities were performed in space by onboard specialists, though none has proved transformative (Hersch). Claims about the commercial opportunities of producing new materials at zero g have proven exaggerated. Indeed, the shuttle has a mixed record in terms of doing outstanding research that became one of its main justifications before the ISS was authorized. Ironically, its unique features were often exploited as regards satellites only when things went wrong. Astronauts saved the $2 billion HST by installing correcting optics to its main mirror, and they extended its research capabilities and its lifespan by repeated technological upgrades (Foster, Hersch). Two stranded commercial communication satellites were recovered in space walks, returned to the shuttle's payload bay, and brought back to Earth. The shuttle is best seen as opening up a new range of opportunities beyond those provided by Expendable Launch Vehicles (ELVs), rather than as a replacement for them. Wide-ranging and ambitious space programs that include human spaceflight require a mixed fleet if they are to succeed (McCurdy). To think otherwise, to phase out ELV production when the shuttle became operational, was a serious policy blunder, rectified just in time by the *Challenger* accident.

In the immediate aftermath of the Apollo moon landings, NASA promoted the shuttle as one essential component of a two-part system, the other being a space station (McCurdy). It took 15 more years before President Reagan authorized a station in 1984. NASA originally planned to use Saturn V rockets for most of the assembly, saving the shuttle for the far larger number of crew transfer and refurbishment missions. In the event, it was used primarily for assembly, transforming that stage of the process from a rendezvous and docking challenge (which the Soviets automated with their Mir station) into a construction project of what became the equivalent of a 35-story office building. The assembly phase lasted for more than 12 years and required 36 shuttle flights and 155 spacewalks totaling almost 1000 hours (McCurdy). International partners played a major role during this phase, conducting the majority of the missions for crew supply and refurbishment. The space station could not have been built without the shuttle, even if that was not its primary role when the post-Apollo program was promoted, and even if assembly was not the most cost-effective use of the orbiter.

When President Nixon authorized the shuttle in January 1972, he suggested that it would become a platform for international collaboration. He particularly mentioned flying astronauts from the Soviet bloc: In the shuttle's operational life, it actually hosted astronauts from 16 countries. The first Israeli astronaut, Ilan Ramon, perished in the *Columbia* accident, as did the Asian-American Kalpana Chawla, the first Indian woman in space. In 1972, Nixon also hinted that there might be foreign contributions to the shuttle's hardware, which led NASA Administrator Tom Paine to seek major contributions from Australia, Canada, Western Europe, and Japan (Krige). Canada provided the robotic Canadarm, an invaluable tool for maneuvering hardware in space, whereas the Europeans contributed Spacelab, a shirt-sleeve environment for performing scientific experiments that fit in the shuttle's cargo bay. More integrated packages—European contributions to parts of the orbiter itself and to a space tug to ferry satellites from the LEO to the geostationary orbit—were withdrawn to reduce technology transfer to the minimum. This was in line with a traditional policy of keeping technological and managerial interfaces between NASA and its international partners as clean as possible. By virtue of this approach, Canadarm, Spacelab, and flying foreign astronauts all satisfied one of NASA's core missions (to collaborate internationally) without jeopardizing the other (to maintain America's scientific and technological leadership in space) (Krige).

The shuttle was not just a technological marvel. It was a symbol of American global leadership, the embodiment of American technological prowess, a source of American pride, an inspiration to the hundreds of thousands who watched its launches live in Florida and the millions who followed its fortunes on television, and a statement about the ability of a government agency, along with hundreds of American contractors and their workforces, to develop, build, and maintain a complex space transport system (Billings). Today, no national alternative is available to replace the shuttle's main mission of carrying people and hardware to LEO. The way forward is not clear

(Logsdon). Should a public-private partnership driven by commercial opportunity take over the shuttle's traditional role? Should NASA and the federal government concentrate on developing a crew carrier to go back to the moon and to explore other bodies in the solar system? Should taxpayer dollars be spent at all on human spaceflight?

Certain technological choices seem to have been made already. The spacecraft of the future may well be spaceplanes, not capsules, but this is unclear as of the end of 2012 (Launius). The fascination with this concept persists and is reinforced by all that we have learned from the huge technical, industrial, and financial effort that has been vested in the spaceplane over the past 40 years. The cooling system will likely be metallic, the option rejected for the shuttle (Jenkins). The craft could be reusable, though it will be commercially successful only if its operator can sustain flight rates (which will still be very sensitive to weather constraints) and can cut refurbishment and inspection costs after each trip (Hale).

The outcome of this debate will involve more than making technological choices. It will also express social and political values. The history of the Space Shuttle, and of human space flight in general, is not simply a history of technology. It is a history rich with multiple meanings. Indeed, it mirrors the history of the kind of society that America wants to be, of how it balances private investment with public interest, and of how—if at all—it sees the human conquest of space as defining its role in the world.

About the Authors

Robert E. Biggs was a member of the Rocketdyne Engineering staff for 55 years, including the entire 40 years of the Space Shuttle program as a member of the SSME management team. As the systems development manager and as the chief project engineer, he directed the SSME ground test program from the first engine test in 1975 through the first 1100 engine tests in 1985. Throughout the program, he served as Chairman, Rocketdyne SSME Engineering Change Board; and Chairman, SSME Rocketdyne Software Change Board. In addition, he was the Rocketdyne representative on the MSFC SSME Problem Review Board. Mr. Biggs is the author of *Space Shuttle Main Engine: The First Twenty Years and Beyond* (AAS History Series, Volume 29, Univelt, Inc., 2008).

Linda Billings is a Fellow at the George Washington University School of Media and Public Affairs in Washington, D.C., and Director of Science Communication with the Center for Integrative STEM Education at the National Institute of Aerospace. She does communication research for NASA's astrobiology, near-Earth object, and planetary protection programs in the Science Mission Directorate. Dr. Billings earned her Ph.D. in mass communication from the Indiana University School of Journalism, M.A. in international transactions from George Mason University, and B.A. in social sciences from the State University of New York at Binghamton (now Binghamton University). Her research interests and expertise include mass communication, science communication, risk communication, rhetorical analysis, journalism studies, and social studies of science. Her research has focused on the role that journalists play in constructing the cultural authority of scientists, the rhetorical strategies that scientists and journalists employ in communicating about science, and the rhetoric of space exploration.

James I. Craig is Emeritus Professor of Aerospace Engineering at the Georgia Institute of Technology. His research is in the areas of structural mechanics, structural dynamics, and design. Applications include active structural control, integrated aeromechanical design of rotor blades, earthquake engineering structural analysis and design, and probabilistic response and damage estimation for populations of structural systems and infrastructure systems. He is responsible for teaching undergraduate engineering courses in structures, experimental methods, and software engineering, and graduate courses in vibration measurement and analysis, and computer-aided engineering and design. He is coauthor of *Structural Analysis* (Springer, 2009) and *Mechanics of Failure Mechanisms in Structures* (Springer, 2012).

Amy E. Foster is an historian of technology and gender. She has published *Integrating Women into the Astronaut Corps: Politics and Logistics at NASA, 1972–2004* (Johns Hopkins, 2011), wrote a chapter on American women in aviation in the 1990s in *American Women and Flight Since 1940* (University of Kentucky Press, 2004), and has written multiple articles on gender and science fiction. She has also presented work on women astronauts and women in engineering. Her research interests include the U.S. space program, gender and the space program, gender issues in engineering, aviation history, and the history of technology. She has been a recent Guggenheim Fellow at the Smithsonian's National Air and Space Museum and held a fellowship sponsored jointly by NASA and the American Historical Association. She is an Associate Professor of History at the University of Central Florida in Orlando.

N. Wayne Hale Jr. was the manager of the Space Shuttle program for NASA at the Lyndon B. Johnson Space Center, Houston, Texas, a position he held from September 2005 until his retirement in 2010. In this capacity, he was responsible for overall management, integration, and operations of the Space Shuttle program. He also served as deputy manager from July 2003. Hale began his career with NASA in 1978 as a propulsion officer in the Propulsion Systems Section, Flight Control Division of Flight Operations at the JSC. From May to November 1985, Hale was head of the Integrated Communications Section, Systems Division, Mission Operations, and head of the Propulsion Systems Section, Systems Division, Mission Operations, from November 1985 to March 1988. Between March 1988 and January 2003, Hale served as a flight director in Mission Control for 41 Space Shuttle missions. He also served as deputy chief of the Flight Director Office for Shuttle Operations from June 2001 to January 2003.

Matthew H. Hersch is a lecturer in science, technology, and society in the Department of History and Sociology of Science at the University of Pennsylvania, where he received his Ph.D. During his doctoral studies, he held a HSS-NASA Fellowship in the History of Space Science and a Guggenheim Fellowship at the Smithsonian Institution's National Air and Space Museum. Most recently, he served as the Postdoctoral Teaching Fellow for the Aerospace History Project of the Huntington-USC Institute on

California and the West. His book, *Inventing the American Astronaut*, was published in 2012 by Palgrave Macmillan.

Dennis R. Jenkins, a consulting engineer in Cape Canaveral, Florida, works on various aerospace projects. He spent more than 30 years as a contractor on the Space Shuttle program, including stints at Vandenberg AFB, the Dryden Flight Research Center, and KSC. Jenkins served as an investigator on the CAIB, and as technical staff to the President's Commission on Implementation of United States Space Exploration Policy and the Return-to-Flight Task Group. His last assignment was as a project manager for planning and coordinating the delivery of the four remaining orbiters to their final display sites. Jenkins is the author of *Space Shuttle: The History of the National Space Transportation System—The First 100 Missions* (Specialty Press, 2001), *Dressing For Altitude: Aviation Pressure Suits, Wiley Post to Space Shuttle* (NASA/ARMD, 2012), and more than 30 other works on aerospace history.

John Krige has a Ph.D. in physical chemistry from the University of Pretoria (South Africa) and a Ph.D. in the history and philosophy of science from the University of Sussex (Brighton, U.K.). He joined the Georgia Institute of Technology in 2000 as Kranzberg Professor in the School of History, Technology, and Society. Prior to that, he directed a research group in the history of science and technology in Paris and was the project leader of a team that wrote the history of the European Space Agency. Krige's research focuses on the intersection between support for science and technology and the foreign policies of governments. Since being at Georgia Tech, he has expanded his interest beyond the study of intergovernmental organizations in Western Europe to include an analysis of –United States–European relations during the Cold War. He co-edited, with Kai-Henrik Barth (Security Studies Program, Georgetown University), *Global Power Knowledge. Science, Technology and International Affairs* (*Osiris*, Vol. 21, University of Chicago Press, 2006). His most recent monograph is *American Hegemony and the Postwar Reconstruction of Science in Europe* (MIT Press, 2006). His study of *NASA in the World: Fifty Years of International Collaboration in Space,* written along with Angel Long Callahan and Ashok Maharaj, will be published by Palgrave Macmillan.

W. Henry Lambright is professor of Public Administration, International Affairs, and Political Science at the Maxwell School at Syracuse University. His research interests include federal decision-making on space technology, environmental policy, transboundary issues, national security, and the integration of science with policy, ecosystem management, biotechnology, technology transfer, and leadership issues. Lambright has written scores of articles and has written or edited seven books, including *Powering Apollo: James E. Webb of*

NASA (Johns Hopkins, 1995). His most recent book, which he edited, is *Space Policy in the 21st Century* (Johns Hopkins University Press, 2003). His research has been supported by various federal agencies, including NSF, NASA, and the State Department. He has written on NASA administrative leadership under IBM sponsorship. He earned his Ph.D. from Columbia University in 1966. He is a fellow of both the American Association for the Advancement of Science and the National Academy of Public Administration.

Roger D. Launius is Associate Director for Collections and Curatorial Affairs at the Smithsonian Institution's National Air and Space Museum in Washington, D.C. Between 1990 and 2002, he served as chief historian of NASA. He has written or edited more than 20 books on aerospace history, most recently *Exploring the Solar System: The History and Science of Planetary Exploration* (Palgrave Macmillan, 2013); *Coming Home: Reentry and Recovery from Space* (NASA, 2012); *Globalizing Polar Science: Reconsidering the International Polar and Geophysical Years* (Palgrave Macmillan, 2010); *Smithsonian Atlas of Space Exploration* (HarperCollins, 2009); *Robots in Space: Technology, Evolution, and Interplanetary Travel* (Johns Hopkins University Press, 2008), and others. He is a Fellow of the American Association for the Advancement of Science, the International Academy of Astronautics, and the American Astronautical Society, and associate fellow of the AIAA. He also served as a consultant to the CAIB in 2003 and presented the prestigious Harmon Memorial Lecture military at the United States Air Force Academy in 2006. He is frequently consulted by the electronic and print media for his views on space issues, and he has been a guest commentator on National Public Radio and all the major television network news programs.

Nancy Leveson is Professor of Aeronautics and Astronautics and Professor of Engineering Systems at MIT. She is an elected member of the National Academy of Engineering (NAE) and has received many awards, such as the ACM Allen Newell Award for outstanding computer science research, the AIAA Information Systems Award for "developing the field of software safety and for promoting responsible software and system engineering practices where life and property are at stake," and the ACM Sigsoft Outstanding Research Award. She has published over 200 research papers and is author of 2 books, *Safeware: System Safety and Computer*s, published by Addison-Wesley in 1995 and translated into Japanese in 2009, and *Engineering a Safer World: Applying Systems Thinking to Safety*, published by MIT Press in 2012. She served on the NASA Aerospace Safety Advisory Panel, was a consultant to the CAIB, chaired a National Academy Study Committee to Review Oversight Mechanisms for Space Shuttle Flight Software Processes (1993), and was on the National Academy Study Committee for Space Shuttle

Upgrades (1999) and the NASA Software Super Problem Resolution Team (2005–2006). Professor Leveson conducts research on all aspects of system safety, including design, operations, management, and social aspects. Although she started in computer science and software engineering, she has migrated to the larger field of system engineering and subfields of particular importance, such as system safety, system analysis, human factors, human–automation interaction, and organizational safety.

John M. Logsdon is Professor Emeritus of Political Science and International Affairs at George Washington University's Elliott School of International Affairs, where he was the founder and long-time director of GW's Space Policy Institute. He is the author of many articles and essays, including the award-winning study *John F. Kennedy and the Race to the Moon* (Palgrave Macmillan, 2010), *The Decision to Go to the Moon: Project Apollo and the National Interest* (MIT Press, 1970; University of Chicago Press, 1976), and the main article on "space exploration" for the *Encyclopedia Britannica.* Logsdon, a sought-after commentator on space issues, has appeared on all major broadcast and cable networks and in the print media. In 2003, he was a member of the CAIB and formerly was a member of the NASA Advisory Council and its Exploration Committee. In 2008–2009, he held the Charles A. Lindbergh Chair in Aerospace History at the Smithsonian Air and Space Museum.

Howard E. McCurdy is Professor of Public Affairs in the Public Administration and Policy Department at American University in Washington, D.C. Professor McCurdy's teaching and research focus on public management, organization theory, science policy, and financial management. An expert on space policy, he recently completed a second edition of his award-winning *Space and the American Imagination* (second edition, Johns Hopkins University Press, 2011) (Smithsonian Institution Press, 1997). A co-authored book, *Robots in Space* (Johns Hopkins University Press, 2008), explores the human-machine debate, while *Faster, Better, Cheaper* (Johns Hopkins University Press, 2001) provides a critical analysis of cost-cutting initiatives in NASA. An earlier study of NASA's organizational culture, *Inside NASA* (Johns Hopkins University Press, 1993), won the 1994 Henry Adams prize for that year's best history on the federal government. Among his other publications are books on public administration, the space station decision, and the myth of presidential leadership. The media often consults him on public policy issues, and he has appeared on national news outlets such as the Jim Lehrer News Hour, National Public Radio, and NBC Nightly News.

Stephen Waring is Professor of History at the University of Alabama in Huntsville. He grew up on the Nebraska prairie in the Village of Bloomington.

He was educated at Doane College and the University of Iowa where he took courses in European and American history and studied the history of modern governance and bureaucratic organizations. He joined the faculty at the University of Alabama in Huntsville where he remains today. He is the author of *Taylorism Transformed: Scientific Management Theory Since 1945* (University of North Carolina Press, 1991) and, with Andrew J. Dunar, *Power to Explore: A History of Marshall Space Flight Center 1960–1990* (NASA SP-4313, 1999). He is working on a history of the *Challenger* accident investigations.

INDEX

Note: page numbers are followed by *f* or **t** (indicating figures or tables).

SUPPLEMENTAL MATERIALS

To download supplemental material files, please go to AIAA's electronic library, Aerospace Research Central (ARC), and navigate to the desired book's landing page for a link to access the materials: arc.aiaa.org.

A complete listing of titles in the Library of Flight series is available from ARC. Visit ARC frequently to stay abreast of product changes, corrections, special offers, and new publications.

AIAA is committed to devoting resources to the education of both practicing and future aerospace professionals. In 1996, the AIAA Foundation was founded. Its programs enhance scientific literacy and advance the arts and sciences of aerospace. For more information, please visit www.aiaafoundation.org.